XXIV AIVELA Annual Meeting 2016

Journal of Physics: Conference Series Volume 882

Brescia, Italy
27 - 28 October 2016

Editor:

Enrico Primo Tomasini

ISBN: 978-1-5108-4786-6
ISSN: 1742-6588

Printed from e-media with permission by:

Curran Associates, Inc.
57 Morehouse Lane
Red Hook, NY 12571

Some format issues inherent in the e-media version may also appear in this print version.

Printed by Curran Associates, Inc. (2017)

For permission requests, please contact the Institute of Physics at the address below.

Institute of Physics
Dirac House, Temple Back
Bristol BS1 6BE UK

Phone: 44 1 17 929 7481
Fax: 44 1 17 920 0979

techtracking@iop.org

Additional copies of this publication are available from:

Curran Associates, Inc.
57 Morehouse Lane
Red Hook, NY 12571 USA
Phone: 845-758-0400
Fax: 845-758-2633
Email: curran@proceedings.com
Web: www.proceedings.com

Table of contents

Volume 882

XXIV A.I.VE.LA. Annual Meeting
27–28 October 2016, Brescia, Italy

Accepted papers received: 10 August 2017
Published online: 29 August 2017

Preface

Papers

IOP Conf. Series: Journal of Physics: Conf. Series **882** (2017) 011001 doi:10.1088/1742-6596/882/1/011001

PROCEEDINGS OF THE XXIV A.I.VE.LA. ANNUAL MEETING
Brescia, Italy, 27-28 October 2016
http://www.aivela.org/XXIV_Convegno/index.html

EDITORIAL

A.I.VE.LA. – the Italian Association of Laser Velocimetry and non-invasive diagnostics is a non-profit cultural association whose objective is to promote and support research in the field of non-contact or minimally invasive measurement techniques, particularly of electromagnetic-based and optical techniques.

Through its Annual Meeting, AIVELA aims at creating an active and stimulating forum where current research results and technical advances can be exchanged and the development of new systems for laboratory use, field testing and industrial application can be promoted. The techniques addressed include Laser Doppler Anemometry – LDA, Phase Doppler Anemometry – PDA, Image Velocimetry – PIV, Flow visualization techniques, Spectroscopic measurement techniques (LIF, Raman, etc.), Laser Doppler Vibrometry – LDV, Speckle Pattern Interferometry – ESPI, Holographic techniques, Shearography, Digital Image Correlation – DIC, Moiré techniques, Structured light techniques, Infrared imaging, Photoelasticity, Image based measurement techniques, Ultrasonic sensing, Acoustic and Aeroacoustic measurements, etc.

The first Annual Meeting was held back in October 1992 and since then the event has met large consensus among the research and scientific communities worldwide for the high scientific interest of the papers presented.

The XXIV AIVELA Annual Meeting was held at the premises of the Faculty of Engineering in Brescia on 27-28 October 2016 and was organised in collaboration with the University of Brescia.
This volume contains a selection of the papers presented at the event.

The detailed Programme of the Meeting can be found at:
http://www.aivela.org/XXIV_Convegno/thursday_27.html

Trusting our Association and its initiatives will meet your interest, I wish to thank you in advance for your kind attention and hope to meet you soon at one of our events.

EDITOR
Enrico Primo Tomasini
President of A.I.VE.LA.

IOP Conf. Series: Journal of Physics: Conf. Series **882** (2017) 011002 doi:10.1088/1742-6596/882/1/011002

Peer review statement

All papers published in this volume of *Journal of Physics: Conference Series* have been peer reviewed through processes administered by the proceedings Editors. Reviews were conducted by expert referees to the professional and scientific standards expected of a proceedings journal published by IOP Publishing.

IOP Conf. Series: Journal of Physics: Conf. Series **882** (2017) 012001 doi:10.1088/1742-6596/882/1/012001

Vision-based localization of the center of mass of large space debris via statistical shape analysis

G Biondi, S Mauro, and S Pastorelli

Department of Mechanical and Aerospace Engineering, Politecnico di Torino, Torino, 10129, Italy

gabriele.biondi@polito.it

Abstract. The current overpopulation of artificial objects orbiting the Earth has increased the interest of the space agencies on planning missions for de-orbiting the largest inoperative satellites. Since this kind of operations involves the capture of the debris, the accurate knowledge of the position of their center of mass is a fundamental safety requirement. As ground observations are not sufficient to reach the required accuracy level, this information should be acquired in situ just before any contact between the chaser and the target. Some estimation methods in the literature rely on the usage of stereo cameras for tracking several features of the target surface. The actual positions of these features are estimated together with the location of the center of mass by state observers. The principal drawback of these methods is related to possible sudden disappearances of one or more features from the field of view of the cameras. An alternative method based on 3D Kinematic registration is presented in this paper. The method, which does not suffer of the mentioned drawback, considers a preliminary reduction of the inaccuracies in detecting features by the usage of statistical shape analysis.

1. Introduction

Since the launching of Sputnik, the first artificial satellite, by the USSR in 1957, a very large number of satellites have been launched into space by several countries and with several purposes. In particular, from the second half of the twentieth century up to the present, the number of artificial satellite in Low-Earth Orbit (LEO) has dramatically increased. Some of them are now inactive and uncontrolled. As the number of failed spacecraft increases significantly every year, also the risk of very dangerous collisions between relatively large space objects increases. A very famous collision that happened in 2009 between Cosmos 2251 and Iridium 33 caused an environmental disaster due to the fragmentation of the two bodies [1].

In this context major space agencies such as NASA and ESA are interested in the possibility of removing relatively large space debris and they planning the first missions. Under this spur, several concepts of debris removal systems were developed [2]. It is common in particular the idea of docking passive spacecraft with active chasers spacecraft.

For a good planning of the docking between spacecraft, it is important to estimate the complete relative dynamic state between the chaser and target spacecraft. In particular, the precise estimation of the position of the center of mass of the target leads to the possibility of predicting torques coming from the interaction forces between spacecraft.

This estimation is particularly challenging because of the absence of helpful sensors that allow the retrieval of information about the kinematics and dynamics of the target spacecraft.

Astrometric measurements from ground-stations are essential to determinate approximately the orbit of the targets, but the accuracy of those measures is only sufficient for the purpose of planning the rendezvous with the chaser spacecraft. Some examples of methods to make an initial orbit determination of a space object are presented in [3] and in [4]. Once the chaser reaches an orbit that is close to the one of the target, a proximity tracking of its natural features become possible if the chaser equips vision-based sensors.

There are in literature some works regarding the complete state estimation of passive targets that rely on 3D sensors. Tracking is pursued in real-time with different techniques but the possession of a 3D CAD model of the target is required. For instance, in [5] an active laser ranging device is adopted to pursue the estimation of the complete state of a known but non-cooperative spacecraft. The ICP algorithm and a Kalman filter are combined together in a closed-loop fashion to make the method tolerant of potential failed detections of the objects. Active sensors represent a fairly reliable solution for tracking the motion of space debris but, as a drawback, they slightly add complexity in the chaser design due to increasing energy consumption and, secondarily, costs.

Against, looking to the state of the art, it is possible to find examples that demonstrates that passive sensors maintain comparable performances to active ones. One of the most significant works is illustrated in [6]: the dynamic relative state of two non-cooperative satellites is estimated through the application of an iterated extended Kalman filter, which is fed with measurements from a stereo-rig system on the chaser spacecraft. Moreover, no prior assumptions about the shape of the target are required by this estimation technique

However, the last mentioned solution, as any other in the state of the art, do not consider the possibility of losing temporary the availability of measurements or having modifications of the measured quantities. To better explain this last concept, imagine that one or more features disappear suddenly from the field-of-view of the stereo-rig system; at the same time, the tracking sensors could detect completely new features. Besides, also occlusions can happen, compromising totally the acquisition of the measurements. In these conditions, no known methods are able to recover the complete dynamic state of the non-cooperative target.

Some previous works of the authors of this paper ([7], [8], and [9]) contemplates the mentioned tracking conditions and provides methods for recovering the attitude dynamics of space debris, but not the orbital one. Therefore, the literature lacks of methods that consider difficult tracking conditions for estimating the location of the center of mass (CoM) of space debris.

The aim of the present paper is to introduce an approach that does not rely on typical recursive Bayesian observers, but considers only the registration of the target kinematics to locate the CoM, assumed coincident with the center of gravity of the passive object of interest. In particular, this approach relies on the identification of the axes and poles of the successive finite rotations of the target in several different time intervals.

Considering no relative translation between the chaser and the target, all the axes intersect in a point of the body that coincides with the CoM. When a relative translation exists, the axes do not pass any more for the CoM but its distance from the axes is positively related to the relative translational velocity. Nevertheless, if the entity of the translation is within certain bounds, it is possible to find an efficient geometric estimator of the CoM location.

Because of the evident unavailability of real measured data, the proposed method is validated on simulated data. These data include the presence of a noise that is coherent to the uncertainty associated to plausible stereo-vision systems for space applications.

The problem of finding one screw axis between two poses has not an exact solution when some noise corrupts the coordinates of the points representing the poses. This issue is often faced by estimating an optimal screw axis in the least-squares sense. On the contrary, in this paper the authors propose to filter the data in a pre-processing stage to obtain series of rigid clusters of homologous points belonging to the target. This approach corresponds to infer from raw data the original shape of the examined body. In this way, a unique screw axis underlies the finite movement between two filtered poses. While the

IOP Conf. Series: Journal of Physics: Conf. Series **882** (2017) 012001 doi:10.1088/1742-6596/882/1/012001

common least-squares estimation techniques lead to a total failure in localizing the CoM, the proposed method is useful for a reliable recovery of the position of the searched point.

The remainder of this paper is organized as follow: section 2 describes how to simulate the measurements for the method validation. Section 3 illustrates a simple technique to identify screw axes from the coordinates of three couples of homologous points. Then, it shows a possible estimator of the CoM location. The proposed estimator is tested on ideal data (not corrupted by noise or missing samples due to occlusions) and its accuracy is briefly discussed. Section 4 presents one of the most famous techniques for the statistical shape analysis of homologous point clusters: the general Procrustes analysis (GPA). Section 5 shows the results of the proposed methodology applied on simulated data. Finally, section 6 provides the conclusions of this work.

2. Tracking simulation

The methods that are presented in this work consider as input the positions of several features of the target in many time samples. These data are assumed to be measurable by using a stereo-rig system that is mounted on a chaser spacecraft. Thus, to create a realistic set of input data it is necessary to simulate the relative orbital motion between two close satellites. Moreover, also the attitude dynamics of the target have to be modeled. Indeed, considering the ideal case in which no relative orbital dynamics exists between the satellites, the chaser would see the features of the target rotating around the searched CoM. Otherwise, the trajectories of the features would be the result of the combination of both the target rotation and translation from the chaser point of view.

The attitude dynamics of a satellite is well-approximated by the Euler's equation for torque-free rigid bodies:

$$J\,{}^{b}\dot{\boldsymbol{\omega}} + {}^{b}\boldsymbol{\omega} \times J\,{}^{b}\boldsymbol{\omega} = \mathbf{0} \tag{1}$$

where J is the principal inertia matrix of the target and ${}^{b}\boldsymbol{\omega}$ is the angular rate of the target expressed in a principal body-fixed coordinate system b. In the entire paper, the upper prescript on a vector indicates the coordinate system in which the vector is expressed.

The resulting angular rate from equation (1) determines the variation of the attitude of the target according to the following Darboux's equation:

$${}^{b}\dot{A}_{I} = {}^{b}\omega^{\times}\,{}^{b}A_{I} \tag{2}$$

where ${}^{b}A_{I}$ is the orthonormal direction cosines matrix that maps a vector expressed in an inertial frame I into the same vector expressed in the coordinate system b. The notation ${}^{b}\omega^{\times}$ indicates the following skew-symmetric matrix:

$${}^{b}\omega^{\times} = \begin{bmatrix} 0 & \omega_3 & -\omega_2 \\ -\omega_3 & 0 & \omega_1 \\ \omega_2 & -\omega_1 & 0 \end{bmatrix} \tag{3}$$

Equations from (1) to (3) are useful for the simulation of the attitude dynamics of the target. For simulating the formation of chaser and target, it is convenient to consider the motion of the chaser, assumed as a material point, about the CoM of the target. This mentioned relative dynamics is given by the following equation:

$${}^{\ell}\ddot{\boldsymbol{\rho}}_{c} = {}^{\ell}\begin{bmatrix} 2\dot{\vartheta}\dot{\rho}_{c2} + \ddot{\vartheta}\rho_{c2} + \dot{\vartheta}\rho_{c1} - \sigma(r+\rho_{c1}) + \mu r^{-2} \\ -2\dot{\vartheta}\dot{\rho}_{c1} - \ddot{\vartheta}\rho_{c1} + \dot{\vartheta}\rho_{c2} - \sigma\rho_{c2} \\ -\sigma\rho_{c3} \end{bmatrix} \tag{4}$$

where ${}^{\ell}\boldsymbol{\rho}_{c}$ is the position of the chaser in the local-vertical-local-horizontal (LVLH) ℓ coordinate system of the target, ϑ is the true anomaly of the target, μ is the planetary constant, r is the distance between the CoM of the target and the Earth's center. Finally, σ has the following expression:

$$\sigma = \mu\left[\left(r + \rho_{c1}\right)^2 + \rho_{c2}^2 + \rho_{c3}^2\right]^{-3/2} \tag{5}$$

The position of the target CoM can conveniently be expressed in a coordinate system that is placed on the chaser and whose axes are aligned to the ones of the inertial Earth centered system I:

$${}^{c}\boldsymbol{\rho}_g = -{}^{I}A_{\ell}{}^{\ell}\boldsymbol{\rho}_c \tag{6}$$

Once the initial conditions for equations (1), (2), and (4) are set, it is possible to evaluate at any sample time the position ${}^{c}\boldsymbol{\rho}_i$ of any feature whose known constant position in the body-fixed reference frame is ${}^{b}\boldsymbol{\rho}_i$. Hence:

$${}^{c}\boldsymbol{\rho}_i = {}^{c}\boldsymbol{\rho}_g + {}^{b}A_I^{T}{}^{b}\boldsymbol{\rho}_i \tag{7}$$

For a better comprehension of the mentioned quantities and coordinate systems refer to figure 1. The computation of the second member of the equation (7) is the last step of the data generation procedure, which includes the solution of all the mentioned differential and algebraic equations. To guarantee the robustness of the presented approach, we split the simulation time in several intervals in which different cluster of features are considered. In addition, for some time intervals, we considered a complete unavailability of data simulating the occurrence of occlusions. Figure 2 illustrates this concept. Besides, figure 3 shows an example of the computed trajectories for five particular features during the period of one orbit of the target. From that figure it is possible to evidence the joint effect of the attitude and orbital dynamics.

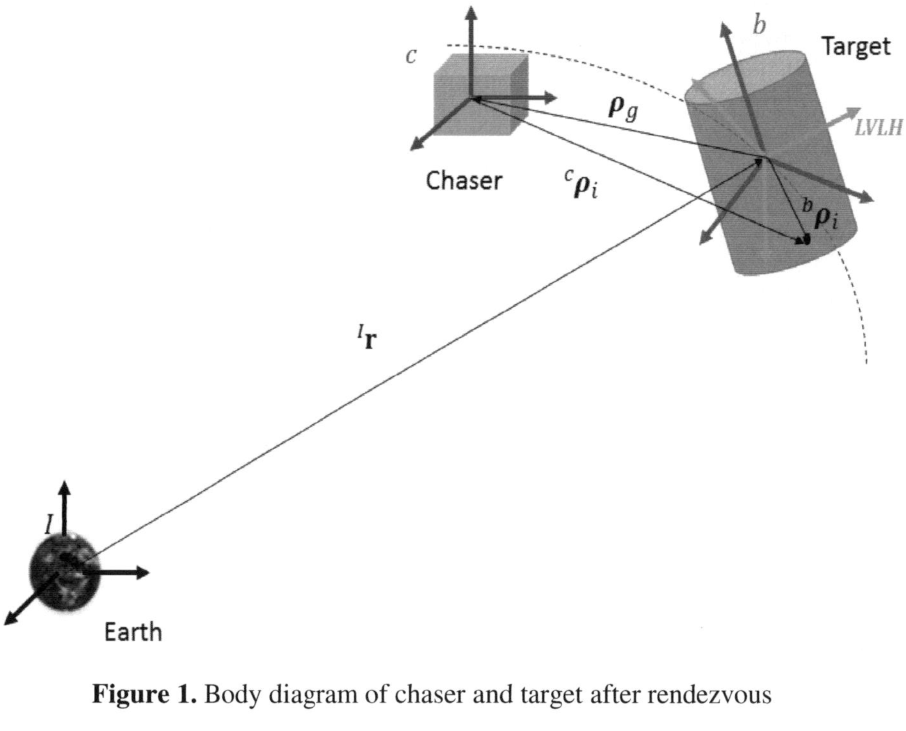

Figure 1. Body diagram of chaser and target after rendezvous

IOP Conf. Series: Journal of Physics: Conf. Series **882** (2017) 012001 doi:10.1088/1742-6596/882/1/012001

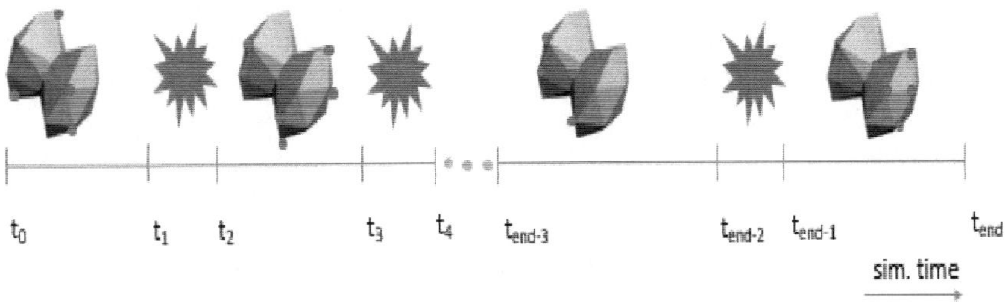

Figure 2. Different clusters of features considered during the simulation

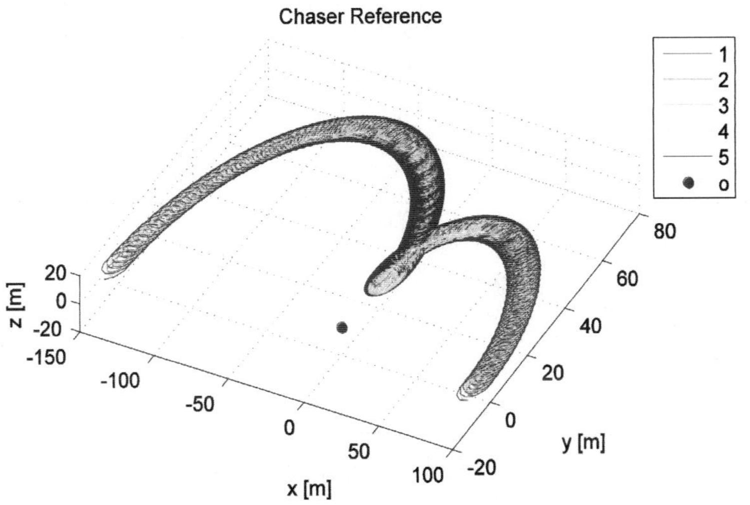

Figure 3. Trajectories of five different features for one orbital period of the target

3. 3D kinematic registration of rigid poses

The kinematic registration is a procedure that consists in finding one synthetic representation of a possible motion that happened between the current pose of a rigid body a past one of the same body. Thus, it is a procedure that produces an alternative representation of the motion of a rigid body. Actually, performing the kinematic registration of a body does not mean to estimate the real evolution of the posture of the body. However, it will be shown that performing such registration leads to information that is useful to estimate the unknown location of the CoM of the target.

The very well-known Mozzi-Chasles' theorem states that two poses of a rigid body can be mapped one with respect to the other through one translation and one rotation about the direction of the translation. Thus, it is possible to evidence the existence of one axis, whose direction and location characterize the displacement of a rigid body. If this displacement happens during an interval of time whose duration tends to zero, then this axis is identified as the *instantaneous screw axis.* Otherwise, if the duration is finite, it is named as the finite *screw axis.*

To characterize the pose of one object it is sufficient to know exactly the Euclidean coordinates of three points that belong to it. Indeed, it is possible to build a reference triad of orthogonal axes by making appropriate cross products of vectors connecting the points. Thus, the knowledge of the positions of three couples of homologous points at two separate instants is sufficient for the computation of the finite screw axis. There are several methods in the literature that address this computation (see [10] for instance). This paper considers a previously developed method consisting of the 3D generalization of the planar Reauleaux's rule [11]. This rule individuates the pole of planar rotations by intersecting the bisecting lines connecting two couples of homologous points. Therefore, according to this rule, two poses of a rigid body overlaps through a single planar rotation. Actually, if the real motion between the poses was a pure rotation, the found pole would be exactly the CoM of the body. On the contrary, the existence of a translational motion leads the found pole distancing from the actual CoM. In particular, in the instantaneous case, the distance is proportional to the ratio between the linear and the angular speed of the body.

In the 3D case, two poses overlaps through one translation and one rotation. However, the rotation develops on a plane, which is perpendicular to the direction of the translational displacement. This means that the Reuleaux's rule remains applicable once the effect of the translation is removed in some way. For instance, given three vectors $\boldsymbol{g}_i$ connecting a couple of homologous point (P_i and P_i'), the vectors $\boldsymbol{c}_1 = \boldsymbol{g}_2 - \boldsymbol{g}_1$ and $\boldsymbol{c}_2 = \boldsymbol{g}_3 - \boldsymbol{g}_1$, span a plane which is perpendicular to the screw axis $\boldsymbol{a}$. Thus, the direction of the screw axis is given by the following expression:

$$\boldsymbol{a} = \frac{\boldsymbol{c}_1 \times \boldsymbol{c}_2}{\|\boldsymbol{c}_1 \times \boldsymbol{c}_2\|} \tag{8}$$

Equation (8) provides the direction along with projecting two of the three couples of homologous points on the plane spanned by $\boldsymbol{c}_1$ and $\boldsymbol{c}_2$. This projection eliminates the translation, so the mentioned Reulaeaux's rule holds on the projected points for the computation of the pole U of the remaining rotation. This pole is one of the infinite points belonging to the screw axis.

Differently from the pure Reauleaux's rule, its 3D generalization admits a translational displacement to overlap two different poses of a rigid body. However, this translation must be along $\boldsymbol{a}$. Actually, if the real motion between the poses involved a misaligned translational component, the screw axis would not pass for the CoM of the body. In particular, in the instantaneous case, the distance from the axis to the CoM is proportional to the component of the linear speed that is orthogonal to the axis.

Furthermore, even if no translations occur, the generalization does not individuate the CoM but only a screw axis. Thus, even in the case of pure rotations, it is necessary to compute two screw axes considering three different poses. The axes will intersect in the searched CoM. However, in the general case, two successive screw axes are non-coplanar; so, the CoM cannot be individuated.

Assuming that the translational motion is relatively slow, as it should be the relative orbital motion of two satellites after rendezvous, the authors of this paper proposed to perform a *pseudo-intersection* between successive screw axes [12]: the CoM is estimated as the middle point of the shortest line segment connecting the non-coplanar axes (see figure 4).

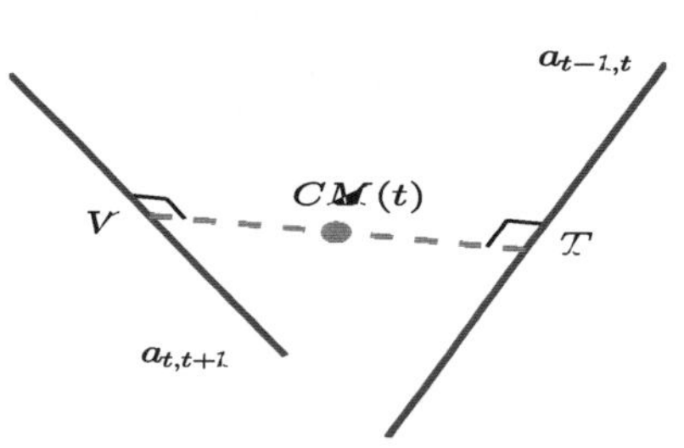

Figure 4. Pseudo intersection at time t between two successive finite screw axes

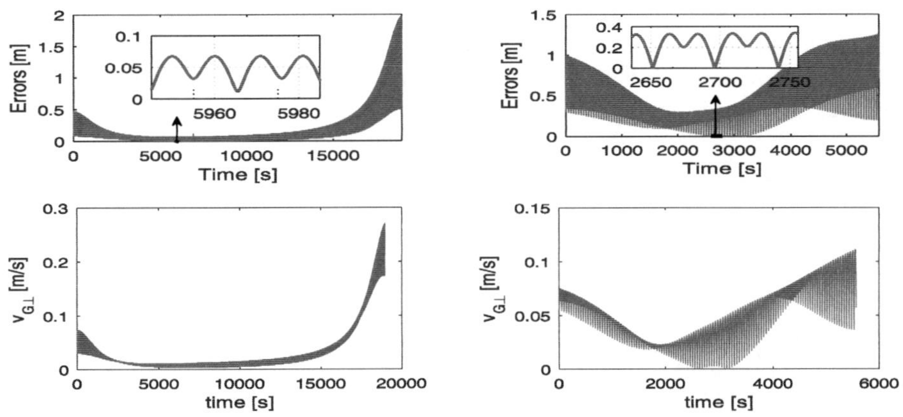

Figure 5. Estimation errors after performing pseudo-intersection on ideal data and comparison with the normal component of the target velocity to the instantaneous screw axis

The work in [12] showed that if two satellites remain close, the line segments computed from ideal data are extremely short (less than 2 mm). Moreover, their length do not depend by the sampling time of the input data but depend only on the simulated dynamics of the target and chaser. On the other hand, the main source of error in the estimation of the location of the CoM is the value of the linear velocity of the target relative to the chaser. In particular the most influencing value is the magnitude of the component that is orthogonal to the instantaneous screw axis, whose direction is the same of the angular velocity vector. The last statement is confirmed by application examples. For instance figure 5 illustrates the comparison between the mentioned component of the velocity and the error in localizing the CoM by performing the pseudo-intersection on two different sets of ideal trajectories (no noise and no missing samples).

4. General Procrustes analysis to regularize noise on trajectories

The results that are shown in figure 5 refers to data that do not suffer the corruption provided by uncertainties due to the imperfections of the sensors used for the data acquisition. The same procedure applied to corrupted data produces unbounded error. The main reason of this sensibility of the procedure to noise is mainly attributed to an apparent loss of stiffness of the acquired shape of the target. Indeed, the distance between two generic points of a rigid body have to remain constant during the motion of the object. The randomness of the noise on the acquired trajectories causes instead a violation of this important property. Moreover, the presented solution to find screw axes is radically based on the non-deformability of the moving body.

Therefore, to maintain bounded the error it is absolutely needed to remove some of the noise while the distance between the acquired features is constrained to remain constant. However, to perform this filtering, the various poses should be comparable such to determinate one particular set of features, which is optimally mappable to all the poses through simple isometries, i.e. combinations of rotations and translations.

Given two different poses of an object, each characterized by a corrupted set of points outlining the surface of the object, a first operation that is useful to make comparisons is the overlap of the centroids of the two sets. The overlapping is obtained by simply subtracting the vector $\boldsymbol{t}_2$ that connects the centroids to all the points characterizing the second pose. If the positions of those points are the columns of a *configuration matrix* $\boldsymbol{X}_2$, this operation corresponds to subtracting to $\boldsymbol{X}_2$ the matrix $\boldsymbol{T}_2$, whose columns are all equal to the vector $\boldsymbol{t}_2$.

IOP Conf. Series: Journal of Physics: Conf. Series **882** (2017) 012001 doi:10.1088/1742-6596/882/1/012001

Hence, after filtering-out the translational information, one has two configurations matrices $\boldsymbol{X}_1$ and $\boldsymbol{X}_2'$, which should be separated by one rotation about the coincident centroid of the two sets of points. Actually, due to the corruption, there are no rotations that exactly maps the two matrices. Thus, the best possible operation to compare the poses consists in finding the orthonormal rotation matrix $\boldsymbol{R}_2$ that most closely maps $\boldsymbol{X}_2'$ to $\boldsymbol{X}_1$

This last statement correspond to the so-called Orthogonal Procrustes problem, whose formulation is the following:

$$\boldsymbol{R}_2 = \underset{R}{\operatorname{argmin}} \|\boldsymbol{R}\boldsymbol{X}_2' - \boldsymbol{X}_1\|_2 \quad \text{s.t.} \quad \boldsymbol{R}\boldsymbol{R}^T = \boldsymbol{I}_3 \tag{9}$$

This problem is a convex constrained quadratic optimization whose closed-form solution is shown in [13]. The solution comes from the singular value decomposition of the matrix $\boldsymbol{X}_1\boldsymbol{X}_2'^T$ to find its nearest orthogonal matrix. In particular it holds:

$$\boldsymbol{X}_1\boldsymbol{X}_2'^T = \Sigma V D^T \tag{10}$$

where Σ and D are two orthogonal matrices that represent two following rotations, while V is a diagonal matrix that represent an intermediate scaling. Thus, it holds:

$$\boldsymbol{R}_2 = \Sigma D^T \tag{11}$$

The same applies to the configuration matrices $\boldsymbol{X}_3', \boldsymbol{X}_4' \ldots, \boldsymbol{X}_n'$ with respect to $\boldsymbol{X}_1$ for the entire duration of the time interval in which a certain set of features is detected. Once a new set is detected, the comparison procedure restart using a new reference configuration matrix (see figure 2).

Obviously, the transformed sets of homologous points do not matches exactly but they form a group of samples of a random variable, which is called *shape*. The mean of these samples can be considered as the *optimal shape,* which is represented by the resulting mean configuration matrix $\overline{\boldsymbol{X}}_1$.

The estimation of the original shape of a moving body from the detection of its natural features or landmarks takes the name of *statistical shape analysis.* In particular the described procedure is better known as *general Procrustes analysis* (GPA) [14]. Figure 6 illustrates this technique.

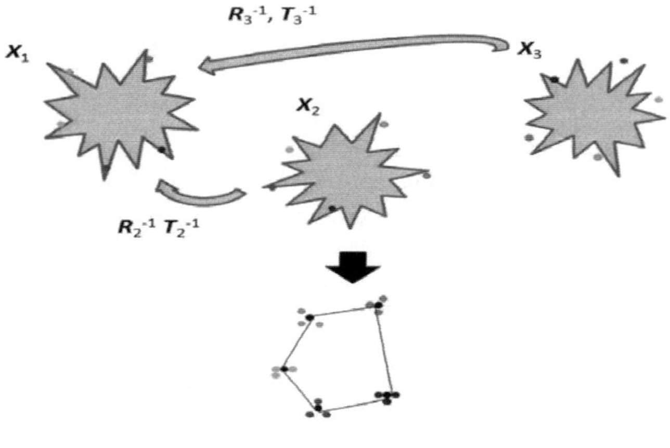

Figure 6. The general Procrustes analysis: the clusters of features are all optimally mapped to the first one in time order. The mean of the resulting positions for each point gives the mean shape of the body

Note that after the computation of the mean configuration matrix $\bar{\boldsymbol{X}}_1$, it is possible to perform the following operation:

$$\bar{\boldsymbol{X}}_i = \boldsymbol{R}_i \bar{\boldsymbol{X}}_1 + \boldsymbol{T}_i \tag{12}$$

where $i = 2, 3, \dots, n$.

The new obtained positions of the features are such to preserve the values of the distances between the features itself. This regularization of the raw acquired data allows the efficient determination of the screw axes through the mentioned 3D generalization of the Reuleax's rule.

5. Estimation of the trajectory of the CoM of the target

Although the GPA has a significant positive effect on the capacities of the method exposed in section 2, the uncertainties on the determination of the screw axes remain non-negligible. An example of the results of such determination after the GPA of raw simulated data is in figure 7: the directions $\boldsymbol{a}$ and the positions of the poles U are plotted as signals in the time domain; moreover, they are compared to their reference values.

The typical regularity of the attitude dynamics of space debris leads one to understand that the irregular behaviour shown by the screw axes in figure 7 is due to residual uncertainties. A low-pass filter is then exploited to remove these residuals. Phase delays are not acceptable because they would provide bias to the final estimated trajectory of the CoM. Therefore, zero-phase filtering is considered.

The pseudo-intersection of the filtered screw axes provides, as shown in section 2, the estimated trajectory of the CoM of the target. However, as visible from the example in figure 8, the resulting signal contains a significant high-frequency noise. Moreover, the signal presents missing samples due to the introduced occlusions.

The final trajectory is recovered by polynomial curve-fitting. The error related to the example in figure 8 is shown in figure 9. This error has a non-random behaviour but it remains quite bounded. Considering the previous methods listed in the introduction, the one proposed in this paper showed a competitive accuracy. Notwithstanding, the level of required assumptions to let the method working is drastically lowered. In particular, there is no need to detect always the same features. Moreover, the proposed method is robust to occlusions or any kind of sudden loss of data.

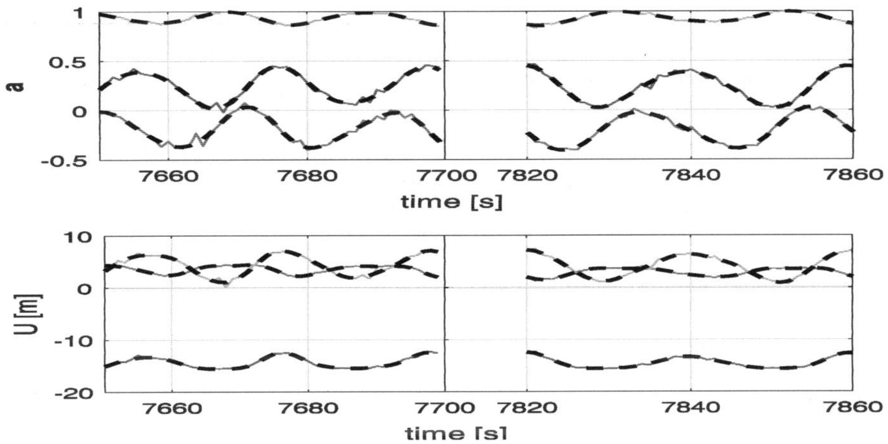

Figure 7. Raw (solid line) and filtered (dashed line) signal expressing a possible time behaviour of the direction and of the position of the screw axes found via 3D kinematic registration

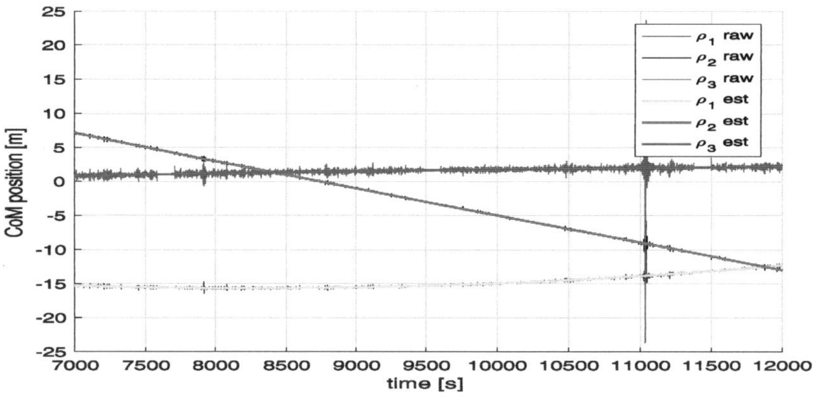

Figure 8. Raw and fitted estimated trajectory of the CoM of the target

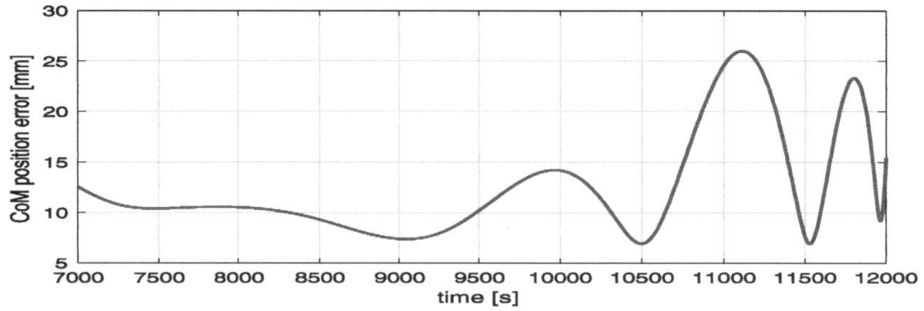

Figure 9. Estimation errors corresponding to the trajectory in Figure 8.

6. Conclusions

The goal of this work was the investigation of methods and algorithms for localizing the CoM of a space debris during an active removal mission. Space debris mitigation is one of the most important topics that are currently discussed within the space community. An accurate knowledge of the CoM location helps the capture and the removal of the object significantly. Indeed, the contact between two non-cooperative systems severely undermines the stability of the joint system.

The underlying assumption under which the methods were developed regards the possibility of tracking the positions of several features of the passive object. However, part of the research work was spent to overcome one of the most common but restrictive assumptions in the literature, that is to consider the possibility of tracking one fixed set of features of the objects.

In this work, a pure kinematic estimation approach was developed. The approach is based on the computation of the finite screw axes that represent the motion between the various detected poses of a space debris. In particular, the combined usage of statistical shape analysis and low-pass filters was decisive for guaranteeing high accuracy in case of corrupted input data. The errors obtained, although distributed in a non-catalogued way, resulted bounded. The found accuracy is similar to one of the best methods known; however, the level of the necessary hypothesis was significantly lowered.

References

[1] Tan A, Zhang X and Dockanian M 2013 Analysis of the Iridium 33 and Cosmos 2251 collision using velocity perturbations of the fragments *Adv. Aerosp. Sci. Appl.* **3** 13-25

[2] Shan M, Guo J and Gill E 2016 Review and comparison of active space debris capturing and removal methods *Prog. Aerosp. Sci.* **80** 18-32

[3] Der G J 2012 New angles-only algorithms for initial orbit determination *AMOS* Maui 53
[4] DeMars K J, Jah M K and Shumacher P W Jr 2012 Initial orbit determination using short-arc angle and angle rate data *IEEE T. Aero. Elec. Sys.* **48** 2628-37
[5] Aghili F, Kuryllo M, Okuneva G and English C 2011 Fault-tolerant position/attitude estimation of free-floating space objects using a laser range sensor *IEEE Sens. J.* **11** 176-85
[6] Segal S, Carmi A and Gurfil P 2014 Stereovision-based estimation of relative dynamics between noncooperative satellites: theory and experiments *IEEE T. Contr. Syst. T.* 568-84
[7] Biondi G, Mauro S, Mohtar T, Pastorelli S and Sorli M 2017 Attitude recovery from feature tracking for estimating angular rate of non-cooperative spacecraft *Mechanical Systems and Signal Processing* **83** 321-36
[8] Biondi G, Mauro S, Mohtar T, Pastorelli S and Sorli M 2016 Feature-based estimation of space debris angular rate via compressed sensing and Kalman filtering *MetroAeroSpace 2016* Firenze IEEE 215-20
[9] Biondi G, Mauro S, Pastorelli S and Sorli M 2016 Fault-tolerant feature-based estimation of space debris rotational motion during active removal missions *IAC Proceedings 2016*
[10] Page Á, de Rosario H, Mata V and Atienza C 2009 Experimental analysis of rigid body motion. A vector method to determine finite and infinitesimal displacements from point coordinates *J. Mech. Design* **131**
[11] Eberharter J K and Ravani B 2006 Kinematic registration in 3D using the 2D Reuleaux method *J. Mech. Design* **128** 349-55
[12] Biondi G, Mauro S, Mohtar T, Pastorelli S and Sorli M 2015 A geometric method for estimating space debris center of mass position and orbital parameters from features tracking *MetroAeroSpace 2015* IEEE 265-70
[13] Schönemann P H 1996 A generalized solution of the orthogonal Procrustes problem *Psychometrika* **31** 1-10
[14] Dryden I L and Mardia K V 1993 Multivariate shape analysis *Sankya Ser. A* **55** 460-80

IOP Conf. Series: Journal of Physics: Conf. Series **882** (2017) 012002 doi:10.1088/1742-6596/882/1/012002

Non-contact measurement of tremor for the characterisation of Parkinsonian individuals: comparison between Kinect and Laser Doppler vibrometer

L Casacanditella[1], G Cosoli[1], MG Ceravolo[2], EP Tomasini[1]
[1]Department of Industrial Engineering and Mathematical Sciences, Università Politecnica delle Marche, via Brecce Bianche, 60033 Ancona, Italy
[2]Department of Experimental and Clinical Medicine, Università Politecnica delle Marche, via Tronto 10/a, 60126 Ancona, Italy

E-mail: g.cosoli@univpm.it

Abstract. Parkinson's disease is a progressive neurodegenerative disorder affecting the central nervous system. One of its main and most evident symptoms is the tremor, which usually manifests at rest with varying intensity during time. An important diagnostic challenge is the differential diagnosis between Parkinson's disease and the other most widely represented tremor syndrome, i.e. Essential (or senile) tremor. At present there are no standard methods for the quantification of tremor and the diagnosis of both Parkinson's disease and Essential tremor is mainly done on the base of clinical criteria and by using rating scales. The aim of this work is to objectively and non-invasively assess the tremor linked to the quoted diseases, using non-contact techniques: Laser Doppler Vibrometer (LDV) and Kinect for Windows device. Two subjects with Parkinson's disease and one with Essential tremor were tested in different conditions: at rest, during a cognitive task, with forward stretched arms and in "Wing position". The results from data processing in terms of tremor frequency seem to be comparable, with a mean deviation of 0.31 Hz. Furthermore, the values computed are consistent with what is stated in the literature (i.e. 4-12 Hz). So, both LDV and Kinect device can be considered suitable to be used as an objective means for the assessment and monitoring of Parkinson's disease tremor, helping the clinician in the choice of the most suitable treatment for the patients.

1. Introduction

According to the World Health Organization (WHO), hundreds of millions of people worldwide are affected by neurological disorders, such as epilepsy, Alzheimer disease or other dementias and Parkinson's disease (1). Parkinson Disease (PD) is a progressive neurodegenerative disorder of the central nervous system (2–4). It is characterised by several features (both motor and non-motor symptoms (5–7) – the motor signs are caused mainly by dopamine deficiency (8)) that appear slowly over time and can impact significantly on functional and cognitive capabilities of the subject (9–11). Furthermore, it has also a substantial economic impact on society (12,13), considering a rising prevalence of the pathology with age. In fact, it is possible to state that the prevalence of PD (per 100.000) is equal to (14):

- 41 in 40-49 years;

- 107 in 50 to 59 years;
- 428 in 60 to 69 years;
- 1087 in 70 to 79 years;
- 1903 in older than age 80.

Some differences in prevalence of PD by geographic location and gender can be detected (15–18).
Among its motor signs, e.g. bradykinesia, rigidity and tremor, the last is the most evident and embarrassing one (19): it is an involuntary rhythmic oscillation, quite regular, of a body part around a fixed point, along a unique plan. This phenomenon is stressed when that body part is at rest condition, in fact rest tremor is the most common and easily recognised symptom (20). However, due to the erratic occurrence of tremor and its varying intensity over time, it can be difficult to either detect or quantify this symptom. Therefore, it would be important to provide the clinicians with objective quantitative measurements of Parkinson's disease symptoms, in order to help the diagnosis and personalise the treatment (21–23). Moreover, when the differential diagnosis between tremor syndromes becomes challenging, like in subjects with atypical tremor, the support of quantitative measures able to characterize the frequency and time course of involuntary movements could be useful.
In the literature, it is possible to find studies related to the measurement of tremor by means of triangulation sensors and accelerometers (24), but also displacement and velocity sensors and electromyography (25), gyroscopes (26) and laser diodes (27).
The aim of this work is to quantitatively measure the tremor frequency in the upper limbs of Parkinsonian individuals by means of two different non-invasive and non-contact sensors: a Laser Doppler Vibrometer (already used for non-invasive measurements of several physiological parameters (28–33)) and the Kinect device. This study is intended to be a pilot step for an objective evaluation that can be useful for a reliable quantitative assessment of tremor in Parkinson's disease.

2. Materials and methods

2.1. Measurement setup

Three subjects complaining for tremor syndrome (2 male, 1 female, mean age = 67 years, average disease duration = 17 years) were enrolled into the present pilot study; they were included because of the manifestation of unilateral or bilateral tremor affecting the upper limbs. The subjects were tested in different conditions: at rest or during the execution of a cognitive task, seated or standing. All the measurements were carried out in the morning, while undergoing their daily treatment. Two of the tested patients (subjects 2 and 3 in the Results section) were affected by Parkinson's disease, while subject 1 suffered from essential tremor (34,35), the most common "movement disorder" (36), which is more evident during voluntary actions, both in postural and in kinetic situations. In the literature, the frequency of the essential tremor is in the range 4-12 Hz (37).
Two non-invasive and non-contact measuring instruments were used:

- A Laser Doppler Vibrometer, LDV (PDV 100 by Polytec (38), sensitivity: 0.2 V/(mm/s)), measuring the velocity of the surface displacement;
- A Kinect for Windows device (39) (frame rate: 30 Hz; image resolution: 480 pixels x 640 pixels; depth measurement distance: 0.5-4.5 m), measuring both a colour image (by means of an RGB camera) and a depth image (by means of a depth sensor, composed of an IR projector and an IR camera).

The subject was asked to relax and sit on a chair with armrests. The LDV laser beam (HeNe, Class II, visible (38)) was directed perpendicularly on the hand of the patient, at a distance of 1 m; the signal was sampled with a sampled frequency of 1 kHz. At the same time, the RGB and the depth cameras of the Kinect device were used to record videos during the tests.
The measurement setup is reported in 'figure 1'.

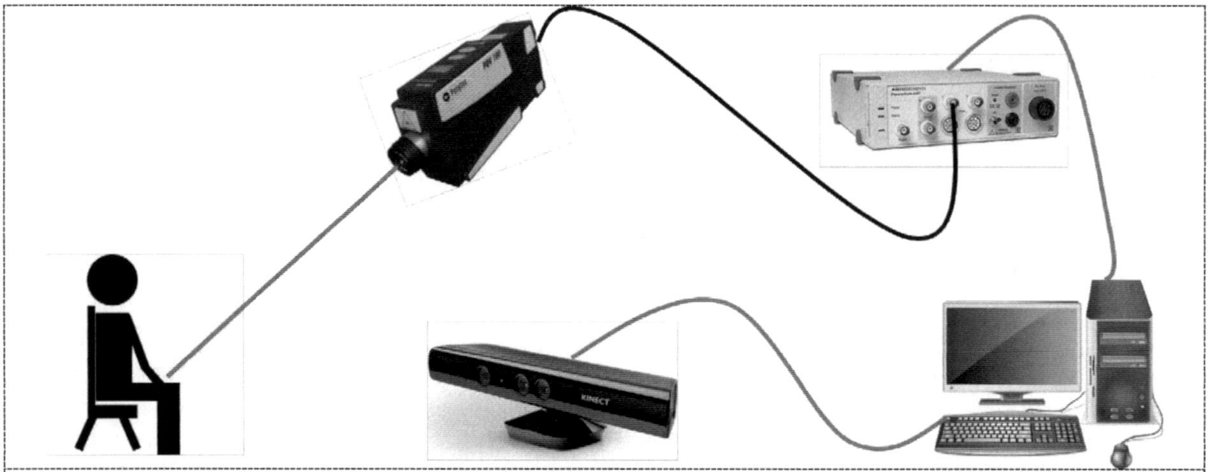

Figure 1. Test bench: LDV beam is directed perpendicularly on the subject's hand; Kinect device is used to record the video during the test

Different kinds of test were performed:

1. At rest: the subject was seated, with the back leaned against the seatback of the chair, the elbows at 90°, the forearms laid on the armrests and the hands out of the supports;
2. At rest with a consequent cognitive task: the subject was seated and he was asked to do a cognitive task (e.g. to count backwards out loud);
3. "Wing position": the subject was asked to stand up and stay with the arms lifted on a plane parallel to the floor, the elbows bended with an acute angle, the hands in front of the chest, with the palms downwards;
4. With the arms stretched out: the subject was asked to stand up and stay with the arms lifted in front of the shoulders, on a plane parallel to the floor, with the palms downwards.

The second test is intended to increase the subject's stress, highlighting the tremor (even when it is latent). In the third and the fourth tests, the assumption of a determined position is useful to evaluate the so-called re-emergent tremor *(40)*.

The four tests are represented in 'figure 2'. Each kind of test had a duration of 20 s.

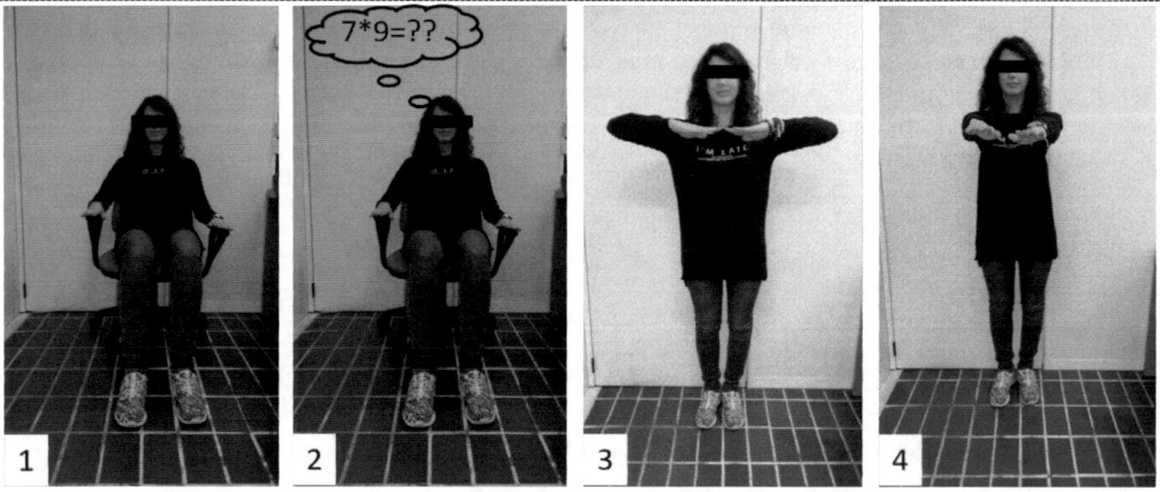

Figure 2. Test kinds: 1) at rest; 2) at rest with a consequent cognitive task; 3) "Wing position"; 4) Arms stretched, with palms downwards

IOP Conf. Series: Journal of Physics: Conf. Series **882** (2017) 012002 doi:10.1088/1742-6596/882/1/012002

2.2. Data processing

The signals extracted from LDV and Kinect device were processed in order to compute the tremor frequency. The two sensors required the use of different processing techniques in order to obtain the frequency value. Both the signals were processed in MATLAB® (41) environment.

In particular, with regard to LDV signal, the processing procedure consisted in the following steps:

- Removal of the mean value of the signal;
- Filtering with a 3rd order Butterworth low-pass filter (cutoff frequency = 20 Hz);
- Fast Fourier Transform (FFT) of the signal, in order to evaluate its spectral components.

An example of such a processing procedure is reported in 'figure 3'.
On the other hand, two different approaches were applied to the signals from the Kinect depth camera:

- One approach to extract the tremor in the direction orthogonal to the sensor field of view. In this case, the colour image was used; the processing steps were the following:
  - Selection of the Region of Interest (ROI);
  - Conversion into black and white pixels, depending on the intensity;
  - Computation of centroids;
  - Fast Fourier Transform;
- The other approach to individuate the tremor in the direction parallel to the sensor field of view. In this case, the depth map was used; the processing stages were the following:
  - Selection of the Region of Interest (ROI);
  - Computation of the mean value;
  - High pass filtering (cutoff frequency = 0.5 Hz);
  - Mean removal;
  - Fast Fourier Transform of the signal.

An example of signals from Kinect sensor is reported in 'figure 4'.

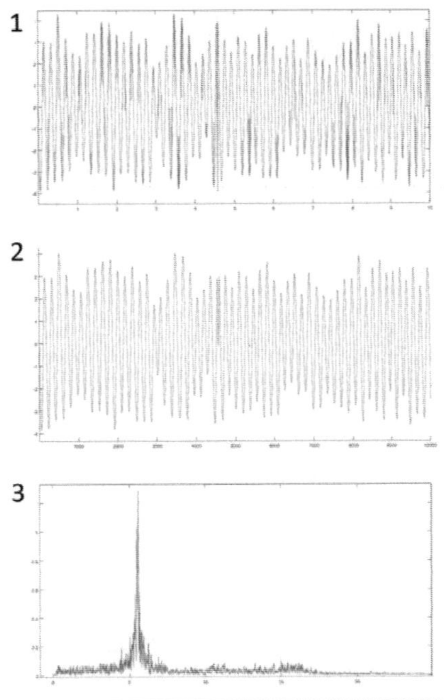

Figure 3. LDV signal processing: 1) raw signal after mean removal; 2) filtered signal; 3) frequency content of the signal

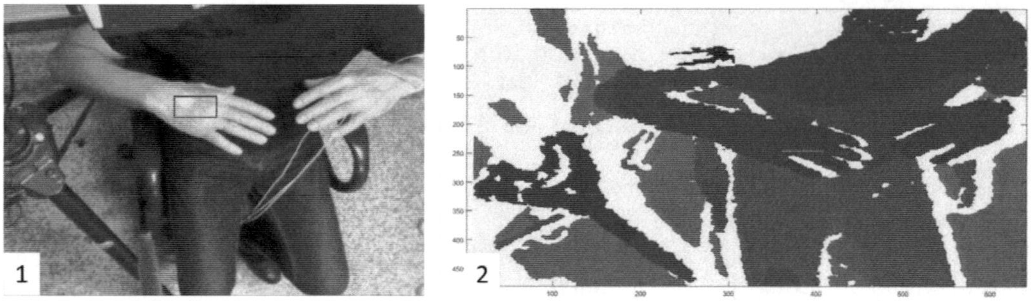

Figure 4. Images from Kinect device: 1) colour image; 2) depth map

3. Results

The results obtained from LDV and Kinect sensors (in particular, the frequency of the tremor) were compared.
The results in terms of tremor frequency, measured by means of LDV and Kinect sensor, for the 3 subjects in the different kinds of test, are reported in 'table 1'. It is possible to observe that they are comparable, with an average deviation of 0.31 Hz.
No significant differences in terms of tremor frequency can be identified between subjects with essential tremor (subject 1) and the other two (subjects 2 and 3).

Table 1. Tremor frequency measured by means of LDV and Kinect sensor in the three tested subjects during the different kinds of test

Tremor frequency [Hz]					
Subject	**Sensor**	**At rest**	**Cognitive task**	**Stretched arms**	**Wing position**
1	**LDV**	4.7	4.3	2.1	5.6
	Kinect	4.5	4.3	2.1	5.5
2	**LDV**	3.7	4.9	5.6	5.5
	Kinect	3.4	N.A.	5.2	5.6
3	**LDV**	4.4	4.6	2.3	5.6
	Kinect	4.1	4.7	1.9	4.2

4. Discussion and conclusions

The results from this preliminary study allow the authors to state that both Laser Doppler Vibrometer and Kinect device are able to detect and characterize upper limb tremor in terms of frequency, providing a suitable measurement tool towards a quantitative assessment of this condition. They are comparable in terms of performance in quantifying the tremor frequency; however, it is possible to say that, for sure, Kinect device is preferable in terms of cost (for this reason, it would be possible to use it even in home environments), while LDV requires simpler data post-processing techniques.
Moreover, the obtained results are in line with those reported in literature, both referring to rest tremor (3-7 Hz) (42) and to essential tremor (4-12 Hz) (37).

In the future, it would be interesting to further exploit the videos recorded by the Kinect camera and in particular its ability to detect anatomical landmarks, which are useful to reconstruct the trajectory of a specific movement.

5. References

1. WHO | Neurological Disorders: Public Health Challenges [Internet]. WHO. [cited 2017 Feb 23]. Available from: http://www.who.int/mental_health/publications/neurological_disorders_ph_challenges/en/
2. Schapira AHV. Neurobiology and treatment of Parkinson's disease. Trends Pharmacol Sci. 2009 Jan;30(1):41–7.
3. Grosset D, Fernandez H, Grosset K, Okun M. Parkinson's Disease: Clinican's Desk Reference. CRC Press; 2009. 179 p.
4. Schapira A. Parkinson's Disease. Oxford University Press; 2010. 131 p.
5. Brooks DJ. Parkinson's disease: Diagnosis. Parkinsonism Relat Disord. 2012 Jan;18, Supplement 1:S31–3.
6. Shulman LM, Taback RL, Rabinstein AA, Weiner WJ. Non-recognition of depression and other non-motor symptoms in Parkinson's disease. Parkinsonism Relat Disord. 2002 Jan;8(3):193–7.
7. Aarsland D, Larsen JP, Karlsen K, Lim NG, Tandberg E. Mental symptoms in Parkinson's disease are important contributors to caregiver distress. Int J Geriatr Psychiatry. 1999 Oct 1;14(10):866–74.
8. Lang AE, Obeso JA. Time to move beyond nigrostriatal dopamine deficiency in Parkinson's disease. Ann Neurol. 2004 Jun 1;55(6):761–5.
9. Plotnik M, Dagan Y, Gurevich T, Giladi N, Hausdorff JM. Effects of cognitive function on gait and dual tasking abilities in patients with Parkinson's disease suffering from motor response fluctuations. Exp Brain Res. 2011 Jan 1;208(2):169–79.
10. Garber CE, Friedman JH. Effects of fatigue on physical activity and function in patients with Parkinson's disease. Neurology. 2003 Apr 8;60(7):1119–24.
11. Antonini A, Barone P, Marconi R, Morgante L, Zappulla S, Pontieri FE, et al. The progression of non-motor symptoms in Parkinson's disease and their contribution to motor disability and quality of life. J Neurol. 2012 Dec 1;259(12):2621–31.
12. Schenkman M, Wei Zhu C, Cutson TM, Whetten-Goldstein K. Longitudinal evaluation of economic and physical impact of Parkinson's disease. Parkinsonism Relat Disord. 2001 Sep;8(1):41–50.
13. Noyes K, Liu H, Li Y, Holloway R, Dick AW. Economic burden associated with Parkinson's disease on elderly Medicare beneficiaries. Mov Disord Off J Mov Disord Soc. 2006 Mar;21(3):362–72.
14. Pringsheim T, Jette N, Frolkis A, Steeves TDL. The prevalence of Parkinson's disease: a systematic review and meta-analysis. Mov Disord Off J Mov Disord Soc. 2014 Nov;29(13):1583–90.
15. Wright Willis A, Evanoff BA, Lian M, Criswell SR, Racette BA. Geographic and Ethnic Variation in Parkinson Disease: A Population-Based Study of US Medicare Beneficiaries. Neuroepidemiology. 2010 Apr;34(3):143–51.
16. Newman EJ, Grosset KA, Grosset DG. Geographical difference in Parkinson's disease prevalence within West Scotland. Mov Disord Off J Mov Disord Soc. 2009 Feb 15;24(3):401–6.
17. Haaxma CA, Bloem BR, Borm GF, Oyen WJG, Leenders KL, Eshuis S, et al. Gender differences in Parkinson's disease. J Neurol Neurosurg Psychiatry. 2007 Aug;78(8):819–24.
18. Gillies GE, Pienaar IS, Vohra S, Qamhawi Z. Sex differences in Parkinson's disease. Front Neuroendocrinol. 2014 Aug;35(3):370–84.
19. Gelb DJ, Oliver E, Gilman S. Diagnostic criteria for Parkinson disease. Arch Neurol. 1999 Jan;56(1):33–9.

20. Jankovic J. Parkinson's disease: clinical features and diagnosis. J Neurol Neurosurg Psychiatry. 2008 Apr 1;79(4):368–76.
21. Pastorino M, Arredondo MT, Cancela J, Guillen S. Wearable sensor network for health monitoring: the case of Parkinson disease. J Phys Conf Ser. 2013;450(1):12055.
22. Synnott J, Chen L, Nugent CD, Moore G. WiiPD--an approach for the objective home assessment of Parkinson's disease. Conf Proc Annu Int Conf IEEE Eng Med Biol Soc IEEE Eng Med Biol Soc Annu Conf. 2011;2011:2388–91.
23. Rigas G, Tzallas AT, Tsipouras MG, Bougia P, Tripoliti EE, Baga D, et al. Assessment of tremor activity in the Parkinson's disease using a set of wearable sensors. IEEE Trans Inf Technol Biomed Publ IEEE Eng Med Biol Soc. 2012 May;16(3):478–87.
24. Beuter A, de Geoffroy A, Cordo P. The measurement of tremor using simple laser systems. J Neurosci Methods. 1994 Jul;53(1):47–54.
25. Norman KE, Edwards R, Beuter A. The measurement of tremor using a velocity transducer: comparison to simultaneous recordings using transducers of displacement, acceleration and muscle activity. J Neurosci Methods. 1999 Oct 15;92(1–2):41–54.
26. Salarian A, Russmann H, Wider C, Burkhard PR, Vingerhoets FJG, Aminian K. Quantification of tremor and bradykinesia in Parkinson's disease using a novel ambulatory monitoring system. IEEE Trans Biomed Eng. 2007 Feb;54(2):313–22.
27. Chang R-S, Chiu J-H, Chen F-P, Chen J-C, Yang J-L. A Parkinson's Disease Measurement System Using Laser Lines and a CMOS Image Sensor. Sensors. 2011 Jan 26;11(2):1461–75.
28. Pinotti M, Paone N, Santos FA, Tomasini EP. Carotid artery pulse wave measured by a laser vibrometer. In: Third International Conference on Vibration Measurements by Laser Techniques: Advances and Applications [Internet]. International Society for Optics and Photonics; 1998 [cited 2015 Oct 12]. p. 611–616. Available from: http://proceedings.spiedigitallibrary.org/proceeding.aspx?articleid=954179
29. Morbiducci U, Scalise L, Melis MD, Grigioni M. Optical Vibrocardiography: A Novel Tool for the Optical Monitoring of Cardiac Activity. Ann Biomed Eng. 2007 Jan 1;35(1):45–58.
30. Scalise L, Morbiducci U. Non-contact cardiac monitoring from carotid artery using optical vibrocardiography. Med Eng Phys. 2008 May;30(4):490–7.
31. Scalise L, Ercoli I, Marchionni P, Tomasini EP. Measurement of respiration rate in preterm infants by laser Doppler vibrometry. In: 2011 IEEE International Workshop on Medical Measurements and Applications Proceedings (MeMeA). 2011. p. 657–61.
32. Cosoli G, Casacanditella L, Tomasini EP, Scalise L. Evaluation of Heart Rate Variability by means of Laser Doppler Vibrometry measurements. J Phys Conf Ser. 2015;658(1):12002.
33. Cosoli G, Casacanditella L, Tomasini EP, Scalise L. The non-contact measure of the heart rate variability by laser Doppler vibrometry: comparison with electrocardiography. Meas Sci Technol. 2016;27(6):65701.
34. Benamer HTS, Patterson J, Grosset DG, Booij J, de Bruin K, van Royen E, et al. Accurate differentiation of parkinsonism and essential tremor using visual assessment of [123I]-FP-CIT SPECT imaging: The [123I]-FP-CIT study group. Mov Disord. 2000 May 1;15(3):503–10.
35. Geraghty JJ, Jankovic J, Zetusky WJ. Association between essential tremor and Parkinson's disease. Ann Neurol. 1985 Apr 1;17(4):329–33.
36. Movement Disorders - Wiley Online Library [Internet]. [cited 2017 Feb 23]. Available from: http://onlinelibrary.wiley.com/journal/10.1002/(ISSN)1531-8257
37. Louis ED. Essential tremor. Lancet Neurol. 2005 Feb;4(2):100–10.
38. GmbH P. Polytec: PDV-100 Portable Digital Vibrometer [Internet]. [cited 2016 Nov 30]. Available from: http://www.polytec.com/it/products/vibration-sensors/single-point-vibrometers/complete-systems/pdv-100-portable-digital-vibrometer/
39. Kinect. In: Wikipedia [Internet]. 2017 [cited 2017 Feb 24]. Available from: https://en.wikipedia.org/w/index.php?title=Kinect&oldid=766578983

IOP Conf. Series: Journal of Physics: Conf. Series **882** (2017) 012002 doi:10.1088/1742-6596/882/1/012002

40. Jankovic J, Schwartz KS, Ondo W. Re-emergent tremor of Parkinson's disease. J Neurol Neurosurg Psychiatry. 1999 Nov;67(5):646–50.
41. MATLAB - MathWorks [Internet]. [cited 2017 Feb 23]. Available from: https://www.mathworks.com/products/matlab.html
42. Duval C. Rest and postural tremors in patients with Parkinson's disease. Brain Res Bull. 2006 Jun 15;70(1):44–8.

IOP Conf. Series: Journal of Physics: Conf. Series **882** (2017) 012003 doi:10.1088/1742-6596/882/1/012003

Stereo particle image velocimetry set up for measurements in the wake of scaled wind turbines

Gabriele Campanardi[1], Donato Grassi[1], Alex Zanotti[1], Emmanouil M Nanos[2], Filippo Campagnolo[2], Alessandro Croce[1], Carlo L Bottasso[1,2]

[1] Politecnico di Milano, Dipartimento di Scienze e Tecnologie Aerospaziali, via La Masa 34, I-20156, Milano, Italy
[2] Technische Universität München, Lehrstuhl für Windenergie, Boltzmannstraße 15, D-85748 Garching bei München, Germany

E-mail: gabriele.campanardi@polimi.it

Abstract. Stereo particle image velocimetry measurements were carried out in the boundary layer test section of Politecnico di Milano large wind tunnel to survey the wake of a scaled wind turbine model designed and developed by Technische Universität München. The stereo PIV instrumentation was set up to survey the three velocity components on cross-flow planes at different longitudinal locations. The area of investigation covered the entire extent of the wind turbines wake that was scanned by the use of two separate traversing systems for both the laser and the cameras. Such instrumentation set up enabled to gain rapidly high quality results suitable to characterise the behaviour of the flow field in the wake of the scaled wind turbine. This would be very useful for the evaluation of the performance of wind farm control methodologies based on wake redirection and for the validation of CFD tools.

1 Introduction

In the recent years particle image velocimetry (PIV) for measurements on scaled wind turbines models was extensively used with various purposes. In particular, the use of a stereoscopic PIV set up enables to survey the wake extent at different cross-sections in order to study wake instabilities and the development of the flow in the near and the far wake of the wind turbine scaled model [1]. Moreover, this measurement technique was also useful to capture flow structures developed from the blade tip and root [2].

In the present paper, a new stereo PIV set up, purposely designed to survey the wake of scaled wind turbine models in the large wind tunnel of Politecnico di Milano (POLIMI), is described. In particular, the present set up was used to survey the three velocity components on cross-flow planes at different longitudinal locations behind a scaled wind turbine model designed and developed by Technische Universität München (TUM). The main scope of the present tests was to evaluate the capabilities of the present set up to gain rapidly high quality measurements for the characterisation of the behaviour of the flow field in the wake of the scaled wind turbine model. This result was considered very interesting for several goals. The former is to study the performance of wind farm control methodologies developed by TUM based on wake redirection. Indeed, this strategy is currently one of the most promising wind farm control methodologies which may lead to improved power capture and reduced structural loading. The current research on wake redirection is very active and covers high-fidelity numerical simulations [3] in addition to scaled experiments in the wind tunnel [4]. Thus the present test results provided also an interesting data base useful for the validation of CFD tools purposely developed for these studies.

2 Experimental Set Up

2.1 The wind turbine model

The scaled wind turbine model used in the present tests has a 1.1 m rotor diameter and was designed and developed by TUM. The layout of the wind turbine model is presented in Fig. 1.
The model is characterized by a realistic aerodynamic performance, both at the airfoil and rotor levels, and its wake is in good accordance in shape, deficit and recovery with those of full-scale machines. Moreover, the model features active individual pitch, torque and yaw control that, together with a comprehensive onboard sensorization of the machine (including measures of shaft and tower loads), enables the testing of modern control strategies. Indeed, the model is controlled by a `M1 Bachmann` module that executes, similarly to what is done on real scale wind turbines, hard-real-time control laws similar to the ones described in Bossanyi [5].

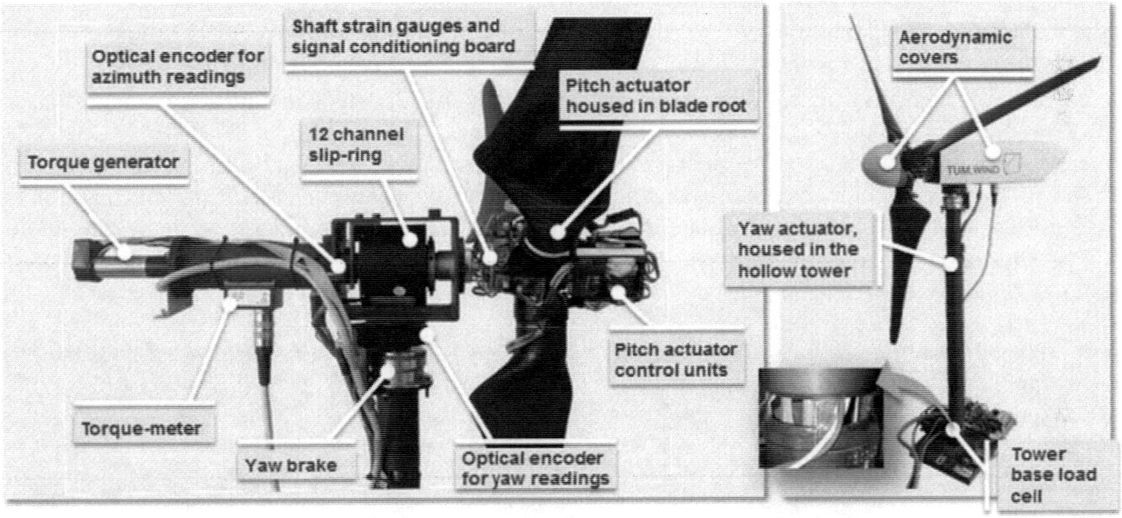

Figure 1: Layout of the scaled wind turbine model designed and developed by Technische Universität München

2.2 The wind tunnel

The Politecnico di Milano large wind tunnel presents a closed-circuit configuration, arranged in a vertical layout with two test sections in the loop (see the layout in Fig. 1). This facility has two different test chambers: a 4 m x 3.84 m aeronautical test section located downstream a settling chamber and a contraction and characterized by a very good flow quality (freestream turbulence level less than 0.1%) and a 13.84m x 3.84m wind engineering test chamber located in the return duct characterized by a stable floor boundary layer thickness in the order of 0.4 m and a mean turbulence level in the order of 2%. The experimental study of scaled wind turbine performance was carried out in the latter test chamber (boundary layer test section). Indeed, due to the larger dimensions, this choice enables to reduce to the minimum the blockage and the interference with the boundaries so allowing tests also for wind farm configurations. For further details about the wind tunnel facility please see Gibertini *et al.* [6].

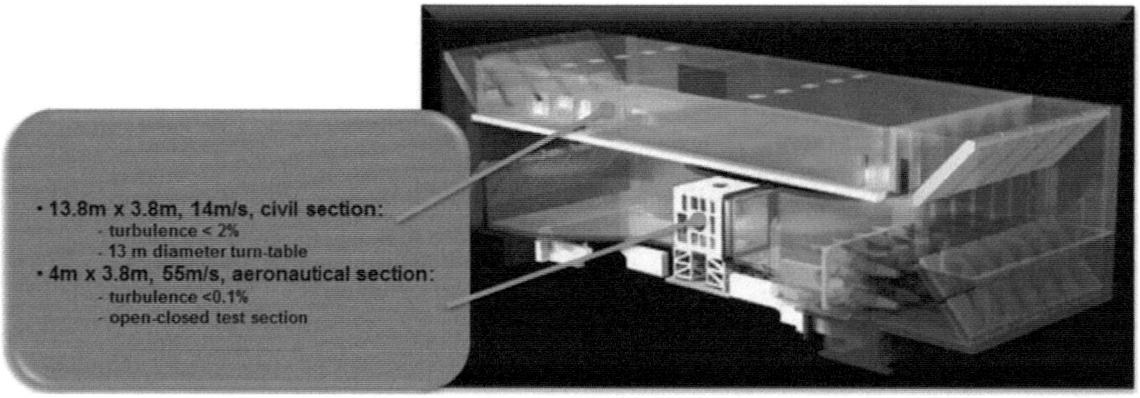

Figure 2: Layout of the Politecnico di Milano large wind tunnel

2.3 *Stereo particle image velocimetry set up*

The PIV instrumentation was set up to survey the three velocity components (u,v,w) on cross-flow $(x\text{-}z)$ planes at different longitudinal locations (along x-axis) behind the wind turbine model rotor disk. In particular, in the here described test campaign two PIV measurement planes were considered, respectively located at 0.56 D and 6 D past the rotor disk.
The stereo PIV instrumentation comprised a Litron NANO-L-200-15 Nd:Yag double pulsed laser with a 200 mJ output energy and a wavelength of 532 nm, and two Imperx ICL-B1921M CCD cameras with a 12 bit, 1952×1112 pixel array [7]. Each camera was equipped with a Nikkor 50 mm lens and tilting type lens for correct focusing of the measurement window. The tilting lens mountings were adjusted in order to achieve the Scheimpflug condition. The camera separation angle was set to 40°. The layout of the instrumentation set up in the boundary layer test section of the POLIMI wind tunnel is shown in Fig. 3.

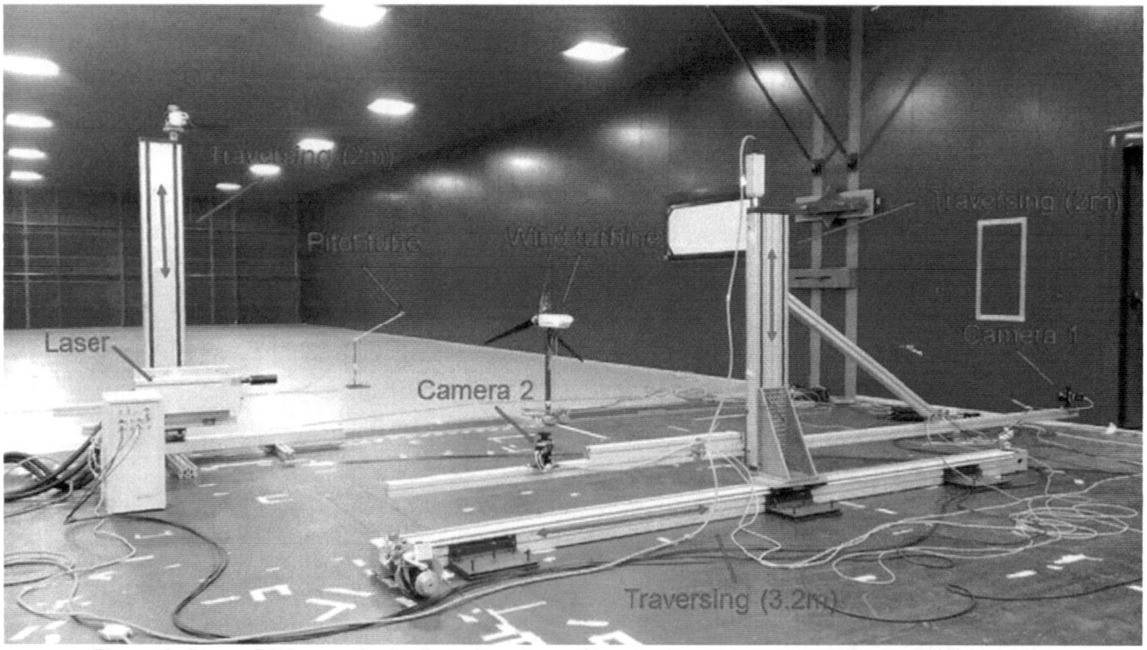

Figure 3: Stereo PIV set up in the Boundary Layer Test Section (14m x 4m) of POLIMI wind tunnel

IOP Conf. Series: Journal of Physics: Conf. Series **882** (2017) 012003 doi:10.1088/1742-6596/882/1/012003

The measurement planes covered the entire extent of the wind turbines wake and in order to achieve a higher resolution of the velocity survey, the entire measurement area was composed by several windows with a certain overlapping band between them. In particular, the area of investigation on the measurement plane at 0.56 D was 1586 mm × 1032 mm and was composed by twelve measurement windows (arranged as four rows and three columns, see the layout in Fig. 4a) covering 786 mm × 427 mm each. On the other hand, the area of investigation on the measurement plane at 6 D was 1786 mm × 830 mm and was composed by fifteen measurement windows (arranged as three rows and five columns, see the layout in Fig. 4b) with the same dimensions (786 mm × 427 mm each).

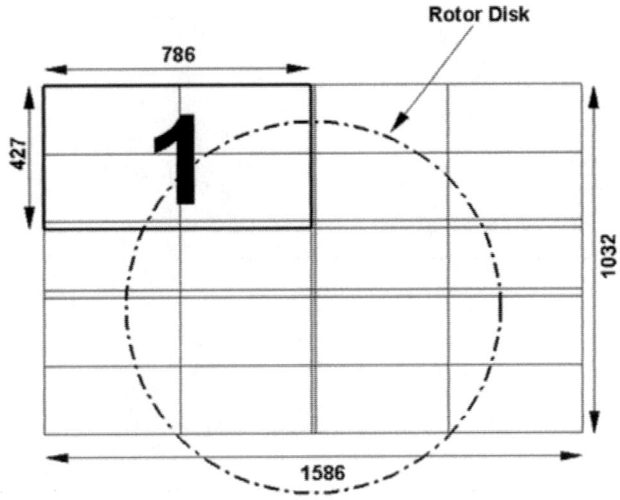

(a) PIV plane at 0.56 D past the rotor disk (dimensions in mm)

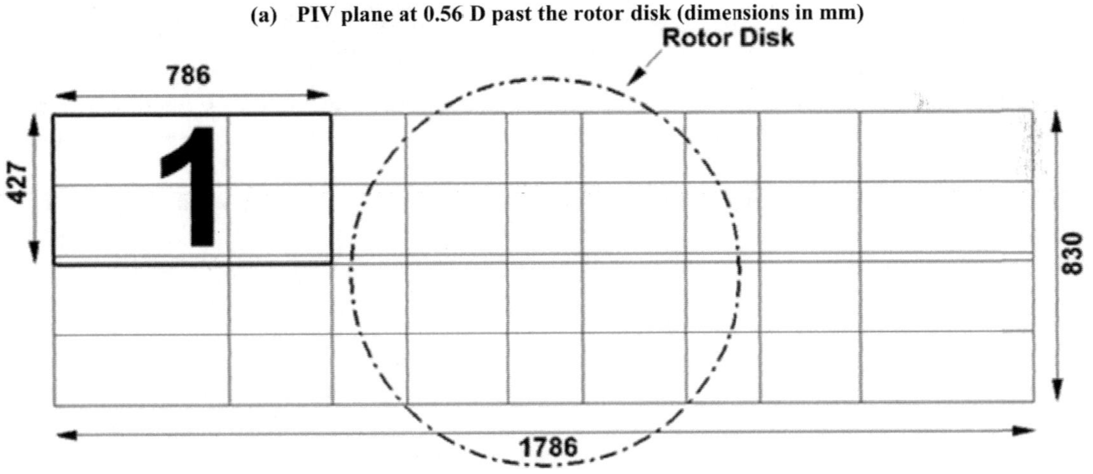

(b) PIV plane at 6 D past the rotor disk (dimensions in mm)

Figure 4: Layout of the PIV measurement planes, the dimensions of the measurement windows are shown for the upper-left window denoted with bold 1)

A rapid survey of the entire measurement area was obtained by the use of two separate traversing systems for both the laser and the cameras. In particular, as can be observed from the layout in Fig. 3, the two cameras were connected by a metallic arm and mounted to a 3.2m

IOP Conf. Series: Journal of Physics: Conf. Series **882** (2017) 012003 doi:10.1088/1742-6596/882/1/012003

x 2m double axis traversing system to move the cameras along the *y* and *z*-axis. The laser was also mounted on a traversing system to move the laser sheet along *z*-axis simultaneously with the measurement window. This solution enables to scan rapidly the large measurement area allowing a proper lightning of each window to ensure high quality results on the whole area.
The synchronization of the two laser pulses with the image pair exposure was controlled by a 6-channels Quantum Composer QC9618 pulse generator. The digital images were captured using a GigaEthernet EBus connection. The calibration of the stereo PIV system was performed using a unique calibration grid covering the entire measurement area positioned in correspondence of the measurements planes (see Fig. 5). Thus, the calibration was performed for each measurement window moving the cameras with the traversing system.

Figure 5: Calibration of the stereo PIV system

The test chamber was fulfilled with the seeding using two particle generators with Laskin atomizer nozzles (PIVpart30 by PIVTEC) located downstream the wind turbine model. The particles consisted of small DEHS oil droplets with a diameter in the range of 1-2 μm.
The image pair analysis was carried out using PIVview 3C software [8], developed by PIVTEC in close cooperation with the PIV-Groups of the German Aerospace Center (DLR) in Gottingen and Koln. The image pairs were post-processed using the multigrid interrogation algorithm [9] with a final interrogation window of 96 pixels × 96 pixel. In particular, two hundreds image pairs were acquired for each measurement window. This choice was considered a fair compromise to obtain a reliable ensemble average for the velocity field and a contained run time taking into account the large area to be scanned during the test. The accuracy of the present PIV measurements can be estimated considering a maximum displacement error of 0.1 pixels [7]. Thus, taking into account the employed pulse-separation time and the optical magnification [10], the maximum out-of-plane velocity components error is about 0.1 m/s.

3 Test Results

The PIV surveys were performed with a rated rotor rotational speed of 850 RPM and a wind tunnel freestream velocity (U_∞) of 5.95 m/s. In the present paper sample results of the test campaign obtained with the wind turbine model operating without wake redirection control are presented. In particular, Figs. 6 and 7 show the contours of the ensemble average of the streamwise velocity component (*u*) respectively for the measurement planes at 0.56 D and 6 D past the rotor disk. In particular, on the plane at 0.56 D past the rotor disk measurements are missing in two areas on the left and on the right of the rotor disk (depicted in white on Fig. 6), where, due to the close proximity of the measuring plane with the wind turbine, part of the nacelle was visible on the background of the image pairs, leading to bad correlation between the images.

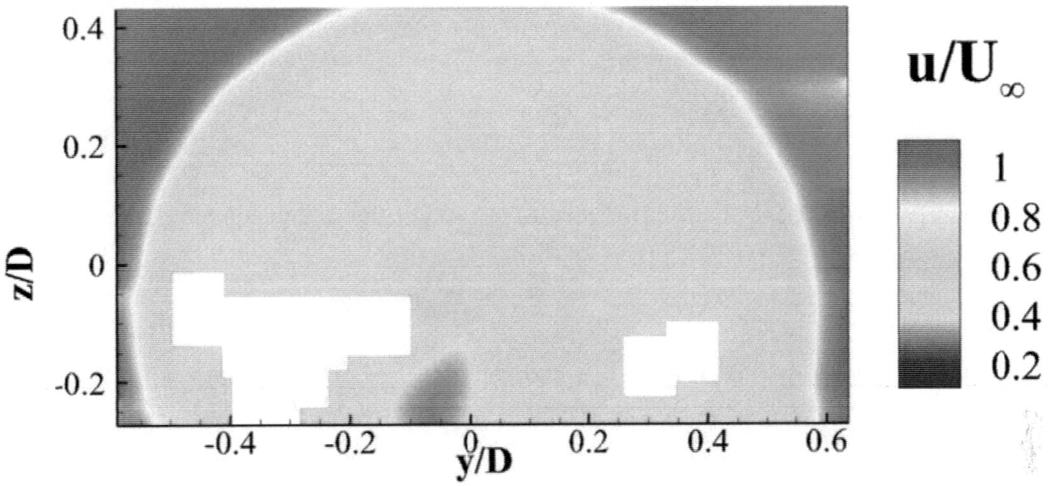

Figure 6: Normalized averaged streamwise velocity component on the yz-plane at 0.56D past the rotor disk

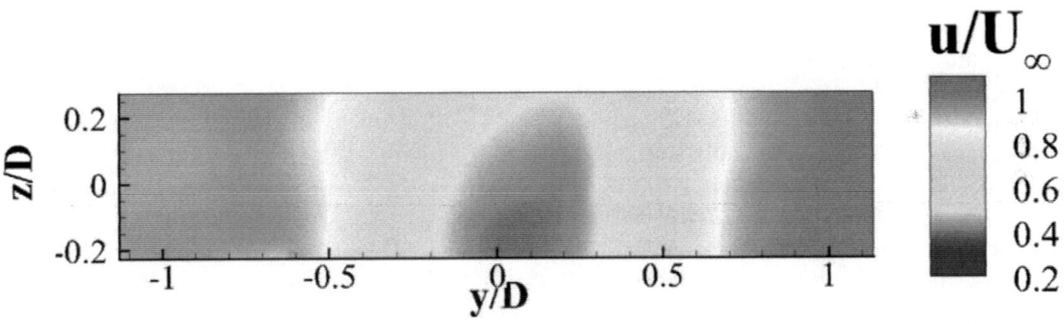

Figure 7: Normalized averaged streamwise velocity component on the yz-plane at 6D past the rotor disk

The overall flow fields shows very high quality results obtained for both the measurement planes. The sample results presented in this paper show clearly the streamwise evolution of the wind turbine wake. In particular, on the measurement plane at 0.56 D past the rotor disk, an apparent gradient of the streamwise velocity is visible close to the blade tip passage region, while inside the rotor disk wake the velocity field is quite uniform except in the region of the nacelle wake (see Fig. 6).
On the other hand, an apparent increase of the nacelle wake is appreciable from the measurement results obtained on the plane at 6 D past the rotor disk (see Fig. 7).

IOP Conf. Series: Journal of Physics: Conf. Series **882** (2017) 012003 doi:10.1088/1742-6596/882/1/012003

4 Conclusions

A new stereo PIV set up was developed and tested at the large wind tunnel of POLIMI for the application of wake surveys on scaled wind turbine models. The set up enables to rapidly scan the entire extent of the wind turbine wake by the use of two separated traversing systems for both the laser and the cameras. The wind tunnel campaign provided high quality results that enabled to investigate the behaviour of the flow field in the wake of the scaled wind turbine useful for the evaluation of the performance of wind farm control methodologies based on wake redirection developed by TUM.
Moreover, the present experimental data base was used for the validation of CFD tools purposely developed by TUM for these studies.

Acknowledgements

This project was partly funded by the EU Horizon 2020 research and innovation programme under the Marie Sklodowska-Curie grant agreement No. 642108.

References

[1] Naumov I V, Mikkelsen R F , Okulov V L and Sørensen J N 2014 PIV and LDA measurements of the wake behind a wind turbine model *The Science of Making Torque from Wind 2014 (TORQUE 2014) IOP Publishing Journal of Physics: Conference Serie* vol 524

[2] Meyer K E, Naumov I V, Kabardin I, Mikkelsen R and Nørkær Sørensen J 2013 PIV in a model wind turbine rotor wake *10th International Symposium on Particle Image Velocimetry* (PIV13) Delft, The Netherlands, July 1-3, 2013

[3] Fleming P, Gebraad P, Lee S, van Wingerden J W, Johnson K, Churchfield M, Michalakes J, Spalart P and Moriarty P 2014 Evaluating techniques for redirecting turbine wakes using sowfa *Renewable Energy* vol 70 pp 211-218

[4] Campagnolo F, Petrovi_c V, Bottasso C L and Croce A 2016 Wind tunnel testing of wake control strategies *Proceedings of the American Control Conference* pp 513-518

[5] Bossanyi E 2000 The design of closed loop controllers for wind turbines *Wind Energy* vol 3 pp 149-163

[6] Gibertini G, Gasparini L, and Zasso, A 1997 Aerodynamic design of a civil-aeronautical low speed large wind tunnel *AGARD 79th Fluid Dynamics Panel Symposium* Moscow, Russia, 30 Sept.-3 Oct. 1996, pp 6.1-6.10

[7] Zanotti A, Ermacora M, Campanardi G and Gibertini G 2014 Stereo Particle Image Velocimetry Measurements of Perpendicular Blade-Vortex Interaction over an Oscillating Airfoil *Experiments in Fluids* vol 55 (9) pp 1-13

[8] PIVview 2C/3C, User Manual, PIVTEC, www.pivtec.com

[9] Raffel M, Willert C, Wereley S and Kompenhans J 2007 *Particle Image Velocimetry - A Practical Guide* Springer Verlag, Berlin

[10] De Gregorio F, Pengel K and Kindler K 2012 A comprehensive PIV measurement campaign on a fully equipped helicopter model *Experiments in Fluids* vol 53 pp 37-49

IOP Conf. Series: Journal of Physics: Conf. Series **882** (2017) 012004 doi:10.1088/1742-6596/882/1/012004

Advanced Imaging Techniques for Multiphase Flows Analysis

A. Amoresano, G. Langella, M. Di Santo, P. Iodice
University of Naples "Federico II" - Via Claudio 21 80125 Naples

amedeo.amoresano@unina.it

Abstract. Advanced numerical techniques, such as fuzzy logic and neural networks have been applied in this work to digital images acquired on two applications, a centrifugal pump and a stationary spray in order to define, in a stochastic way, the gas-liquid interface evolution. Starting from the numeric matrix representing the image it is possible to characterize geometrical parameters and the time evolution of the jet. The algorithm used works with the fuzzy logic concept to binarize the chromatist of the pixels, depending them, by using the difference of the light scattering for the gas and the liquid phase.. Starting from a primary fixed threshold, the applied technique, can select the 'gas' pixel from the 'liquid' pixel and so it is possible define the first most probably boundary lines of the spray. Acquiring continuously the images, fixing a frame rate, a most fine threshold can be select and, at the limit, the most probably geometrical parameters of the jet can be detected.

1. Introduction

The study of the phenomena of two phase systems is an important topic in the academic research [1]. It regards a lot of application in several fields like Mechanical Industry, Biology [3], Chemical [4] Aerospace [5] etc...The development of objective detection criteria for multiphase flow regimes has been the central research subject among many researchers during the recent past. One of the first research works on this subject is the one by Hubbard and Dukler (1966) [6] in which different flow regimes were characterized through their spectral signatures of temporal pressure signals. An interesting short review of the technique to analyses the dynamic of a two phase system by using imaging technique has been reported by has been done by F. L. Kleine et. Al in the introduction of their paper. [7]Many are the approaches used to analyses how two systems of different phase interact each other. In the last 10 years, with the coming of new sensors, the investigation of the two phase phenomena by images has grown considerably. The acquisition of data by using images technique has be done more carefully, because the acquired image couldn't represent the expected results. The acquisition can be affect by several mistake and uncertain depending on several other parameters. It is possible to sentence that digital images are subject to a wide variety of distortions during acquisition, processing, compression, storage, transmission and reproduction, any of which may result in a degradation of visual quality [8]. The quality image affect strongly the experimental results expected in special way if the measurements need several devices lying between the lens and the measuring region and if it the aim is to define the boundary lines between two or more phases. In this case can be useful to work with particular system that increase the contrast between the two fluids. For example the determination of the individual droplet diameter during the intake and the compression strokes was based on the intensity of the light scattered from the droplet which was subjected to the monochromatic irradiation of He-Ne laser with uniform intensity profile [9][10].

The determination of the boundary layers able to define the correct limit between the gas and the liquid phase can be complex and couldn't give the correct information especially if the interested phenomena developing themselves in very short time[11]. Due to all the cited parameters the correct pattern or other types of information extractable from the images can be treated with more sophisticated technique like statistical approach [12] and by using the wavelet transform to locate sub images with details coefficient able to define important particulars not always visible [13]. The data extracted from the recorded images hold several information can be used in different way. In this paper the data are considered like solutions of the Lagrangian equations [14] and define the work points of the dynamic system considered, i.e. the diagram of the phases. Because of the data representing the phenomena are images, their extraction have to be taken from them carefully. In other words the images and, consequently the associated matrix, can be treated with algorithms able to extract the wanted parameters. In this paper different measurements are presented like a two phase flow in a centrifugal pump where, by applying fuzzy method, bubbles, cluster of bubble and their trajectories have been calculated [15]. Measurements in a stationary spray where the lagrangian variable characterizing the spray is the half cone angle obtained from the images applying the Big Data approach [16].

2. Experimental set up

Two experimental different setup one for each experiment are presented. All the three cases use the same camera to record the images.

2.1. Centrifugal pump setup

Experimental analysis that follows numerical calculations has pointed out two-phase flow behavior by means of optical techniques. In Figure 1 a schematic layout experimental system has shown of the. A CCD image acquisition device has been used with high rate of frame per second. In order to visualize the two phase flow rate, the aspiration duct and the external flange of the impeller are transparent (fig 2). In this way it is possible to observe the behavior of the two phase flow outside and inside the impeller. The flow is visualized in the transparent duct (Fig.2) to study the flow field characteristics in the stationary case to understand the distribution of the gas phase inside the impeller. For the recording of the frames, a high speed CCD camera is used. The dissipation phenomena are quantified by the electrical balanced motor. The amplitude of the rotation around the zero value of the rotation axis is strongly related with the quantitative presence and permanence of the gas phase inside the impeller

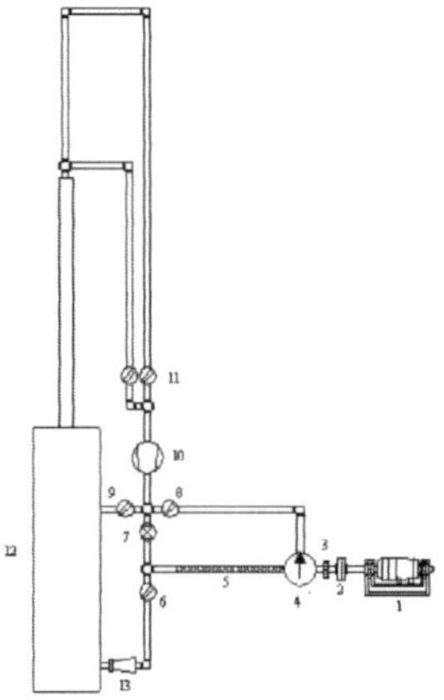

1 Triphase electrical balance motor
2 Double Hooke's joint
3 Magnetic Pick - up
4Centrifugal pump
5 Transparent Duct
6 Ball valve
7 By - Pass Valve
8 Lamination valve
9 Ball valve
10 Aero meter
11 Ball valve
12 water tank
13 water flowrate

Figure 1. Layout of the experimental setup

Figure 2 Pump working with two phase flow. Particular of the transparent flange

2.2. A Stationary spray experimental setup

The atomizer is set up on a vertical test (Fig.3) stand which allows the atomizer to translate both in vertical and in horizontal directions with steps of 1 mm, allowing to measure at different points in the vertical plane. A laser source (λ = 514.5 nm) generates a laser line that becomes a laser beam due to a divergent lens intercepting the spray to a centerline of his projection on the axial plane [17]. The laser sheet intercepts the spray exactly in the centerline and produces a planar image of the spray like shown in fig 4.

The laser beam induces the light scattering phenomenon when in contact with the liquid and the formed image (Fig. 4) can be acquired by using an optical device. In this case has been used a fastcam Photron Fastcam PCI. It is placed with its optical axis with an angle of 90° with respect to the laser sheet. The image sequences are recorded and stored in raw or bitmap format to acquire all the information.

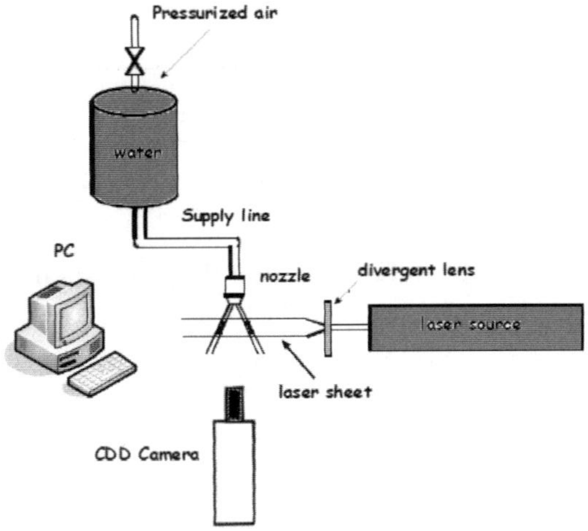

Figure 3- Experimental setup of data acquisition of stationary spray

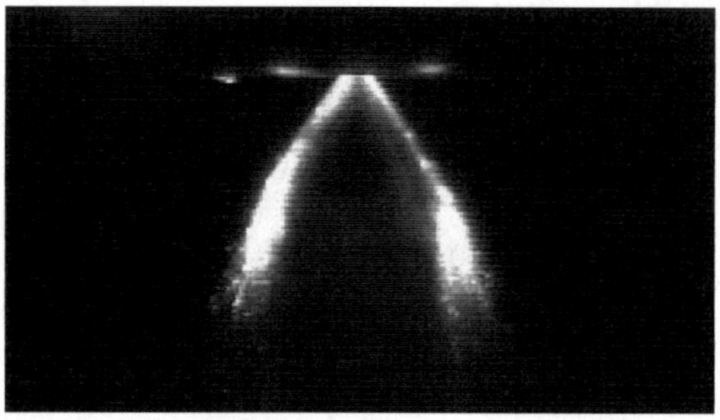

Figure 4- Planar image generated by a laser beam with a hollow cone spray

3. Data processing by fuzzy logic approach

In this paper, we had neither target images nor information about the final result of segmentation and classification, so we went towards the realization of a Multilayer Perceptron (MLP) not supervised but self-organized. Such multilayer is structurally and functionally different from a traditional MLP. Unlike supervised learning networks [18-21], in which the operator selects the examples for the training by assigning both the input arrays and output ones, the not supervised learning networks receive only arrays in the input stage. The final configuration of the weights of individual neurons allows to subdivide the elements supplied to the input into "similarity" groups which represent a classification. The Self Organizing Maps, therefore, can be considered as classifiers whose classes are arranged on a two-dimensional grid. Thanks to this pattern the Self-Organizing Maps keeps the topology of the input space. This means that if two inputs are neighbors, they are grouped in the same class or in closer classes on the two-dimensional grid. Then, the map tries to represent all the available observations by using a restricted set of arrays. Such arrays are ordered on a two-dimensional grid in such a way that similar arrays are disposed closer to each other while, conversely, the different ones are disposed far from each other. The greater is the difference larger will result the distance on the map.

The node output of a level (Fig. 5) is transmitted to the nodes of another layer by "links" which can amplify or attenuate or suppress these outputs through "weighted factors." Except for the nodes belonging to the first layer (input), the overall input to each node is the sum of the weighted outputs of nodes belonging to the previous level.

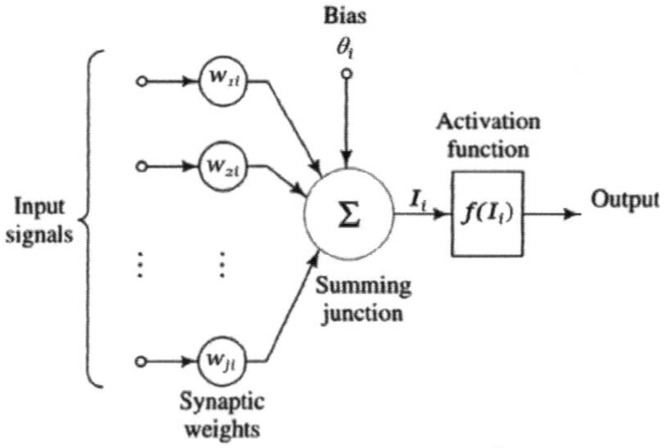

Figure 5. Information flow through the ith network node.

Each node is activated depending on input and activation function related to the node itself.

Each node is enabled depending on the input and on the activation function of node itself. The total input (I_i) to the ith unit (node) of each layer is given by

$$I_i = \sum_j w_{ij} o_j \qquad (1)$$

o_j being the output of the jth neuron of the previous layer and w_{ij} the connection, weighed between the ith node of a layer and the jth one of previous layer.
Each node output is evaluated as:

$$o_j = f(I_i) \qquad (2)$$

wheref is the activation function, sigmoidal in our case:

$$f(x) = \frac{1}{1 + e^{-(x-\theta)/\theta_0}} \qquad (3)$$

Using the sigmoidal function to calculate the output of a network rather than a step function allows the classification of complex objects rather than divide the input pattern with simple lines.

3.1. Network structure

A three layers network was developed in the present work (Fig. 6). Each level (*layer*) is made of MxN neurons (MxN imagine). Each neuron corresponds to a single pixel. Between the input and the output layer there must be a number of layers greater than zero. Neurons belonging to a same layer do not have connections between each other. Each neuron of a layer is connected to the corresponding neuron of the previous layer as well as to the neurons belonging to the chosen neighborhood (N^d). In this way, each neuron on the layer *i* (*i* >*1*) will have $|N^d| + 1$ (dove $|N^d|$ is the number of pixels in N^d) links of the *(i - 1)th* layer.
For the case N^1 a neuron has 5 links, while for the N^2 it has 9 links, and so on. For border nodes the number of links may be less than$|N^d| + 1$. Each neuron belonging to the output layer is also connected to the corresponding neuron on the input layer.

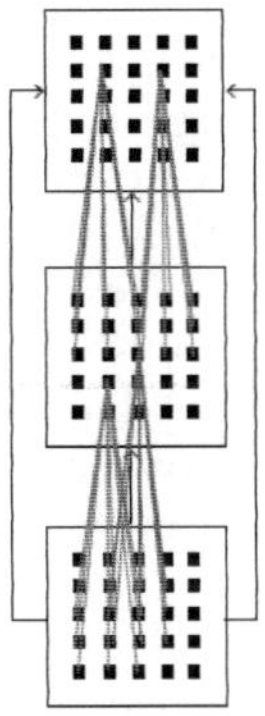

Figure 6. Scheme of a self-organizing multilayer neural network.

The proposed structure differs from a traditional *MLP* by the following points:

- link distribution;
- connection (feedback) from output layer to the input one.

3.2. Learning

The guiding principle allowing the network to learn is to let the network itself to learn from its mistakes. The input to a neuron belonging to the input layer is a real number in the interval [0,1] proportional to the value of the gray level of the corresponding pixel. Since the interest was in the elimination of noise and in the extraction of spatially compact regions, all weights were initially set

IOP Conf. Series: Journal of Physics: Conf. Series **882** (2017) 012004 doi:10.1088/1742-6596/882/1/012004

equal to 1. No external bias is imposed on weights. In this case random assignment of the weights could act as a pseudo - noise, altering accordingly the compactness of the regions extracted. Since all weights are set equal to 1, initially the total input for each node lives in $[0,n_e]$ where n_e is the number of links of each neuron. From here, we can also set the threshold value of θ (transport function for input / output eq. (3)) by

$$\theta = \frac{n_e}{2} \tag{4}$$

half total of the input rang
The input value I_i for each neuron of the ith layer (except the input layer) is calculated by Eq. (1). The state of the neurons in this layer is calculated through the transport function $f(x)$, eq. (3). These outputs are used as inputs to the next layer. The image is provided by MxN pixels; each pixel has a value in the range [0,1]. Hence the fuzzy aspect of the problem: each pixel is an element of the fuzzy-set more or less clear or more or less dark. The support of this fuzzy-set is equal to the number of neurons of the output.
The degree of blur (fuzziness) of this set, as a whole, can be considered the error as well as the instability of the whole system since this will reflect the deviation from the "state" of the network desired. In this way there are not target a-priori as output values. We can take the blur value to measure the error of the system and back- propagate it in order to "tune" the weights, reducing it with time (epochs), at least down to zero. Therefore, the measure of the error E, can be adopted as a function capable of measuring the degree of blur, that is to say

$$E = g(I) \tag{5}$$

where I is the blur level of the fuzzy set (index of fuzziness).
After adopting the weights, the output of the neurons belonging to the output layer is "passed" to the corresponding neurons of the input layer. The second step goes on taking these as input. The iteration (update of weights) is continuous until the entire system stabilizes, that is to say that the error (measure of blur) becomes small. In this situation most of the neurons belonging to the output layer will assume the values "0" or "1". Therefore, neurons with output value "0" will constitute a group while those having "1" will constitute another one. In this way the final image should present only compact regions in well-defined edges. In the cases where there is a no a-priori knowledge of the image (clustering problems) such a system is very instable.
In the case of a fuzzy set, we know that the index of indeterminacy (or blur) is minimum when the membership value of all the elements is "0" or "1" and is maximum when all the values are 0.5. Consequently, a neural network is stabilized when all the values of the output layer are "0" or "1" and is maximum (network unstable) when all values are equal to 0.5.

4. Results
The first step to have a good response applying the technique above described is to record a large number of frames. In this way the fuzzy logic and the related statistical approach give results strongly reliable. It want to emphasize that the found contour between gas and liquid phase are not the real contours but the most probably contours.

4.1. Centrifugal Pump
By using the CCD camera Photron Fastcam the sequence of the behaviour of air bubbles inside the impeller has been carried out. A typical sequence of the air bubbles behaviour has been rebuilt and it is shown in Fig.7

IOP Conf. Series: Journal of Physics: Conf. Series **882** (2017) 012004 doi:10.1088/1742-6596/882/1/012004

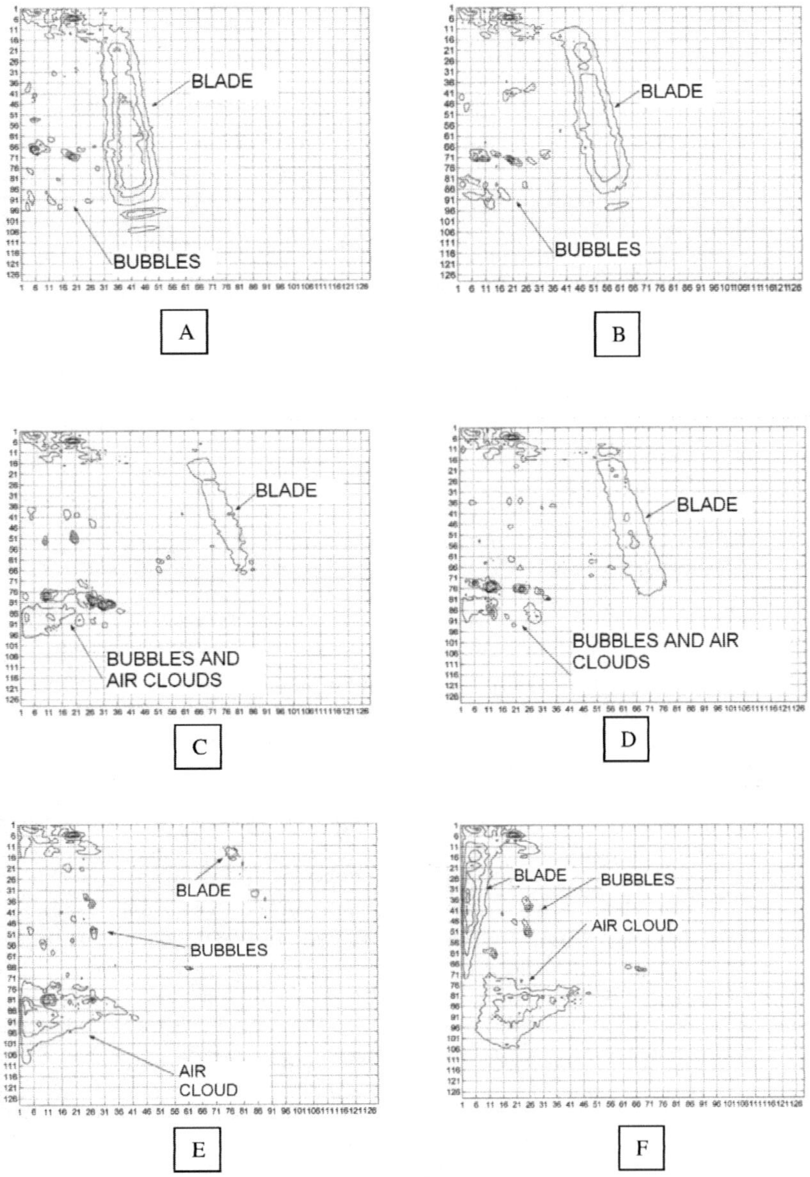

Fig. 7 a,b,c,d,e,f. Trajectory and behavior of air bubble inside a centrifugal pump

Particles that start from the low-pressure blade, or near this region, collide with the blade, slide on its surface, and go to the upper zone in the outlet zone of the rotor. The particles that crowd this region will change the blade profile and also the components of the velocity. Furthermore, experimental analysis showed that air bubbles move inside the inlet duct on the upper side of the blade and crowd near the inlet flange. This is a very important wake point. In fact, if the air flow rate is relatively high in this region the bubbles can coalesce and for bubbles that the impeller is not able to break in to smaller sizes. It is well documented that bubbles with diameter in the range [0.1-0.2] mm pass easily from the inlet region to the outlet region following, with a good approximation, the water stream lines inside the impeller. Bubbles with $0.3 < d < 0.5$mm can detach from the coalescence region and move inside the blade to the blade duct from the high pressure region to low pressure region until the impact with the blade. These bubbles inside the impeller are governed by the balance of the pressure gradient, momentum, and drag force and can find the equilibrium point inside the impeller. Growing the range of diameter of the bubbles $0.6 < d < 0.8$–1mm, the equilibrium point moves to the eye of the pump and the bubbles can coalesce with the bubbles near the inlet flange. These conditions can also cause the total block of the water flow rate. By contour extraction software it is possible to evaluate the extension of the gas phase with respect to rotor vane, as reported in Figure 12.The cited sequence, from (a) to (f), has been acquired at 1 kHz and so the behavior of bubbles can be investigated duct by duct, from bubble motion to coalescence.

In Figure 7 (a), by contour analysis software, small bubbles starting from the inner radius, according to numerical results, move towards the blade side. In Figure 7(b) all bubbles coming from the rotor inlet, increasing radial coordinate, begin to coalesce, in small clusters. In figures. 7 (c), 9 (d), 9(e), and 9(f), coalescence goes on, generating a gas volume which keeps on invading the whole van. Energy transfer, due to Euler equation, is different between gas and liquid phases. If the coalescence phenomenon goes on, the percentage of gas will be more significant of the liquid phase and momentum variation transferred will be as small as liquid flow which rate decreases to zero. With such analysis a correlation between rotor angular speed, water flow rate, and gas phase extension can be carried out, identifying the limit of the work condition.

4.2. Stationary Spray

As demonstrated in previous research works [21], the single half-angle cone allows to characterize most of the spray. On the base of these results, and considering the difficulty to study both the "moustaches" Fig 4., due to the complexity of the whole system and process that would have led to incomparable values, we decided to analyze only the right spray side, which is the more fare from the laser source, focusing on the internal angle. In fact, the internal angle is much more interested by turbulence compared to the outer angle, due to the complex fluid-dynamics phenomena that take place around the atomizer axis both inside and outside, which lead to irregularities on the internal spray contours, as shown in fig.10. These irregularities influence a lot the spray evolution and need to be understood more. Therefore, to better catch the effective value of the angle, avoiding these irregularities, the AOI (Area of Interest) has been focused only on the first about two millimeter just outside the atomizer instead of the whole illuminated spray area.

The half-cone angle has been defined as the one formed by the spray axis and the segment between the hypothetical nozzle exit, considered as a point, and the middle point of the best fit curve that linearizes the most likely internal spray edge, as shown in Fig.11. The axis can be considered perfectly vertical and central in the image if the system is well regulated. The analysis of the image foresees to individuate the position of the points that belong to the most likely internal spray edge, defining an algorithm based on the brightness gradient, which is related to the grey value of the 8 bit image. As is well known, Matlab assigns to each image a (i,j) pixel matrix, where i is the row index and j the column one. These values depend on the image brightness ranging from 0 to 255; the lowest ones are dark pixels, while the highest are light pixels.

The image has elaborated by using neural network previous described, pixels outside the spray contours are usually black, then dividing the image in two equal parts at the nozzle axis, considering the right side only, where is visualized the right spray "moustache", and choosing a low gradient threshold (about 5 points between the j pixel and j+1 one) it is possible, for each row, starting from the

IOP Conf. Series: Journal of Physics: Conf. Series **882** (2017) 012004 doi:10.1088/1742-6596/882/1/012004

first column j=1, to individuate every pixel that meet the threshold criteria. The first of these is the pixel which most likely correspond to the internal edge, while the last one most likely correspond to the outer edge instead, as shown in Fig. 10.

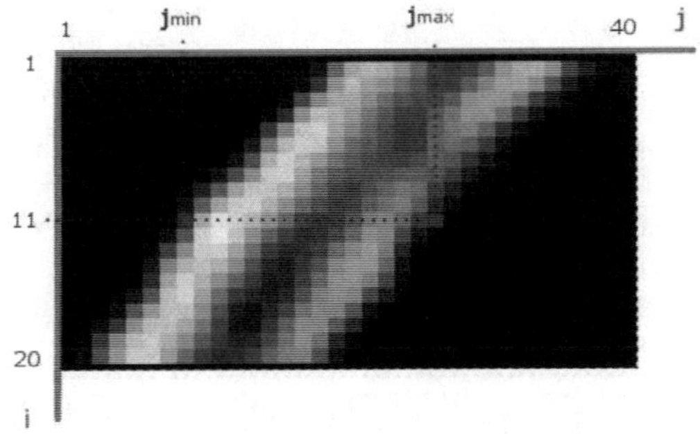

Fig. 10 Boundary region of the spray

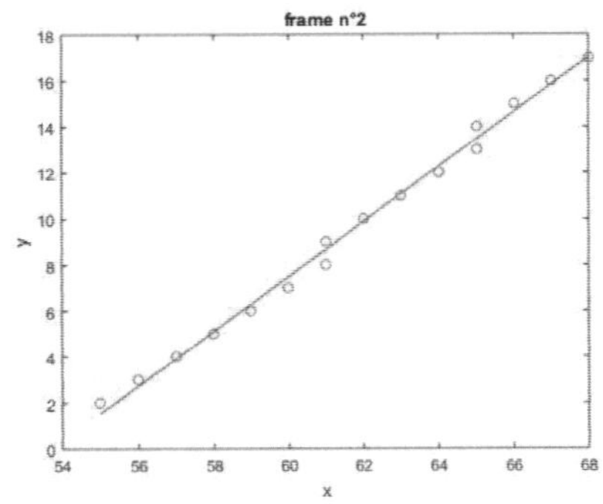

Fig.11 Best fit curve of the frame n°2

Starting from the elaborated images it is possible to calculate with a good confidence the half angle, or better the alfa angles of each fuzzy image and to define its variation band as reported in Fig. 12

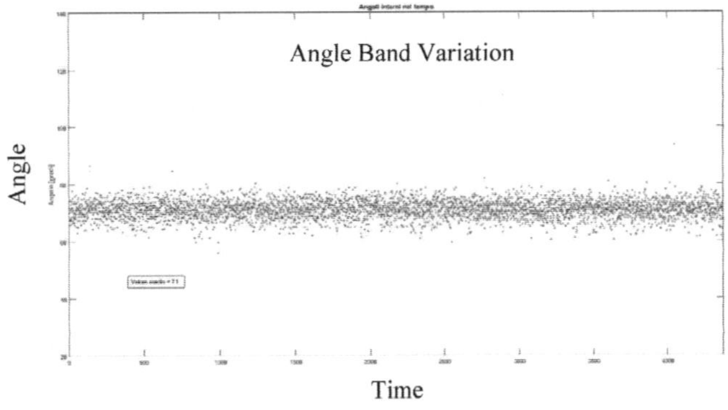

Fig.12 Angle band variation

4. Conclusion

In this paper the methodology to obtain information about a two phase fluid have been discussed. Have been presented two experiments, one related to the air bubbles moving inside an impeller pumping water and the second related to a stationary spray atomizing a liquid fluid. Both the experiments have been analysed by using the images to define their characteristics. To define better for the first the bubble's trajectory and for the spray a characteristic parameter like the half cone angle measures in the first 14mm starting for the tip an innovative approach, based on the use of a Fuzzy Logic coupled to a Neural Network, has been applied for describing their fluid dynamic and morphological behaviour..

The Fuzzy Logic coupled to a Neural Network allowed to objectively define both the interface (boundary layer) of the gas liquid interface and of the liquid-gas jet as a function of a threshold function. This approach, based on analysis of quasi- statistical data (Fuzzy Logic, Multivariate analysis, neural networks), is designed to extract more information from the images obtained from the high definition camera. In fact, assumed that digital images are two-dimensional arrays acquired over time, they define a 3D matrix and therefore it is possible to characterize the space-time evolution of areas at high-density present in liquid.

The digital image processing through the Neuro - Fuzzy Network opens up research on fuel injection systems to new scenarios. A careful investigation of the threshold values to be used in the neural network will lead to the realization of binarized images where you can also distinguish the finely atomized zone from the evaporated one and to well define the boundary layer between gas and liquid phase. In this way it can be a valuable diagnostic tool useful to understand in the centrifugal pump the vacuum fraction and the energy loss pumping a two phase fluid and in a stationary spray how maximize the efficiency of combustion.

References

[1] R. A. S. Brown, G. W. Govier "High-speed photography in the study of two-phase flow" The Canadian Journal of Chemical Engineering Volume 39, Issue 4 August 1961 Pages 159–164

[2] Dongwhi Choi, Donghyeon Lee & Dong Sung Ki "A Simple Approach to Characterize Gas-Aqueous Liquid Two-phase Flow Configuration Based on Discrete Solid-Liquid Contact Electrification" Scientific Reports 5, Article number: 15172 (2015) doi:10.1038/srep15172

[3] Armen R. Kherlopian, Ting Song, Qi Duan, Mathew A. Neimark, Ming J. Po, John K. Gohagan and Andrew F. Laine, "A review of imaging techniques for systems biology" BMC Systems Biology20082:74 DOI: 10.1186/1752-0509-2-74

[4] S. Anweiler, M. Masiukiewicz "Application Of Stereology For Two-Phase Flow Structure Validation In Fluidized Bed Reactors" Thermal Science: Year 2016, Vol. 20, No. 4, pp. 1199-1208

[5] A. Lecuona, P. A. Sosa, P. A. Rodríguez and R. I. Zequeira "Volumetric characterization of dispersed two-phase flows by digital image analysis" Measurement Science and Technology, Volume 11, Number 8

[6] Hubbard, M. G. and Dukler, A. E., 1966, "The Chacacterization of Flow Regimes for Horizontal Two-Phase Flow", Proc. Heat Transfer and Fluid Mech, Eds. M. A. Saad and J. A. Moller, Stanford University Press.

[7] J. Braz. Soc. Mech." Time-frequency analysis of intermittent two-phase flows in horizontal piping"

Journal of the Brazilian Society of Mechanical Sciences and Engineering vol.26 no.2 Rio de Janeiro Apr./June 2004

[8] Zhou Wang, Alan Conrad Bovik, Hamid Rahim Sheikh, Eero P. Simoncelli, "Image Quality Assessment: From Error Visibility to Structural Similarity" Ieee Transactions On Image Processing, Vol. 13, No. 4, April 2004

[9] Kadota, T., Mizutani, S., Wu, C., and Hoshino, M., "Fuel Droplet Size Measurements in the Combustion Chamber of a Motored SI Engine via Laser Mie Scattering," SAE Technical Paper 900477, 1990, doi:10.4271/900477.

[10] Montanaro, A., and Allocca, L., "Flash Boiling Evidences of a Multi-Hole GDI Spray under Engine Conditions by Mie-Scattering Measurements", SAE F&L Congress, Kyoto 2015, JSAE 20159062, SAE 2015-01-1945

[11] Chacon-Murguia, M.I., and Ramirez-Alonso, G. "Visual, Fuzzy-neural Self-adapting Background Modeling with Automatic Motion Analysis for Dynamic Object Detection", Applied Soft Computing 36 (2015) 570–577

Sekoguchi, K., Inoue, K. and Imasaka, T., 1987, "Void Signal Analysis and Gas-Liquid Two-Phase Flow Regime Determination by a Statistical Pattern Recognition Method", JSME International Journal, Vol. 30, No. 266, pp.1266-1273.

[12] C. Sunde, S. Avdic, I. Pafzsit. "Classification Of Two-Phase Flow Regimes Via Image Analysis And A Neuro-Wavelet Approach" Progress in Nuclear Energy, Vol. 46, No. 3-4, pp. 348-358, 2005

Melbourne, I., and Nicol, M., "Statistical properties of endomorphisms and compact group extensions", Preprint, Un. of Surrey, UK, 2003.

[13] Amoresano, A., Langella, G., Niola, V., and Quaremba, G., "Advanced Images Analysis of Two Phase Flow Inside a Centrifugal Pump", Advances in Mechanical Engineering, Volume 2014, Article ID 958320, 11 pages http://dx.doi.org/10.1155/2014/958320

[14] Allocca, L., Montanaro, A., Amoresano, A., Langella, G., Niola, V., Quaremba, G., "Chaos Theory Approach as Advanced Technique for GDI Spray Analysis" SAE Technical Paper 2017-01-0839

[15] T Marchione, C Allouis, A Amoresano, F Beretta "Experimental investigation of a pressure swirl atomizer spray" Journal of Propulsion and Power 23 (5), 1096-1101

[16] Montanaro, A., and Allocca, L., "Flash Boiling Evidences of a Multi-Hole GDI Spray under Engine Conditions by Mie-Scattering Measurements", SAE F&L Congress, Kyoto 2015, JSAE 20159062, SAE 2015-01-1945

[17] Castelli I., Trentin E. "Combination of Supervised and Unsupervised Learning for Training the Activation Functions of Neural Networks", Pattern Recognition Letters 37 (2014) 178–191

[18] Tsoukalas L.H., Uhrig R.E., "Fuzzy and Neural Approaches in Engineering. John Wiley & Sons Inc.", NY, USA (1997)

[19] Iosifidis A., "Extreme Learning Machine Based Supervised Subspace Learning", Neurocomputing 167 (2015) 158–164

[20] Wasserman P.D., "Advanced Methods in Neural Computing", Van Nostrand Reinhold, NY (1993)

[21] A Amoresano, G Langella, V Niola, G Quaremba "A Statistical Method to Identify the Main Parameters Characterizing a Pressure Swirl Spray" International Review of Mechanical Engineering (IREME) 7 (6), 1007-1013

Comparison between two non-contact techniques for art digitalization

F Bianconi[1], S Catalucci[2], M Filippucci[3], R Marsili[4], M Moretti[5], G Rossi[6,8] and E Speranzini[7]

[1] Department of Civil and Environmental Engineering, Università degli Studi di Perugia, Perugia, Italy
[2] Department of Civil and Environmental Engineering, Università degli Studi di Perugia, Perugia, Italy
[3] Department of Civil and Environmental Engineering, Università degli Studi di Perugia, Perugia, Italy
[4] Engineering Department, Università degli Studi di Perugia, Perugia, Italy
[5] Engineering Department, Università degli Studi di Perugia, Perugia, Italy
[6] Engineering Department, Università degli Studi di Perugia, Perugia, Italy
[7] Engineering Department, Università degli Studi di Perugia, Perugia, Italy

[8] Author to whom any correspondence should be addressed: gianluca.rossi@unipg.it

Abstract. Many measurements techniques have been proposed for the "digitalization of objects": structured light 3D scanner, laser scanner, high resolution camera, depth cam, thermal-cam, ... Since the adoption of the European Agenda for Culture in 2007, heritage has been a priority for the Council's work plans for culture, and cooperation at European level has advanced through the Open Method of Coordination. Political interest at EU level has steadily grown cultural and heritage stakeholders recently highlighted in the Declaration on a New Narrative for Europe: "Europe as a political body needs to recognize the value of Cultural Heritage". Photomodelling is an innovative and extremely economical technique related to the conservation of Cultural Heritage, which leads to the creation of three-dimensional models starting from simple photographs. The aim of the research is to understand the full potential offered by this new technique and dedicated software, analysing the reliability of each instrument, with particular attention to freeware ones. An analytical comparison between photomodelling and structured light 3D scanner guarantees a first measure of the reliability of instruments, tested in the survey of several Umbrian heritage artefacts. The comparison between tests and reference models is explained using different algorithms and criteria, spatial, volumetric and superficial.

1. Photomodelling and Cultural Heritage

The innovative technique of photomodelling is a modern frontier of the survey, based on the extraction of information from digital images, which can be considered as experimental [1]. Photomoelling is lexical neologism, combination of the words photography and three-dimensional modelling. Through the common camera, the operator is able to extract spatial information of the elements present in a scene, acquiring multiple frames. This technique introduces a new working environment, which enables the restitution of buildings, objects, archaeological finds, based on the integration between survey, 3D modelling and representation [2]. The incessant improvement and economic progress of tools mean that these new techniques are accessible to an incredible range of users. The digital camera replaces expensive devices such as laser scanners or 3D structured light scanning systems and open source software and freeware ones come up beside the commercial products.

During the last years, these techniques are having great success in the field of conservation and enhancement of cultural heritage, proving to be one of the major breakouts of passive detection. In fact, photomodelling is the result of the digital revolution that influenced all the artefacts of our culture, from images to tools for interpretation and design, and it can be considered as a development of photogrammetry. In fact, one of the most interesting aspects in the field of Cultural Heritage is the possibility to perform three-dimensional surveys, which allows the virtual reconstruction of the objects in a scenario with high level of details. This procedure is recognized not only as a conservation tool, to promote the virtual tourism in museums, and as a teaching resource, but also for purposes of recovery, restoration and enhancement of Cultural and Environmental Heritage. The survey and the representation acquire an essential role in the protection of assets. Digital contents contribute to the development and accessibility of cultural information and three-dimensional representation becomes effective support, in order to document the status of the goods concerned [3].

1.1. Structure from Motion, algorithms and the photographic survey

Photomodelling is based on the methodology of the Structure from Motion [4], the calculation process that allows the geometry reconstruction of an object through the automatic collimation of points from a series of digital images. After the extraction of significant points, named features, well recognizable in three or more frames, the SfM obtains photographic parameters and crosses the homologous points on multiple images, finding the spatial coordinates of each point (figure 1). At first, the dedicated software aligns the various images in three-dimensional space, generating a 3D point cloud. The following step is the reconstruction of a triangular mesh, processed and textured, with a high number of polygons and a clear geometry of the models. The editing step is always necessary, like various elaborations of smoothing, decimation of the faces, closure of gaps, in order to make suitable the generated model [5].

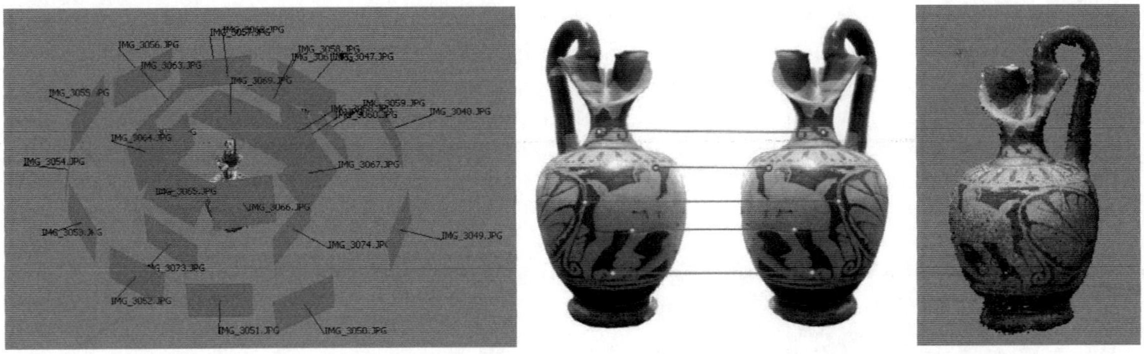

Figure 1. Matching: identification of homologous points (features) between each image.

Photomodelling is different from photogrammetry thanks to the introduction of image or multi-image matching algorithms, which allow the complete three-dimensional reconstruction procedure in automatic [6]. The comparison between the pairs of frames is carried out by decomposing the photographs in a grid of small squares and founding chromatic analogies between them. The calibration of the camera is automatic, as the reconstruction position of each picture in the space; for each key point the real coordinates x,y,z are obtained, materialized in three dimensions in a scattered point cloud, a cloud of low-density points (sparse reconstruction). In the next step the sparse cloud is increased in a dense point cloud (dense reconstruction) thanks to the SfM dense-matching algorithms, developed in photogrammetry: the data relating to the position x,y,z of each key point are used to recognize and extract the coordinates of all the surrounding elements (figure 2). The procedure of SfM, coupled to the algorithms of dense image-matching, allows the reconstruction of the 3D geometry of a scene (Structure), identifying a set of sparse correspondences, and the positions of cameras (Motion).

Figure 2. Sparse reconstruction and dense reconstruction of a bronze artefact of Archaeological Museum of Amelia (TR), Italy.

Photography, the best representation of reality, acquires an innovative role: pixels represent spatial coordinates x,y,z of sparse and dense point clouds, the first shapes of a model. Each step of photomodelling strictly depends on the photographic strategy and on the quality of images. These criteria influence the results obtained following the alignment of frames, manually or automatically. It is really important that users have to evaluate parameters like the sharpness, the brightness and the contrast of the photos, particularly to obtain good results during the orientation of cameras. The lighting of the scene has to be as uniform as possible: overshadowed parts or strongly illuminated ones could cause issues during the reading of information in the images. For these reason it is also important to remove the flash. Users have to analyse the colour characteristics of the scene: objects with particular chromatic variations are better than aseptic scenes. The object must be accessible to the photographic survey entirely, from different angles as much as possible. The ideal plan for photographic survey is to take pictures, going around the vertical axis of the object at 360° for isolated one and 180° for flat plate one, at a constant distance from it and with the camera oriented between 30° / 45° downwards. The number of picture is not fixed, although it is important that the same images are not repeated too many times. In order to avoid the repetition of the same information, the photographic survey should be scheduled at the beginning. The number of images depends on the dimensions of the object to survey, which must be photographed in its entirety as much as possible (figure 3).

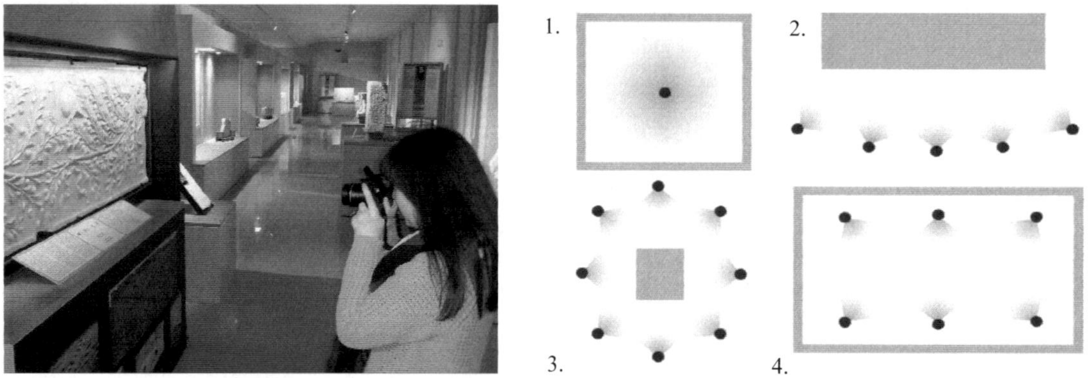

Figure 3. Photographic survey in the Archaeological Museum of Amelia (TR), Italy. Close environment, one point of view, multiple pictures like a 360° panorama (1.); a flat plate object, 180°

pictures in front of the object, multiple point of view (2.); isolated object, 360° pictures around the object, multiple point of view (3.); close environment, multiple point of view and multiple pictures (4.).
The 3D reconstruction steps starting from the analysis of the environment and the object to detect, until the reconstruction and the visualisation of the complete 3D model:

- preliminary study and photographic survey;
- preliminary image processing;
- two stages of point cloud restitution (sparse and dense);
- 3D model definition (mesh/triangular weave);
- editing, texturing and visualisation.

The 3D modelling procedure is always the same: the first operation is the manipulation of the point cloud, created from a set of photographs. Once created the point cloud, models can be imported into a mesh processing and editing software editing. The creation of three-dimensional model is called “triangulation”: starting from the input data vertices, edges and faces are generated. The result obtained is a set of coordinates, which is converted into a polygonal surface [7] [8].

MeshLab [9] and Geomagic Studio [10] are examples of applications in which it is possible to perform manual editing of data, merge, scale, align different surfaces, creating a three-dimensional surface with the aid of different algorithms. The first is an open-source software that provides a simple to use set of tools for editing, cleaning, repair, control, representation and conversion of meshes; the second is a commercial software, with editing and advanced surfacing commands, partially automatic. During the research, Poisson Reconstruction [11] and Ball Pivoting [12] [13] algorithms on MeshLab reproduced the 3D models, at the end scaled on Geomagic Studio with different approaches in relation to the functions available.

2. Dedicated software and instruments

This research involved the analysis of the varied overview of software instruments available today with the aim to understand the potential of these tools. Software were evaluated by their time data processing, difficulty of use, accuracy and precision of results. The applications proposed are also divided into 3D web-service, open-source software and commercial software (figure 4).

Autodesk ReMake	
Time data processing	average
Difficulty	easy
Cost	396,50€ / year

Autodesk 123D Catch	
Time data processing	fast
Difficulty	very easy
Cost	free

Visual SfM	
Time data processing	average
Difficulty	average
Cost	free

Arc3D	
Time data processing	fast
Difficulty	very easy
Costo	free

Agisoft Photoscan	
Time data processing	average
Difficulty	average
Cost	160€ ca.

PhotoModeler Scanner	
Time data processing	slow
Difficulty	difficult
Cost	2240€ ca.

Figure 4. Dedicated software, different from time data processing, difficulty of use and costs.

Agisoft PhotoScan [14], a commercial software, is the most efficient solution for completeness of results; the instrument has provided results during the experiments with the largest number of points. At the same time and with greater accuracy and precision Visual SfM software [15] is the best solution in the reproduction of point clouds, in case of good surrounding conditions. This freeware tool is based on the Structure from Motion technique and it is able to reconstruct elements similar to Laser Scanner products, following a semi-automated workflow, starting from the alignment of frames until the matching between the homologous points. Arc3D [16] is a freeware tool, which allows users to upload images to an online web-server and to obtain a full three-dimensional scene. Arc3D, as Visual SfM, depends on the surrounding conditions of light and space of the scenario. Autodesk Remake [17] and Autodesk 123D Catch [18] are applications, the first low-cost and the second freeware, to get quickly and automatically a borderline result with an average level of details, to print or to be loaded and displayed on an online platform. Photomodeler Scanner [19] software, a commercial tool based on the principles of photogrammetry, permits to follow every step of 3D reconstruction, from the orientation of pictures, until the restitution of a textured mesh.

The purpose of the research is to know and test different types of software and identify among these the best solution, able to return as much as possible complete outcomes.

3. The reliability of instruments

In order to provide scientific results, this research has considered an analytical comparison between photomodelling and structured light 3D scanner Creaform GO!Scan50 [20], tested in the survey of "Doccione di fontana", Galleria Nazionale dell'Umbria in Perugia, Italy. Two different criteria, a spatial one and a surface one, explained the comparison between a reference mesh (the one created with structured light 3D scanner) and testing ones, extrapolating data relating to the distribution of deviations between triangular elements that compose models. Other experimentations have been proposed considering the surveys of several museum artefacts from the collection of the Archaeological Museum of Amelia (TR), Italy. In this case, three objects have been chosen and the test clouds have been compared with the ones recreated on Agisoft Photoscan, which provided results with the highest number of points.

The work was divided into three steps (figure 5):

- acquisition and processing of images relating to dataset samples;
- elaboration and editing of 3D models;
- alignment and comparison of tests with reference models.

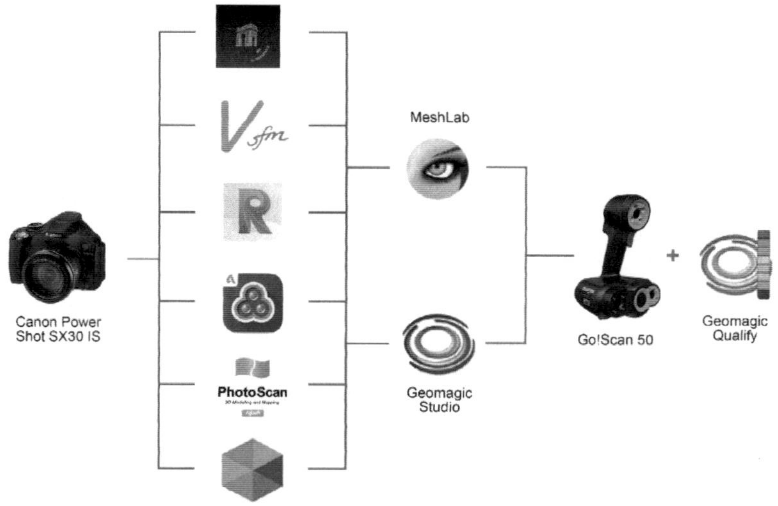

Figure 5. The workflow: from the photographic survey to the comparison between models.

3.1. Measurement, criteria and comparisons

A "measure" is a range of values, acquired with the purpose of controlling a process, perform the calibration of an instrument or allow the physical understanding of a partially known phenomenon [21].

Referring to this assertion, the research considered different methods of analysis (figure 6), a superficial one, a spatial one and a volumetric one:

- comparison of point clouds by a map that identifies the distance of each 3D data point;
- comparison of 2D profiles;
- comparison through a volumetric algorithm.

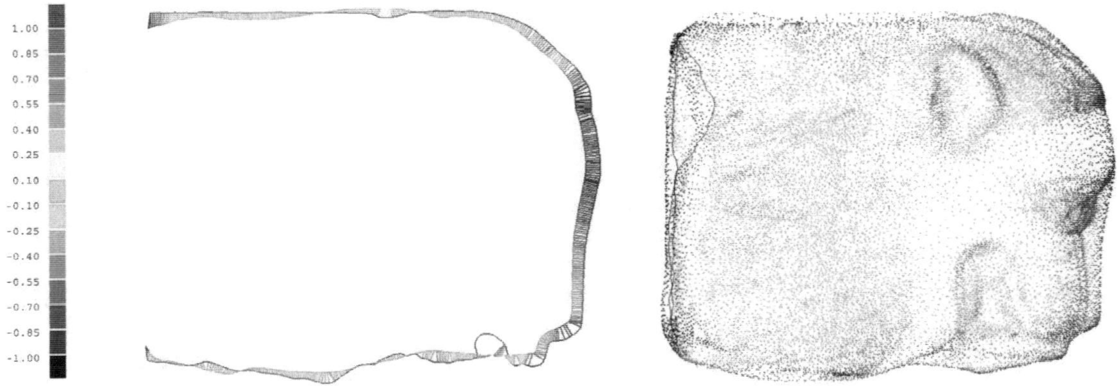

Figure 6. Comparison criteria between reference and test models: comparison between sections and between point clouds.

The tests have been carried out by means of analysis algorithms, which provide variances and deviations between geometric entities in the space, using the software Geomagic Qualify [22]. The analysis was executed by setting a spectrum of 15 intervals of values between -1.0 and +1.0 cm deviation, the range of acceptability; the range between -0.1 and +0.1 cm is the optimum correspondences between the test and reference points. The software applies the colour map of deviations of each coordinate of the point cloud, distributed and projected on the faces of the 3D model. Observing the maps generated, it is possible to notice sometimes areas coloured in grey or missing parts of the models, which the software cannot compare.

The comparison involved the creation of 2D and 3D maps of the deviations, from which it was possible to derive a matrix in .CSV format of the spatial coordinates x,y,z of models points, test and reference ones and the values of the mutual distances for each pair of points. By importing the matrix into MATLAB it was possible to process data in the histograms of frequency, the absolute values of deviations and the curves of the probability density.

3.1.1. Comparison between instruments: case study of "Doccione di fontana", Galleria Nazionale dell'Umbria in Perugia, Italy

Object of the survey is a lion's head in italics marble, dating back to the middle of XIII century.

The reference model was realised using a 3D structured light scanner Go!Scan50 of the Engineering Department in the Università degli Studi di Perugia. This measuring instrument allows through the use of a structured light and the shooting with multiple cameras to capture surface portions of any object. These surface portions are then joined by a dedicated software, which allows to rebuild the three-dimensional object in its entirety, both in its shape, its surface and its microstructure. This

IOP Conf. Series: Journal of Physics: Conf. Series **882** (2017) 012005 doi:10.1088/1742-6596/882/1/012005

instrument can measure with details in the order of tenth of a millimeter on objects of dimensions in a range of 10 cm and a few meters.

The test models were realised using six photomodelling software, starting from the same number of images, captured during the photographic survey in the Galleria Nazionale dell'Umbria (figure 7).

Dataset	Description	Dimension (cm)	Material	Num. of images	Camera	Resolution (pixel)
1	*Doccione di fontana*	15x27x22	Marble	21	Canon PowerShot SX30 IS	4320x3240

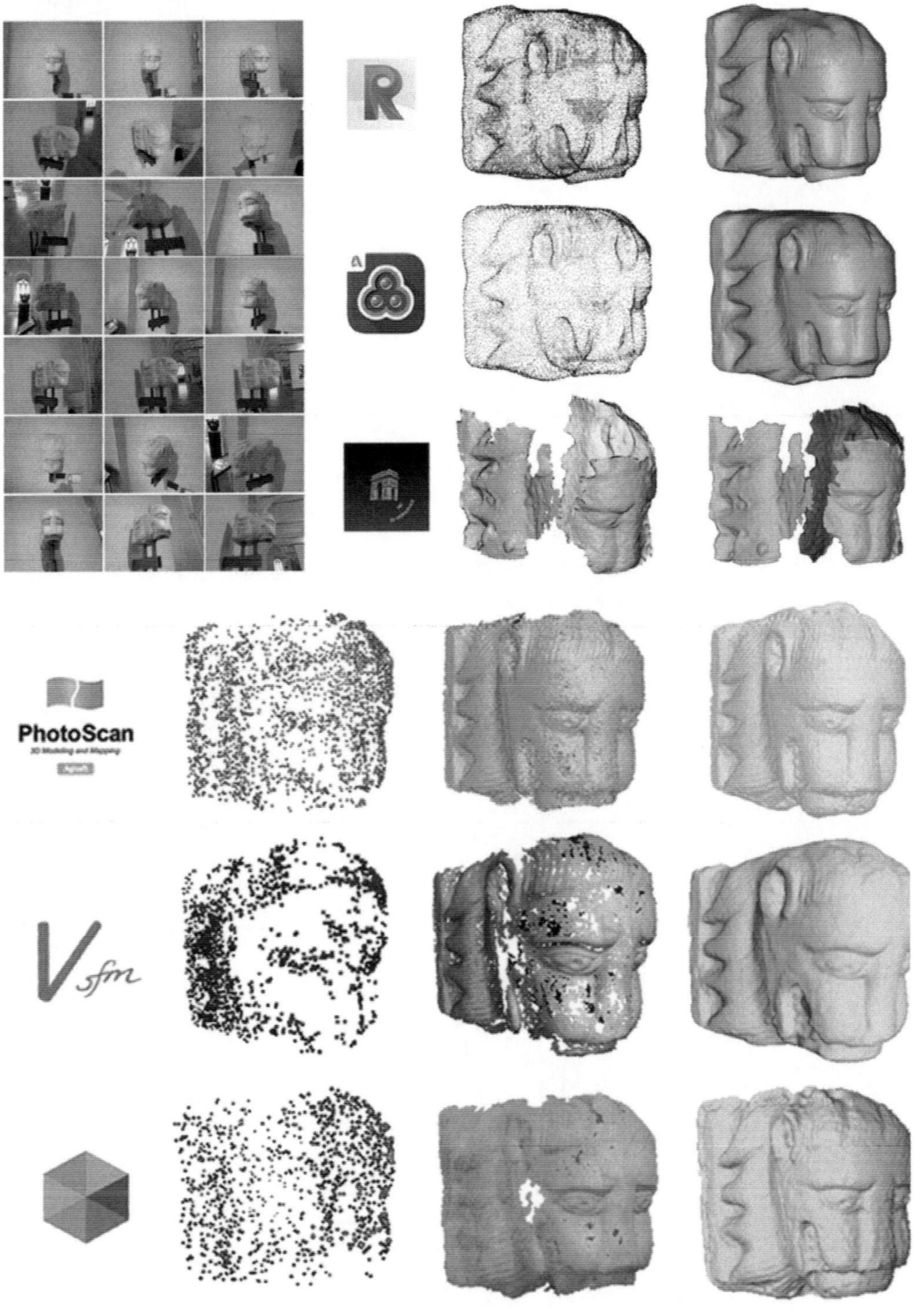

Figure 7. Test models: same number of images, same object, different software.

Considering the graphic representations of 3D and 2D metrics comparisons between test models generated with photomodelling technique and the reference model realized with the structured light GO!Scan50 system, the distribution of the deviations is not uniform.

The colour map projected on the model created with Arc3D shows significant gaps: it appears incomplete and fragmented. An important result is obtained by comparison with the model developed by Visual SfM. Almost the entire front surfaces present deviations within the range of -0.1 and +0.1 cm. The 3D map that presents major deviations is obtained by comparison with Autodesk ReMake. In fact, looking at the map, it is immediately evident that the distances for most of the surface and in correspondence with the front faces are in the range between -1 cm and -0.55 cm. The average value of the deviation between the test and the reference models is the maximum of the results (figure 8,9,10).

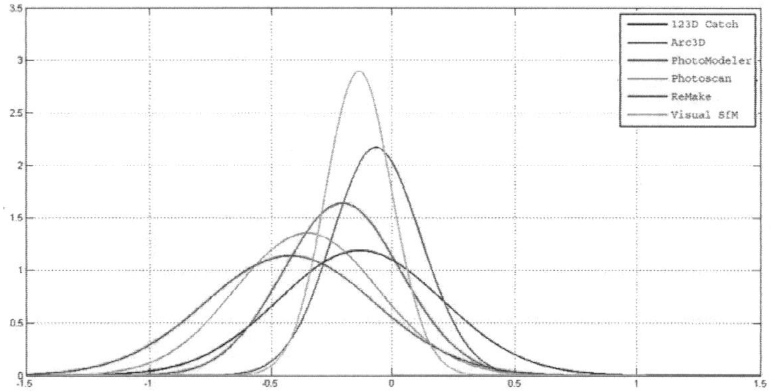

Figure 8. Probability density diagram (left); 3D textured model realized by Agisoft Photoscan (right).

Autodesk 123D Catch

Standard Deviation distribution	Number of point
n>1	105
0.85<n<1	32
0.7<n<0.85	35
0.55<n<0.7	57
0.4<n<0.55	89
0.25<n<0.4	187
0.1<n<0.25	490
-0.1<n<0.1	3764
-0.25<n<-0.1	5311
-0.4<n<-0.25	1851
-0.55<n<-0.4	85
-0.7<n<-0.55	33
-0.85<n<-0.7	27
-1<n<-0.85	24
n<-1	244

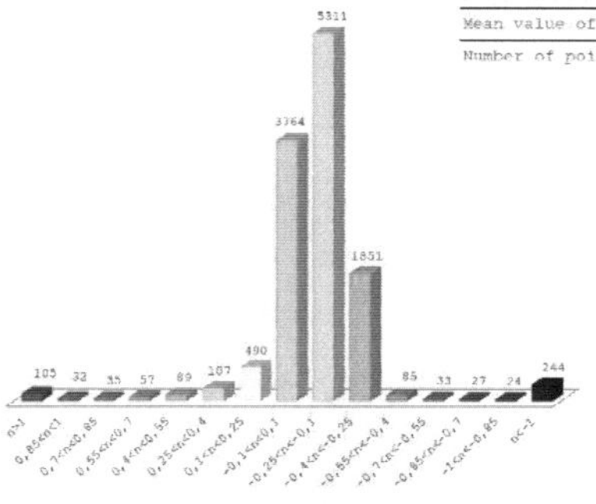

Mean value of distance between models	-0.13402 cm
Standard Deviation	0.3398 cm
Mean value of all differences +- devstd	0.20096 cm, -0.469 cm
Number of points	12334

Arc3D

Standard Deviation distribution	Number of point
n>1	108
0.85<n<1	140
0.7<n<0.85	233
0.55<n<0.7	393
0.4<n<0.55	561
0.25<n<0.4	1349
0.1<n<0.25	4194
-0.1<n<0.1	23510
-0.25<n<-0.1	24113
-0.4<n<-0.25	3466
-0.55<n<-0.4	340
-0.7<n<-0.55	123
-0.85<n<-0.7	95
-1<n<-0.85	43
n<-1	12

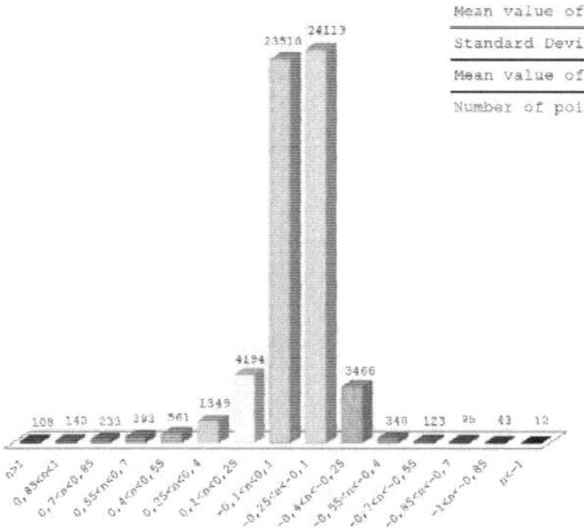

Mean value of distance between models	-0.065572 cm
Standard Deviation	0.18398 cm
Mean value of all differences +- devstd	0.11841 cm, -0.24955 cm
Number of points	58680

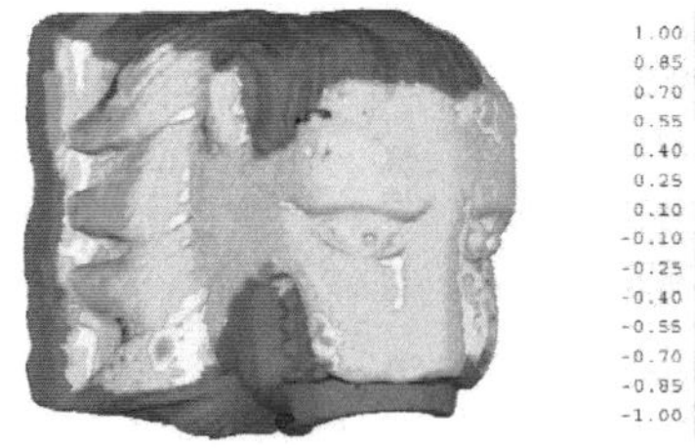

Figure 9. Maps of deviation distribution on surface of modes (Autodesk 123D Catch; Arc3D).

Autodesk ReMake

Standard Deviation distribution	Number of point
n>1	18
0.85<n<1	3
0.7<n<0.85	6
0.55<n<0.7	39
0.4<n<0.55	60
0.25<n<0.4	142
0.1<n<0.25	360
-0.1<n<0.1	1166
-0.25<n<-0.1	3253
-0.4<n<-0.25	7002
-0.55<n<-0.4	3397
-0.7<n<-0.55	2607
-0.85<n<-0.7	1653
-1<n<-0.85	540
n<-1	714

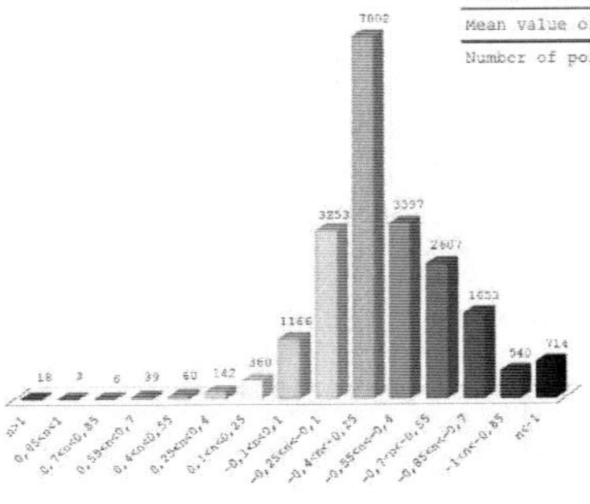

Mean value of distance between models	-0.4229 cm
Standard Deviation	0.35018 cm
Mean value of all differences +- devstd	-0.072719 cm, -0.77309 cm
Number of points	20960

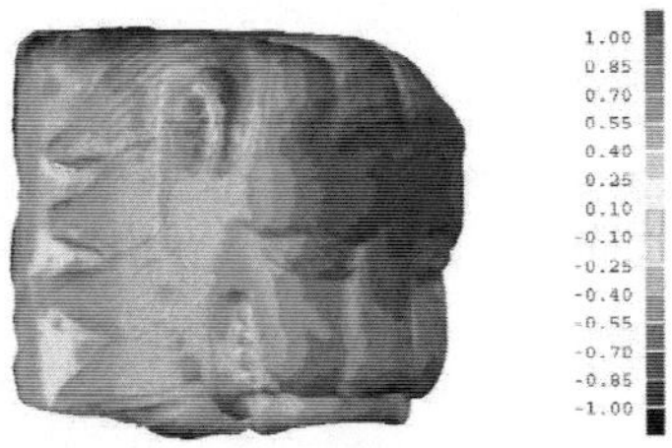

Visual SfM

Standard Deviation distribution	Number of point
n>1	19
0.85<n<1	2
0.7<n<0.85	17
0.55<n<0.7	41
0.4<n<0.55	113
0.25<n<0.4	569
0.1<n<0.25	4155
-0.1<n<0.1	76781
0.25<n<-0.1	66983
-0.4<n<-0.25	34073
-0.55<n<-0.4	4550
-0.7<n<-0.55	580
-0.85<n<-0.7	83
-1<n<-0.85	19
n<-1	54

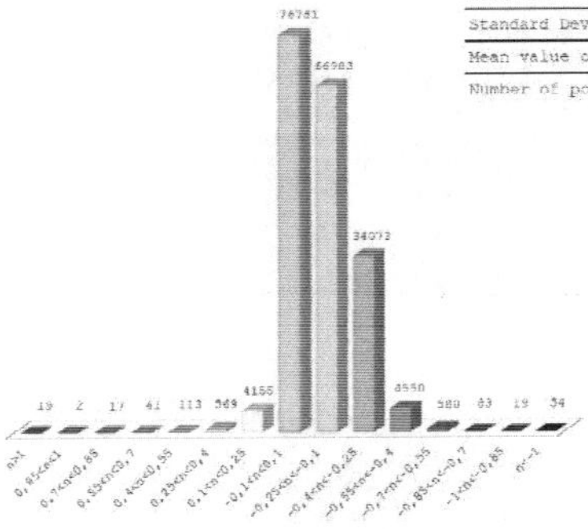

Mean value of distance between models	-0.13778 cm
Standard Deviation	0.13766 cm
Mean value of all differences +- devstd	-0.00011882 cm, -0.27544 cm
Number of points	188039

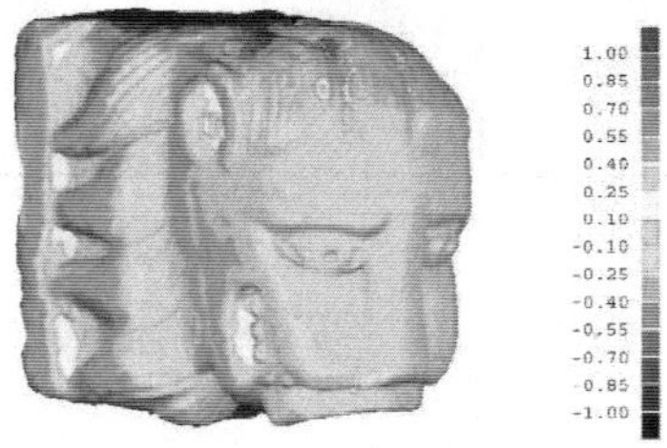

PhotoModeler Scanner

Standard Deviation distribution	Number of point
n>1	132
0.85<n<1	245
0.7<n<0.85	392
0.55<n<0.7	593
0.4<n<0.55	983
0.25<n<0.4	2640
0.1<n<0.25	6959
-0.1<n<0.1	29601
-0.25<n<-0.1	41555
-0.4<n<-0.25	37322
-0.55<n<-0.4	18140
-0.7<n<-0.55	5460
-0.85<n<-0.7	1802
-1<n<-0.85	593
n<-1	419

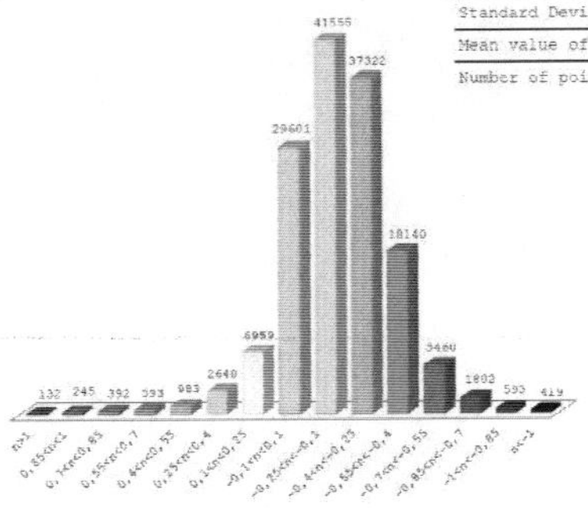

Mean value of distance between models	-0.20708 cm
Standard Deviation	0.24349 cm
Mean value of all differences +- devstd	0.036412 cm, -0.45057 cm
Number of points	146836

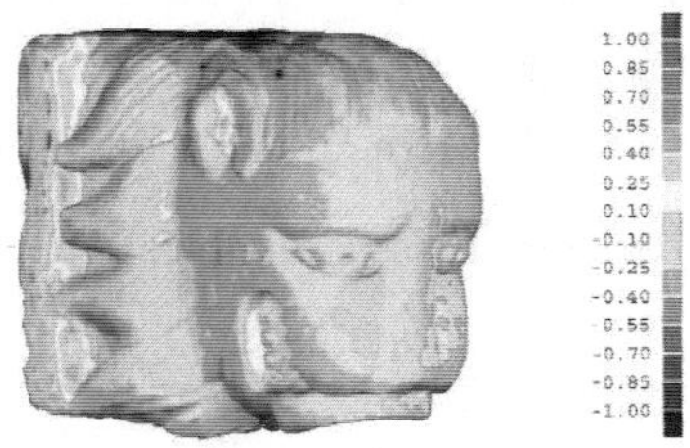

Agisoft Photoscan

Standard Deviation distribution	Number of point
n>1	96
0.85<n<1	64
0.7<n<0.85	194
0.55<n<0.7	520
0.4<n<0.55	941
0.25<n<0.4	1660
0.1<n<0.25	2313
-0.1<n<0.1	13681
-0.25<n<-0.1	22818
-0.4<n<-0.25	27611
-0.55<n<-0.4	26618
-0.7<n<-0.55	14679
-0.85<n<-0.7	6531
-1<n<-0.85	2702
n<-1	2119

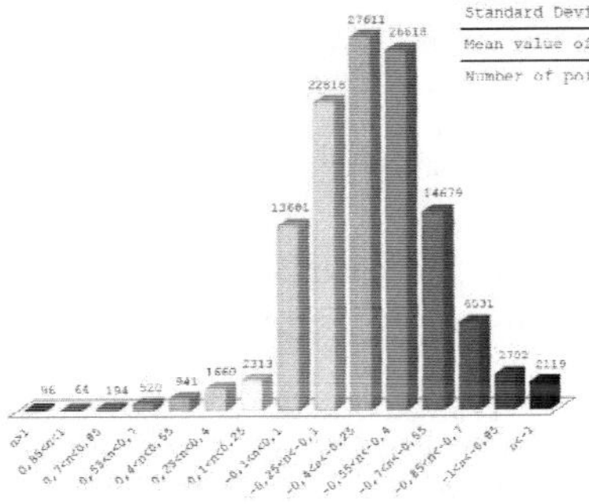

Mean value of distance between models	-0.35253 cm
Standard Deviation	0.29478 cm
Mean value of all differences +- devstd	-0.057747 cm, -0.64731 cm
Number of points	122548

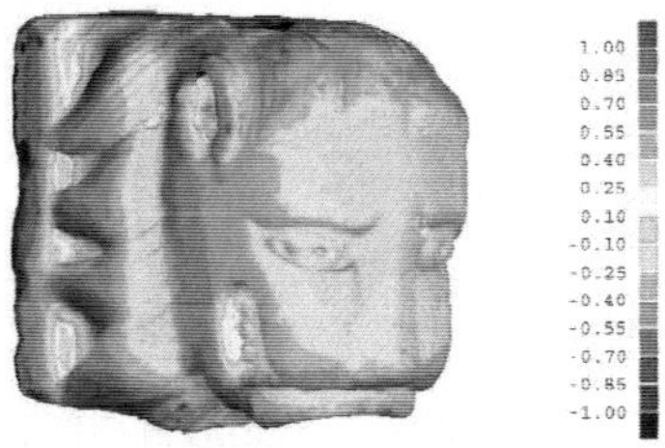

Figure 10. Maps of deviation distribution on surface of modes (Autodesk ReMake; Visual SfM; Photomodeler Scanner; Agisoft Photoscan).

IOP Conf. Series: Journal of Physics: Conf. Series **882** (2017) 012005 doi:10.1088/1742-6596/882/1/012005

3.1.2. Volumetric comparison algorithm

To further support the experimentation, the same obtained 3D models of the "Doccione di fontana" have been imported into the multi-paradigm numerical computing environment MATLAB. Through the elaboration of chromatic maps (figure 11), which identify the volumetric error distribution and the distance between barycentre of the meshes composing the models, it has been possible to compare the reference and tests three-dimensional surfaces through an algorithm. This algorithm is based on a function named ICP (Iterative Closest Point), which identifies the distance between vertices of each triangles that compose mesh test and the surface of triangles that compose reference mesh (figure 12). For each comparison, the algorithm has elaborated box plot statistical diagrams (figure 13) and volumetric error distribution graphs (figure 14).

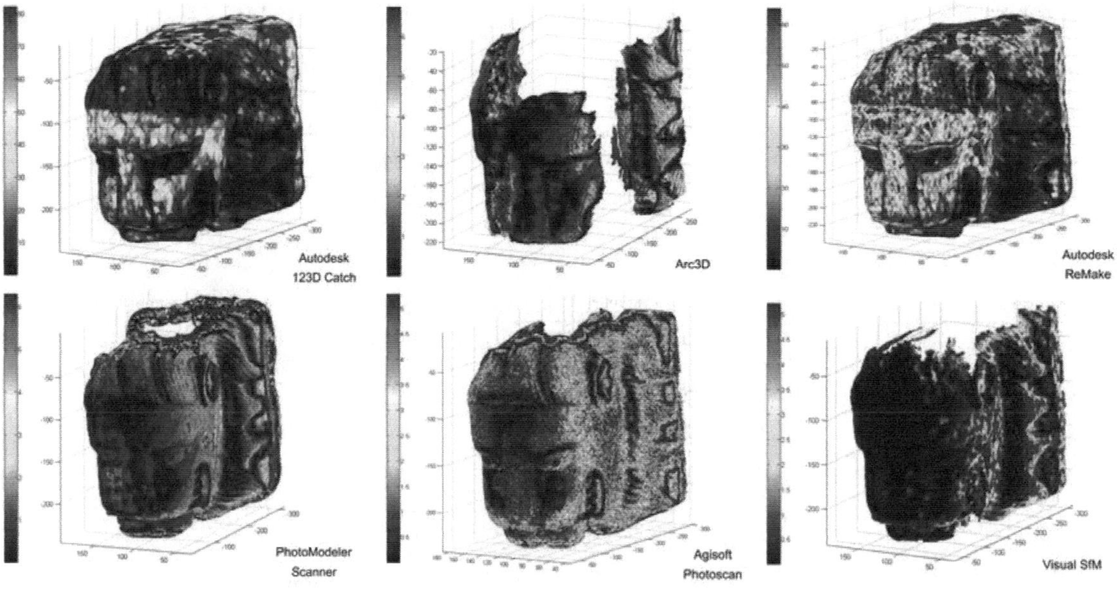

Figure 11. Maps of the volumetric error distribution on surface of modes (MATLAB).

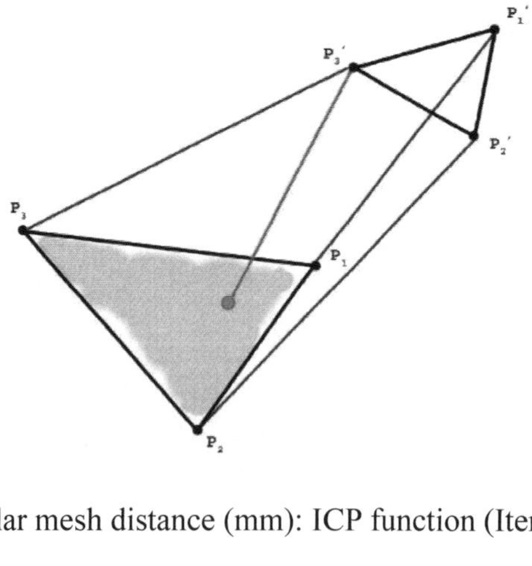

Figure 12. Triangular mesh distance (mm): ICP function (Iterative Closest Point)

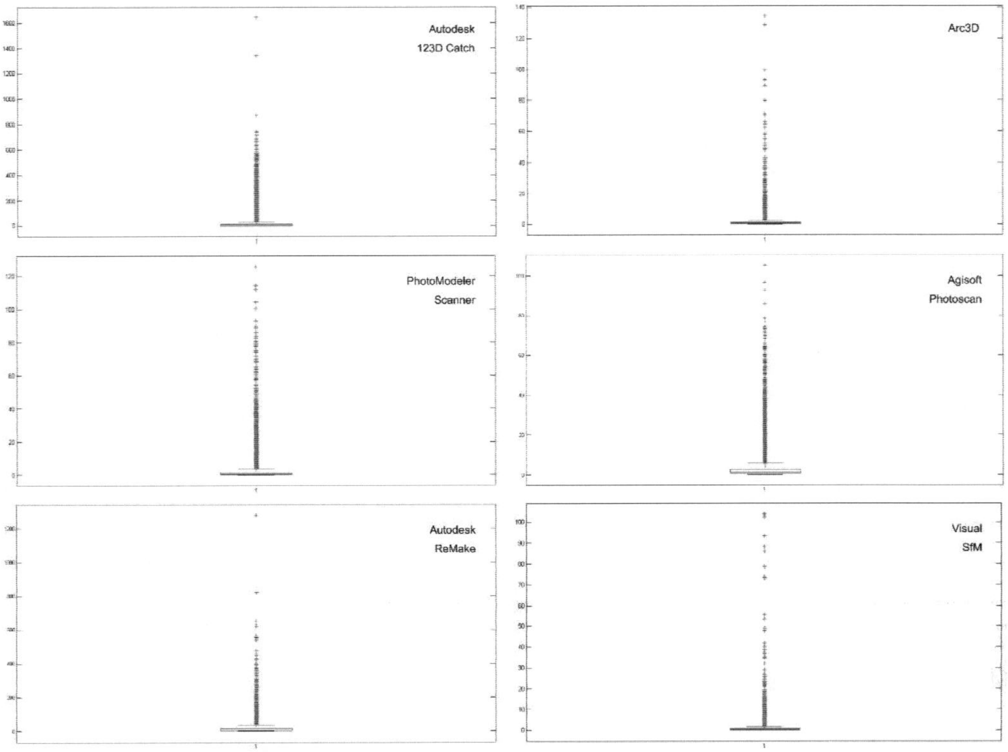

Figure 13. Box plot statistical diagrams (MATLAB).

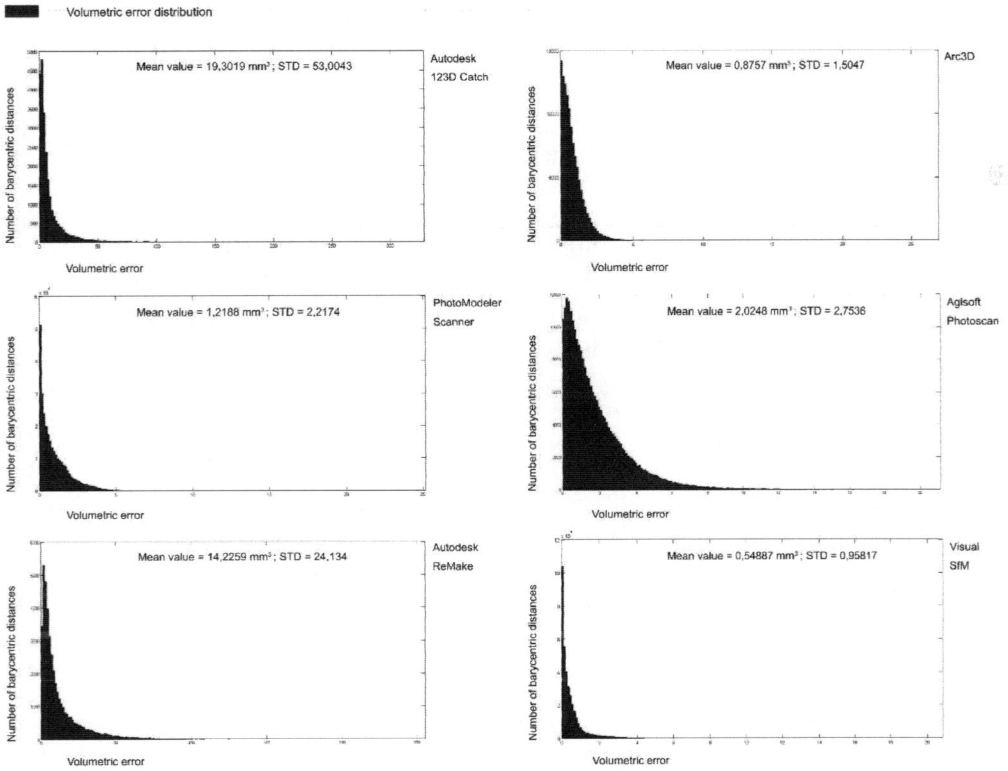

Figure 14. Volumetric error distribution graphs (MATLAB).

IOP Conf. Series: Journal of Physics: Conf. Series **882** (2017) 012005 doi:10.1088/1742-6596/882/1/012005

3.1.3. Comparison between software: case study of Archaeological Museum of Amelia (TR), Italy

Object of comparison have been three artefacts of the Archaeological Museum of Amelia, selected from more than thirty ones, of which was carried out the survey and 3D modelling, with the aim of recreate a real museum multimedia catalogue. The three artefacts are different in material, size and photographic survey. It was used a Canon PowerShot SX30 IS digital camera.

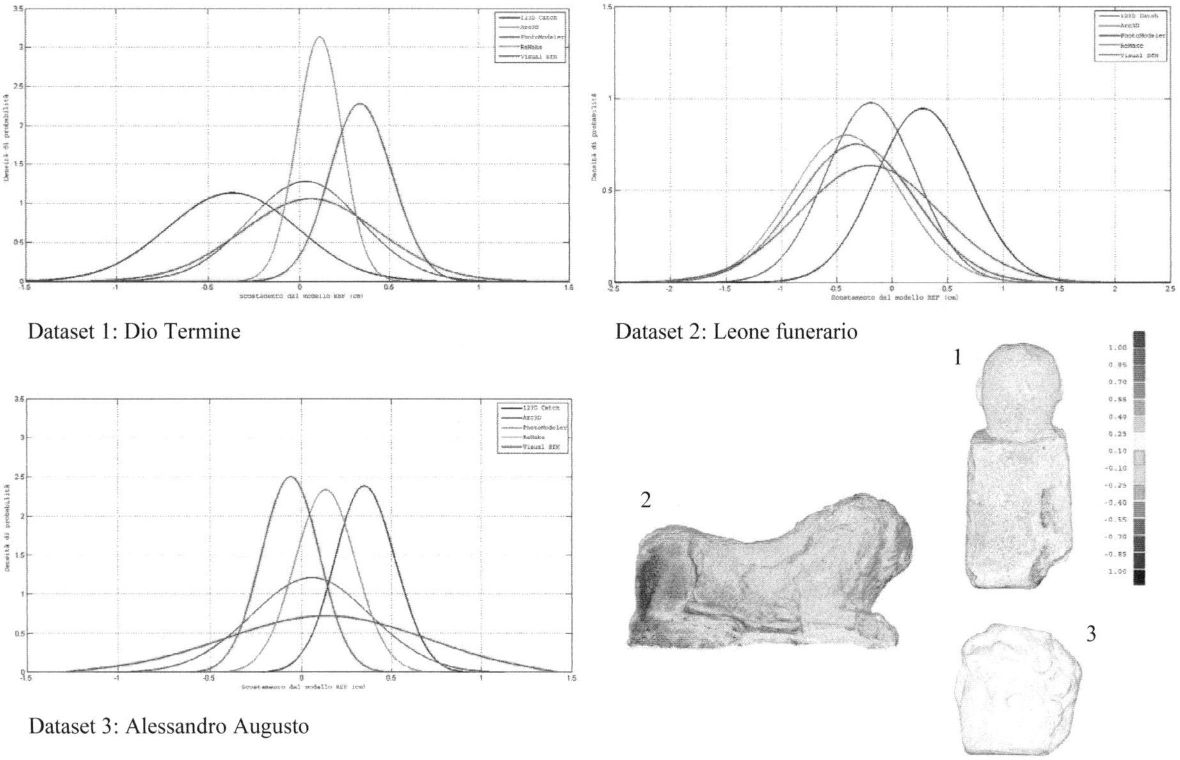

Figure 15. Probability density diagram of each dataset.

From the comparison between geometric models of the first dataset "Dio Termine", it is possible to verify the goodness of the model from Autodesk ReMake, which presents, among all results, the value of the lower Standard Deviation. Compared to tools like Arc3D and PhotoModeler Scanner, which have average values close to zero, Autodesk ReMake, with less uncertainty, ensures a better match with the reference model. The model distinguished by the inaccuracy is the one realized with Autodesk 123D Catch, which is scarcely accurate and precise, largely underestimating the measure and presenting a high dispersion (figure 15,16).

As for the analysis conducted on the third model "Alessandro Augusto", the results are quite consistent with the previous test. The most obvious difference is the greater precision and accuracy of the model generated by Visual SfM and the greater precision of the model developed with Autodesk 123D Catch, which remains in any case poorly accurate. The results obtained by adopting the Arc3D tools, Autodesk ReMake and Photomodeler Scanner is similar to the first case. In this case, it is interesting to notice the results from some software, for example Photomodeler Scanner, which has recreated only a part of the acquired surface, with very accurate results although imprecise (figure 15,16,17).

Very significant case is the second dataset "Leone funerario": comparisons results are closer than the previous tests. This is probably due to the considerable size of the sculpture. The results of all tests are very similar, but the best is the model generated by the 3D web-service Arc3D (figure 15,16,17).

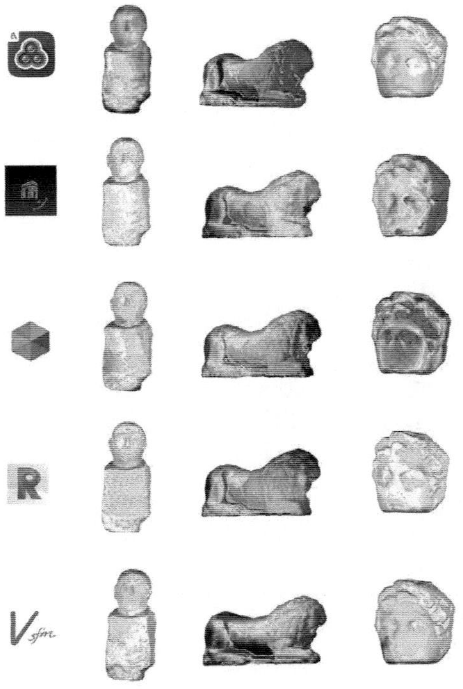

Dataset 1 Dataset 2 Dataset 3

Autodesk 123D Catch

Mean value of distance between models	-0.36856 cm
Standard Deviation	0.3527 cm
Mean value of all differences +- devstd	-0.01586 cm, -0.72127 cm
Number of points	20953

Arc3D

Mean value of distance between models	0.032629 cm
Standard Deviation	0.31241 cm
Mean value of all differences +- devstd	0.34504 cm, -0.27978 cm
Number of points	180320

PhotoModeler Scanner

Mean value of distance between models	0.061274 cm
Standard Deviation	0.37843 cm
Mean value of all differences +- devstd	0.43971 cm, -0.31716 cm
Number of points	104342

Autodesk ReMake

Mean value of distance between models	0.11363 cm
Standard Deviation	0.12733 cm
Mean value of all differences +- devstd	0.24096 cm, -0.013705 cm
Number of points	356725

Visual SfM + MeshLab

Mean value of distance between models	0.34241 cm
Standard Deviation	0.17526 cm
Mean value of all differences +- devstd	0.51766 cm, 0.16715 cm
Number of points	654823

Dataset 1

Figure 16. Maps of deviation distribution on surface of modes and data of Dataset 1.

Autodesk 123D Catch

Mean value of distance between models	0.27629 cm
Standard Deviation	0.4215 cm
Mean value of all differences +- devstd	0.69779 cm, -0.1452 cm
Number of points	151422

Arc3D

Mean value of distance between models	-0.19049 cm
Standard Deviation	0.40824 cm
Mean value of all differences +- devstd	0.21776 cm, -0.59873 cm
Number of points	229329

PhotoModeler Scanner

Mean value of distance between models	-0.3256 cm
Standard Deviation	0.53061 cm
Mean value of all differences +- devstd	0.20501 cm, -0.85621 cm
Number of points	84320

Autodesk ReMake

Mean value of distance between models	-0.41368 cm
Standard Deviation	0.49817 cm
Mean value of all differences +- devstd	0.084494 cm, -0.91185 cm
Number of points	586806

Visual SfM + MeshLab

Mean value of distance between models	-0.20196 cm
Standard Deviation	0.62955 cm
Mean value of all differences +- devstd	0.42759 cm, -0.83151 cm
Number of points	783099

Dataset 2

Autodesk 123D Catch

Mean value of distance between models	0.35488 cm
Standard Deviation	0.16744 cm
Mean value of all differences +- devstd	0.52231 cm, 0.18744 cm
Number of points	40280

Arc3D

Mean value of distance between models	0.056189 cm
Standard Deviation	0.33016 cm
Mean value of all differences +- devstd	0.38635 cm, -0.27397 cm
Number of points	46739

PhotoModeler Scanner

Mean value of distance between models	0.13653 cm
Standard Deviation	0.55322 cm
Mean value of all differences +- devstd	0.68975 cm, -0.4167 cm
Number of points	4195

Autodesk ReMake

Mean value of distance between models	0.13488 cm
Standard Deviation	0.17092 cm
Mean value of all differences +- devstd	0.3058 cm, -0.036042 cm
Number of points	93928

Visual SfM + MeshLab

Mean value of distance between models	-0.059322 cm
Standard Deviation	0.15958 cm
Mean value of all differences +- devstd	0.10026 cm, -0.2189 cm
Number of points	169691

Dataset 3

Figure 17. Data of Dataset 2 and Dataset 3.

3.2. Obtained results

In the totality of the tests, the results of the comparisons highlight the lack of a complete software, suitable to solve every need, while preserving precision and accuracy of the models created (figure 18). The best software is a compromise solution within accuracy and precision: Visual SfM is the one with a good balance between these parameters, with the generation of fairly complete and dense point cloud. Arc3D is a valuable software, referring to the accuracy of the results, but it processes inaccurate models and incomplete point clouds, tied to the conditions of the photographic survey.

Considering the comparison between a reference cloud created with Agisoft PhotoScan and test ones processed through other software, it is clear that, referring for example to the model of Arc3D web-service and toolkit Visual SfM, despite the excellent results obtained in the measurement comparisons, models manifest in almost the totality of the evidence the most significant gaps.

The analysis conducted shows not only the essentiality to undertake an effective campaign of photographic survey, but also the changing of performances, depending on the choice of the acquisition tool and the quality and accuracy of processing software.

Software	Accuracy	Precision
Autodesk ReMake	average	average
Autodesk 123D Catch	low	low
Visual SfM	high	high
Arc3D	high	average
Agisoft Photoscan	low	average
PhotoModeler Scanner	low	average

Figure 18. Accuracy and precision of each software.

4. Conclusions

The research was marked by an experimental approach: starting from the comparison with the structured light instrument Go!Scan50, it was possible to define criteria for assessing the reliability of instruments, using spatial, volumetric and surface analysis, comparing the reference meshes from which values can be extracted such as distribution of the deviations. After the interpretation of the obtained data, it was possible to collect a variety of information regarding the levels of uncertainty of the results, achieved by means of the comparison between instruments and software.

The research shows the limits and the advantages of photomodelling, which appears primarily as a technique able to integrate the operations of detection with excellent results. Compared with other techniques, as the laser scanning and structured light 3D scanning, if the surrounding conditions guarantee a correct process of application, photomodelling will become a valid economic alternative and an important integration tool.

The experience has shown that, regardless of the tools used, a specific software, which can meet the requirements of precision and accuracy of 3D models, doesn't exist. The web-service Arc3D and Visual SfM represent good compromises between these parameters. Both software are freeware, further evidencing the economic efficiency of this new technique.

The broad scope of application considers photomodelling as an efficient technique in case of multimedia reconstructions for digitization and developments related to the virtuality of Cultural and Environmental Heritage, an instrument for any need of restoration after earthquakes and natural disasters. The intent is to make available innovative methods and tools to preserve unaltered the historical memory: it will be possible to reconstruct the objects possibly damaged, by using in advance the technique of photomodelling. At the end of the research, photomodelling is revealed as a suited technique, able to preserve the historical and cultural heritage and guarantee the authenticity of restorative measurements.

References

[1] Filippucci M 2010 Nuvole di pixel. La fotomodellazione con software liberi per il rilievo d'architettura *Disegnarecon* **3(6)** 50-63

[2] De Luca L 2011 *La fotomodellazione architettonica. Rilievo, modellazione, rappresentazione di edifici a partire da fotografie* Dario Flaccovio Editore (Palermo: D. Flaccovio) pp 19-30

[3] De Luca L 2009 Architectural Image-Based surveying, modeling and representation, methodological reflections and research tracks *Architectural Image-Based Modeling Web Portal* Available from http://www.gamsau.map.archi.fr/aibm/Portal_of_Architectural_Image-Based-Modeling/Article-DeLuca2.html

[4] Szeliski R 2010 *Computer vision: algorithms and applications* Springer Science & Business Media (Londra) p 339

[5] Pucci M 2013 Prima che appaia il "divieto di fotorilievo": considerazioni sulla fotomodellazione *Disegnarecon* **7** 1-12

[6] Guidi G and Remondino F 2012 3D Modelling from Real Data *Modeling and Simulation in Engineering* ed C Alexandru (InTech) chapter 3 pp 69-102 Available from: https://www.intechopen.com/books/modeling-and-simulation-in-engineering/3d-modeling-from-real-data

[7] Russo M, Remondino F and Guidi G 2011 Principali tecniche e strumenti per il rilievo tridimensionale in ambito archeologico *Archeologia e calcolatori* **22** Edizioni All'Insegna del Giglio pp 169-198

[8] Marsili R and Garinei A 2014 Thermoelastic Stress Analysis of the Contact Between a Flat Plate and a Cylinder *Measurement: Journal of the International Measurement Confederation* **52(1)** pp 102-110 Available from: http://dx.doi.org/10.1016/j.measurement.2014.03.005

[9] Referring Web Page: http://meshlab.sourceforge.net/

[10] Referring Web Page: http://www.geomagic.com/en/

[11] Kazhdan M, Bolitho M and Hoppe H 2006 Poisson Surface Reconstruction. *Eurographics Symp. on Geometry Processing* ed K Polthier and A Sheffer

[12] Mittleman F B J, Silva H R C and Taubin G 1999 The Ball-Pivoting Algorithm for Surface Reconstruction *IEEE Transactions on Visualization and Computer Graphics* **5(4)**

[13] Marsili R and Garinei A 2013 Design of an optical measurement system for dynamic testing of electrospindles *Measurement* vol 46, Issue 5 (Amsterdam: Elsevier) pp 1715-1721 DOI information: 10.1016/j.measurement.2013.01.006, ISSN: 0263-2241

[14] Referring Web Page: http://www.agisoft.com/

[15] Referring Web Page: http://ccwu.me/vsfm/

[16] Referring Web Page: http://www.arc3d.be/

[17] Referring Web Page: http://remake.autodesk.com/about

[18] Referring Web Page: http://www.123dapp.com/catch

[19] Referring Web Page: http://photomodeler.com/products/scanner/default.html

[20] Referring Web Page: http://www.creaform3d.com/en/metrology-solutions/handheld-portable-3d-scanner-goscan-3d

[21] Rossi G 2010 *Misure meccaniche e termiche. Basi teoriche e principali sensori e strumenti* ed Carocci (Roma: Carocci)

[22] Referring Web Page: http://www.geomagic.com/it/products/qualify/overview

[23] Cardelli E., Faba A., Marsili R., Rossi G., Tomassini R., "Magnetic nondestructive testing", Journal of Applied Physics 117, 17A705 (2015); doi: 10.1063/1.4907180

Identification of calibration and operating limits of a low-cost embedded system with MEMS accelerometer

G D'Emilia[1], D Di Gasbarro[1], A Gaspari[1] and E Natale[1]
[1]University of L'Aquila, Department of Industrial and Information Engineering and of Economics, via G. Gronchi, 18, 67100, L'Aquila, Italy

E-mail (corresponding author): giulio.demilia@univaq.it

Abstract. In this paper some aspects concerning the calibration uncertainty of three-axis low-cost accelerometers for possible use in diagnostics of civil buildings are considered, using a linear slide and a laser vibrometer as the reference.
In order to analyse the principal and cross sensitivity and the offset of the sensor in dynamic conditions, the sensitivity matrix method has been used.
Some considerations about the operating limits of a low-cost embedded system with MEMS accelerometer have been discussed, with reference to the calibration procedure.
In particular, the effects of the non-constant sampling and of the achievable sampling rate, are studied with reference to the calibration uncertainty and to the capability of the calibration procedure to assess the best metrological performances of the system under test.

1. Introduction

Many applications can be found in different fields of engineering, where a network of a significant number of accelerometers, have to be used to monitor the system of interest. Only to cite some relevant examples, a distributed network of accelerometers for Structural Health Monitoring (SHM) of buildings and of civil infrastructures [1], for earthquake monitoring [2], for condition monitoring of high performance tool machines aiming to incipient fault location and classification.

Due to the mechanical complexity of the systems, in most real cases three-dimensional acceleration fields are of interest, therefore requiring a large number of three-axis accelerometers for monitoring and diagnosing purposes. In all these applications accuracy of measured data is obviously a mandatory requirement for reliable diagnosis, together with the need of low cost sensors.

Some solutions based on new sensor technologies (MEMS, FOS, ...), are available for low cost three-axis accelerometers and networking applications of accelerometers already exist for the above-mentioned applications [3], [4]. Today the trend is to design integrated solution systems with several sensors embedded on the same device, in particular low cost accelerometers.

Several low-cost solutions are available on the market [5], based on MEMS accelerometers with integrated ADC system. A new generation of microcontrollers allows us to communicate with these accelerometers with standardized serial buses (for instance I²C). The controller is also very simple to code, as, for example, the Atmega328 controller belonging to the Arduino Uno system, programmed by means of a similar C++ language. Furthermore, this way integrated low cost and low power solutions to monitor accelerations are very simple to implement.

Some aspects have to be considered when such solutions are used, mainly affecting the measurement uncertainty of three axis sensors;

- cross sensitivity;
- frequency rate of data;

- phase control of data coming from different sensors.

In fact, even though many operating examples could be found, nevertheless, these technologies, need to be characterized with respect to many influencing factors, related to both environmental (temperature, humidity, time stability ....) and metrological aspects: in particular, cross sensitivity [6], [7], [8], [9]. Even though a great attention is paid to the indication of cross sensitivity performances by calibration of them, literature indications are often unsatisfactory for many reasons. Some works propose several methods to calibrate accelerometers in dynamic conditions and all are based on sensitivity matrix [6], [7], [8], [9], [10], [11], [12]. Generally the calibration of the embedded solutions is based on non-standard procedures like the mean of data acquired in static conditions [13] or in the best cases is limited to the auto-calibration procedure, available in the integrated accelerometer firmware [2], [14]. Typically, the reference for the auto-calibration procedure is the gravity acceleration. Anyway, in most real applications, the acceleration solicitations are effectively and contemporaneously operating on all the three-axis and mutual interaction between measuring axes should be taken into account accurately; these mutual interaction between axes are often neglected.

Using low cost microcontroller and communication solutions requires that special attention is devoted to both enhance the data rate and the synchronization of data deriving from different sensors especially when modal analyses of monitored structures are useful for the analysis to be carried on [15], [16].

Based on these considerations, this paper describes a procedure for accurate evaluation of operating limits of a low-cost embedded system based on MEMS accelerometer. The frequency range of acceleration is 0 to 10 Hz, realized by linear oscillation according a sinusoidal behavior.

In particular, measurement uncertainty will be evaluated, with reference to main and cross sensitivities, data rate, also in terms of time interval stability as for data acquisition. The effect of settings in terms of coding strategy of the microcontroller and of the transmitting of the data will be also taken into account.

The information gained from the calibration procedure will be used to also improve the settings of the embedded low-cost system, according to a recursive approach.

The effect of reference acceleration will be also studied in order to define a calibration test bench accurate and of reduced cost.

The paper is organized according to the following steps:

– in section 2 the methodology will be discussed, concerning the experimental solutions able to realize the requested reference acceleration and the procedures for data processing;

– in section 3 results of tests are presented which have been carried out according to the proposed methodology. The operating limits and the accuracy improvements of a low cost accelerometer system made by the calibration technique will be discussed. Particular attention will be paid to the reduction of cross sensitivity and to the improvement of the time stability of task times, for uncertainty reduction.

Conclusions end the paper.

2. Materials and methodology

A linear vibrational motion is realized by a linear slide according to a sinusoidal time; frequencies of oscillation in the range 0 to 10 Hz are operated by an actuator driven by the computer. An inclined steel plate is installed on the test bench; the inclination angle is of 45° with respect to the horizontal plane.

Two three-axis accelerometers can be placed on the plate of different characteristics; accurate positioning at different angles of rotation on the inclined plane (0°, 30°, 45° and 60 °) with respect to an horizontal line can be realized by pre-machined holes (figure 1).

The velocity of the slide in the motion direction can be accurately monitored by a Laser Doppler Vibrometer, LDV (figure 2).

IOP Conf. Series: Journal of Physics: Conf. Series **882** (2017) 012006 doi:10.1088/1742-6596/882/1/012006

Figure 1. Test bench.

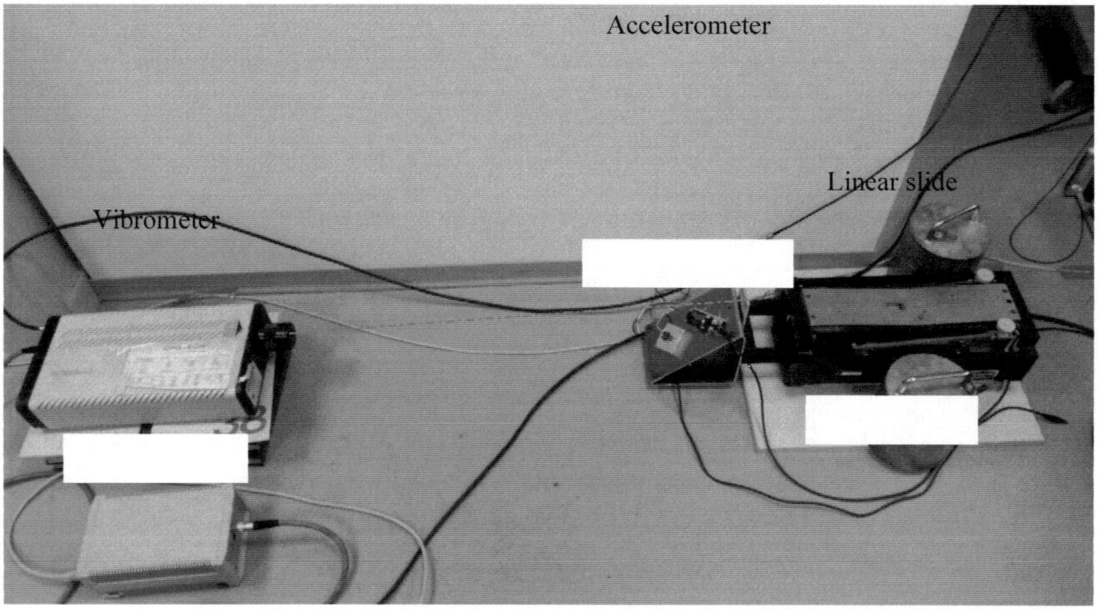

Figure 2. The complete calibration system.

The low-cost system under test is based on Arduino Uno controller and a low-cost MEMS accelerometer (figure 3). Similar solutions are used in [2], [10], [13], [14], [17], [18], [19], [20], [21]. The Arduino Uno software was coded in a similar C++ language and the PC based acquisition software for all the system is coded in NI LabVIEW like the solutions proposed in [2], [19].

The comparing accelerometer is a Sequoia FastTracer, an accelerometer using MEMS technology that incorporates into sensors three-axes sensor technology and the capability of analog to digital conversion. Vibration data are exported directly to the computer via a standard USB port, turning it thanks to an application software.

The velocity signal of the LDV is acquired with a NI c-Rio system coded with Labview RT.

The results are synchronized and elaborated in Matlab.

IOP Conf. Series: Journal of Physics: Conf. Series **882** (2017) 012006 doi:10.1088/1742-6596/882/1/012006

For the general calibration of the sensor the following relationship between the input accelerations and output signals can be set:

$$\begin{pmatrix} V_x \\ V_y \\ V_z \end{pmatrix} = \begin{pmatrix} S_{xx} & S_{xy} & S_{xz} \\ S_{yx} & S_{yy} & S_{yz} \\ S_{zx} & S_{zy} & S_{zz} \end{pmatrix} \begin{pmatrix} a_x \\ a_y \\ a_z \end{pmatrix} + \begin{pmatrix} q_x \\ q_y \\ q_z \end{pmatrix} \quad (1)$$

where:

$V = (V_i)$: output array,

$A = (a_i)$: reference accelerations array,

$S = (S_{ij})$: sensitivity matrix,

$Q = (q_i)$: offset array.

A total of 12 parameters have to be evaluated; a linear Least Squares Optimization will be used to estimate them, as described in references [2], [6] and [7].

It is to be pointed out that this procedure is the most complete approach; in fact, taking into account the possibility of not negligible offset vector means that predefined hypotheses are not considered.

The output from the LDV is a velocity signal and it must be processed to get information about the acceleration. In particular, the velocity signal has to be differentiated and its components, according to the axes of the accelerometer in calibration, should be evaluated. Moreover, the appropriate component of the gravity acceleration g has to be added to each direction.

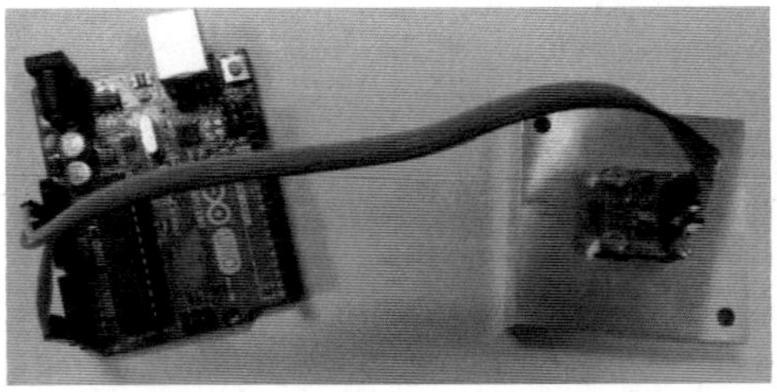

Figure 3. Low cost system based on Arduino Uno controller and an accelerometer ADXL 350Z Eval board.

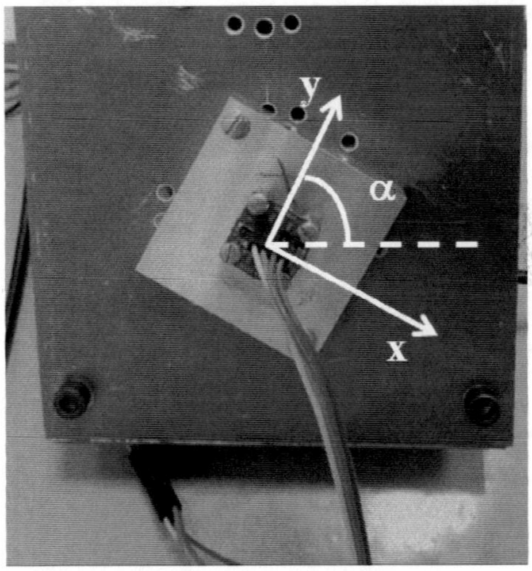

Figure 4. Angle of rotation of the accelerometer on the inclined plate.

Then, if the rotation angle on the inclined plate of the accelerometer is indicated as α (figure 4), the accelerations along the three axes can be obtained as follows:

$$a_x = a_{vib} \cdot \cos(45°) \cdot \cos(\alpha) + g \cdot \cos(45°) \cdot \cos(\alpha) \quad (1)$$

$$a_y = - a_{vib} \cdot \cos(45°) \cdot \text{sen}(\alpha) - g \cdot \cos(45°) \cdot \text{sen}(\alpha) \quad (2)$$

IOP Conf. Series: Journal of Physics: Conf. Series **882** (2017) 012006 doi:10.1088/1742-6596/882/1/012006

$$a_z = a_{vib} \cdot \cos(45°) - g \cdot \cos(45°) \tag{3}$$

where g is the gravity acceleration, and a_{vib} is the acceleration in the motion direction, by differentiating the LDV output.

In order to reduce the numerical noise in acceleration signal from the LDV, the velocity derivative has been computed by the following formula:

$$a_{vib} = \omega \cdot v_{vib} \tag{4}$$

being ω the pulsation of the sinusoidal motion and v_{vib} the measured velocity by the LDV.

The stability of the LDV signal has been evaluated on the base of 30 cycles, as a standard deviation of peaks values, and it is equal to 0.12% of the amplitude.

The LDV has been used as a reference sensor and the two considered acceleration measuring systems, both based on MEMS sensors (the low-cost MEMS and the Sequoia Fast Tracer) have been calibrated with reference to it, in order to highlight the differences.

3. Results of dynamic calibrations and comparison between sensors

The Sequoia accelerometer has been calibrated with reference to the LDV signal.

The Sequoia and the LDV signal have been acquired at a sampling rate of 1024 Hz.

Considering the observations made in previous paragraphs, the accelerometer behavior during a whole oscillation period is considered in comparison with the reference; the test is repeated at different positions of the sensor; for each position the reference accelerations along the x, y and z axes have been calculated, depending on gravity and motion acceleration components and compared with the corresponding outputs of sensors in all directions.

In order to evaluate the variability of the results, 10 different sets of points have been considered, corresponding to different oscillation periods; 10 different matrices S and offsets q have been correspondingly calculated.

Then, the average matrix S and the average offset vector q have been calculated:

$$S = \begin{pmatrix} 0.99 & 0.010 & 0.0033 \\ -0.030 & 1.0 & 0.022 \\ 0.011 & 0.0033 & 1.0 \end{pmatrix}$$

$$q = \begin{pmatrix} 0.16 \\ 0.34 \\ 0.073 \end{pmatrix}$$

and also the standard deviations of them (called dev.st(S) and dev.st(q), respectively) have been calculated:

$$\text{dev.st}(S) = \begin{pmatrix} 0.0019 & 0.0015 & 0.0017 \\ 0.0019 & 0.0012 & 0.0026 \\ 0.0024 & 0.0021 & 0.0016 \end{pmatrix}$$

$$\text{dev.st}(q) = \begin{pmatrix} 0.027 \\ 0.031 \\ 0.031 \end{pmatrix}$$

As can be seen, a maximum relative deviation from the mean values of 0.19% has been evaluated for the coefficients of the main sensitivities, that is a negligible variability.

Considering all causes of uncertainty, these results are in a close agreement with the values on the calibration certificate of the sensor.

IOP Conf. Series: Journal of Physics: Conf. Series **882** (2017) 012006 doi:10.1088/1742-6596/882/1/012006

Also for the low-cost system, the sensitivity matrix S' and the offset vector q' have been calculated on the basis of the reference values, as an average of 10 different matrices and offsets:

$$S' = \begin{pmatrix} 1.1 & 0.012 & 0.019 \\ 0.044 & 0.98 & 0.069 \\ -0.0047 & -0.012 & 0.98 \end{pmatrix}$$

$$q' = \begin{pmatrix} -0.16 \\ -0.74 \\ -0.73 \end{pmatrix}$$

and the standard deviations of them (called dev.st(S') and dev.st(q'), respectively) have been calculated:

$$\text{dev.st}(S') = \begin{pmatrix} 0.021 & 0.012 & 0.017 \\ 0.021 & 0.011 & 0.018 \\ 0.026 & 0.013 & 0.023 \end{pmatrix}$$

$$\text{dev.st}(q') = \begin{pmatrix} 0.26 \\ 0.27 \\ 0.32 \end{pmatrix}$$

Also for low-cost acceleration measuring system the calibration matrix values are according the specification data, if the main sensitivities are evaluated and the cross sensitivity is evaluated as the mean of the matrix data.

The effect of the correction on the raw data of the MEMS accelerometer, by the matrix S and the offset vector q obtained, is shown in figure 5, and appears to be satisfying.

A maximum relative deviation from the mean values of 2.3% has been evaluated for the coefficients of the main sensitivities.

These results confirm that the methodology provides good results.

However, the calibration of the low-cost MEMS accelerometer with reference to the LDV signal, has involved some operational difficulties in applying the methodology and special attention.

The main problems are as in the following:

- the acquisition rate for the low-cost system is not perfectly constant over time;
- a reduced acquisition data rate has to be set, in order to guarantee safe data flow.

Both these aspects increase the variability of results and, therefore, the calibration uncertainty.

In fact, due to variability of the acquisition frequency of data of the low cost acceleration system, there is not time synchronicity between reference system and the one under calibration; this time jitter worsen variability of the results to be processed in order to evaluate the sensitivity matrix.

Furthermore, it is not possible to set an acquisition rate in the order of 1024 Hz, as for the reference, because an average frequency in the order of 200 Hz is the maximum limit beyond which the low-cost measuring system on the whole does not guarantee the reliability of the acquired values. That limit depends, of course, also on the controller and on the acquisition software.

This requires the calculation of the average acquisition rate of the sensor and the setting of the acquisition rate of the reference at that value.

The average acquisition rate of the accelerometer has to be evaluated on the time interval considered for the calibration, because otherwise large synchronization errors with respect to reference data may occur, making calibration results meaningless.

The described calibration procedure highlight the effects of some operating aspects of the low-cost measuring system; in fact, making the data rate high and stable, as much as possible, do not only affect the calibration uncertainty of the system, and this could be considered obvious, but also makes the calibration procedure easy, quick and effective, allowing to better define the system metrological performances.

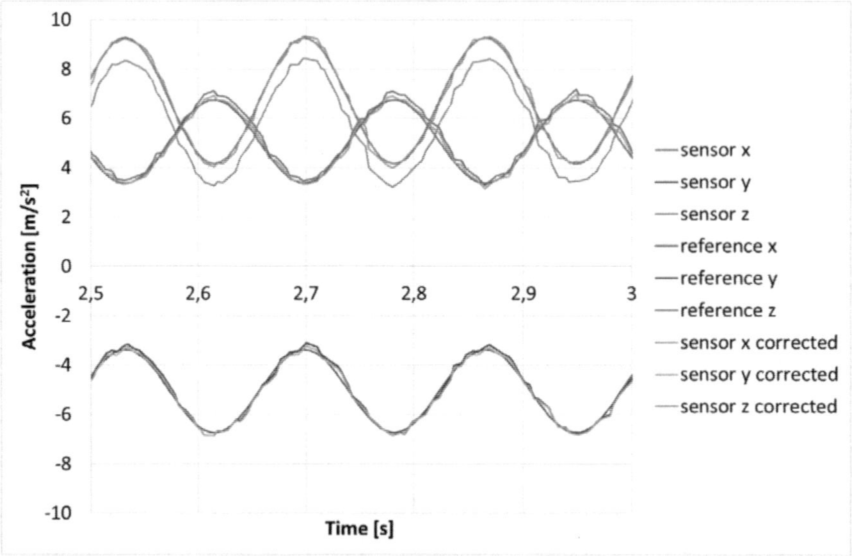

Figure 5. Comparison reference LDV data, low-cost MEMS raw data, and low-cost MEMS data corrected by the sensitivity matrix. a) x-axis acceleration; b) y-axis acceleration; c) z-axis acceleration.

4. Conclusions

In this paper some aspects concerning the calibration uncertainty of three-axis low-cost accelerometers for possible use in diagnostics of civil buildings are considered.

The test bench allows to realize a three-axis acceleration with enough accuracy to show the cross sensitivity and bias effects.

In order to analyze the principal and cross sensitivity and the offset of the sensor in dynamic conditions, the sensitivity matrix method has been used.

A LDV has been used to calibrate an embedded system with low-cost three-axis accelerometer in order to reduce the bias of the sensor and calculate the uncertainty of the system.

Some considerations about the operating limits of a low-cost embedded system with MEMS accelerometer have been discussed.

In particular, some problems in the application of the methodology to the low-cost system have been highlighted: the non-constant sampling, which requires the estimation of an average sampling rate, and the low sampling rate, which requires increasing the acquisition time.

These effects affect the uncertainty of calibration, and this could be considered obvious, but also preclude the calibration procedure to be easy, quick and effective, making difficult to define the best system metrological performances.

References

[1] Rainieri C, Fabbrocino G, and Verderame G M 2013 Non-destructive characterization and dynamic identification of a modern heritage building for serviceability seismic analyses. *NDT & E International* **60** pp 17-31.

[2] Upadhye M Y, Borole P B, and Sharma A K 2015 Real-time wireless vibration monitoring system using LabVIEW *In Industrial Instrumentation and Control (ICIC), 2015 International Conference on* pp 925-928

[3] Liu Y, Wang H, Qin H and Xie Y 2015 Performance enhancement for piezo-resistive micro-accelerometer by geometrical design: a focused review, *Sensor Review* **35**(3), pp. 310-318

[4] Feng D *et al.* 2015 A fiber bragg grating accelerometer based on a hybridization of cantilever

beam, *IEEE Sensors Journal* **15**(3) pp. 1532-1537

[5] Peng C *et al*. 2017 A new type of tri-axial accelerometers with high-dynamic range MEMS for earthquake early warning *Computers and Geoscience* **100**(3) pp.179-187

[6] D'Emilia G, Gaspari A, and Natale E 2016 Evaluation of aspects affecting measurement of three-axis accelerometers. *Measurement* **77** pp 95-104.

[7] D'Emilia G, Gaspari A, and Natale E 2015 Dynamic calibration uncertainty of three-axis low frequency accelerometers *ACTA IMEKO* **4**(4) pp 75-81.

[8] D'Emilia G, Gaspari A, and Natale E 2014 Distributed sensor network for structural characterization of buildings aggregates: Discussion about aspects affecting measurement uncertainty, *13th IMEKO TC10 Workshop on Technical Diagnostics 2014: Advanced Measurement Tools in Technical Diagnostics for Systems' Reliability and Safety* (*Warsaw, Poland*) pp 108-113.

[9] Schiavi A, Mazzoleni F, and Germak A 2015 *Proc. XXI IMEKO World Congress" Measurement in Research and Industry"(Prague, Czech Republic)* vol 1 pp 2161-2164.

[10] Frosio I, Pedersini F, and Borghese N A 2012 Autocalibration of triaxial MEMS accelerometers with automatic sensor model selection *IEEE Sensors Journal* **12**(6) pp 2100-2108.

[11] Nakano A, Hirai Y, Sugano K, Tsuchiya T, Tabata O, and Umeda A 2013 Rotational motion effect on sensitivity matrix of MEMS three-axis accelerometer for realization of concurrent calibration using vibration table *In Micro Electro Mechanical Systems (MEMS), 2013 IEEE 26th International Conference on* pp 645-648

[12] Veldman C S 2015 Accelerometer transverse sensitivity calibration; validation and uncertainty estimation *ACTA IMEKO* **4**(2) pp 52-56

[13] Malatinec T, Popelka V, Huba M, and Hudacko P 2014 Laboratory model helicopter control using a lowcost Arduino hardware *In Control Conference (ICCC), 2014 15th International Carpathian* pp 326-331

[14] Frosio I, Pedersini F, and Borghese N A 2009 Autocalibration of MEMS accelerometers *IEEE Transactions on Instrumentation and Measurement* **58**(6) pp 2034-2041.

[15] Ong Z C, Lim H C, Khoo S Y, Ismail Z, Kong K K, and Rahman A G A 2017 Assessment of the phase synchronization effect in modal testing during operation *Journal of Zhejiang University: Science A* **18**(2) 92-105

[16] Ganguly V, and Schmitz T L 2014 Correction for synchronization errors in dynamic measurements *Transactions of the North American Manufacturing Research Institution of SME* **42** pp. 335-344

[17] Varanis M, Silva A L, Brunetto P H A, and Gregolin R F 2016 Instrumentation for mechanical vibrations analysis in the time domain and frequency domain using the Arduino platform *Revista Brasileira de Ensino de Física* **38**(1).

[18] Köse T, Terzioğlu Y, Azgın K, and Akın T 2015 A single mass two-axis capacitive MEMS accelerometer with force rebalance *In Inertial Sensors and Systems (ISISS), 2015 IEEE International Symposium on* pp 1-4

[19] Hoque R, Hassan S, Sadaf M A, Galib A, and Karim T F 2015 Earthquake monitoring and warning system *In Advances in Electrical Engineering (ICAEE), 2015 International Conference on* pp 109-112

[20] Martin H F S, Groves P D, Newman M, and Faragher R 2013 A new approach to better low-cost MEMS IMU performance using sensor arrays *The Institute of Navigation*

[21] Slyper R, and Hodgins J K 2008 Action capture with accelerometers *In Proceedings of the 2008 ACM SIGGRAPH/Eurographics Symposium on Computer Animation* pp 193-199

IOP Conf. Series: Journal of Physics: Conf. Series **882** (2017) 012007 doi:10.1088/1742-6596/882/1/012007

Accuracy assessment for a multi-parameter optical calliper in on line automotive applications

G D'Emilia[1], D Di Gasbarro[1], A Gaspari[1] and E Natale[1]
[1]University of L'Aquila, Department of Industrial and Information Engineering and of Economics, via G. Gronchi, 18, 67100, L'Aquila, Italy

E-mail (corresponding author): antonella.gaspari@ univaq.it

Abstract. In this work, a methodological approach based on the evaluation of the measurement uncertainty is applied to an experimental test case, related to the automotive sector. The uncertainty model for different measurement procedures of a high-accuracy optical gauge is discussed in order to individuate the best measuring performances of the system for on-line applications and when the measurement requirements are becoming more stringent. In particular, with reference to the industrial production and control strategies of high-performing turbochargers, two uncertainty models are proposed, discussed and compared, to be used by the optical calliper. Models are based on an integrated approach between measurement methods and production best practices to emphasize their mutual coherence. The paper shows the possible advantages deriving from the considerations that the measurement uncertainty modelling provides, in order to keep control of the uncertainty propagation on all the indirect measurements useful for production statistical control, on which basing further improvements.

1. Introduction

The increasing of the performances of automotive components and systems impose the automotive industry to fulfil more stringent requirements for design and tolerances of mechanical components. On the other hand, the need of reducing scrapes to a near zero level and the production costs in the whole asks for more stringent performances of the measuring systems for productive process control [1], [2].

In many cases the improvement of the measurement characteristics can be obtained by the use of complex measuring apparatuses, allowing to automatically and contemporaneously measure many parameters, depending on their setting. Each parameter that can be measured is a different physical quantity and so, although the measurement system is still the same, the measurement process can change, depending on the chosen parameter to be measured. That means the uncertainty evaluation could be also different.

Typically, only the measurement uncertainty is given by the instrument manufacturer concerning quantities which are directly measured [3]. In facts, indirect measurements imply operating and managerial choices, such as, for example the number of measurement points or the working duration of the measurement, making indirect measurement uncertainty depending in a remarkable way on the specific situation.

Many important applications exist, as the case of coordinate measurements, where the assessment of the real measurement accuracy gains key importance; in order to overcome the difficulty of this task some developed methods exist [4], covering the measurements of different geometrical quantities, but they are difficult to extend to other geometrical measuring systems. Furthermore, it has to be pointed out that in the industrial practice, measurement methods previously adopted may influence the choice of the procedure to be implemented, thus limiting the possibility of improvements deriving from the use of new and updated measurement systems.

Anyway, a careful preliminary evaluation of the accuracy is necessary, if effective improvements have to be realized, taking into account the requested high level of quality process control in the automotive sector [5], [6]. Also, the planning of measurement strategies could be influenced by this analysis [2]. Being able to measure different parameters according to the possibilities given by the complex measurement systems, the uncertainty evaluation should be specialized for the particular or the possible parameters which are of interest.

Methods to propagate the uncertainty of length inspection based on the uncertainty propagation of coordinates measurement exist, such as Monte-Carlo Simulation and Taylor Series Method [7]. Anyway, the modelling of the uncertainties causes is often difficult, in particular for on-line indirect measurements, and an experimental validation is required. The theoretical and experimental analysis is organized and realized according to the specific industrial practices and procedures but a prompt action may prevent consequences on the production phases.

A structured and recursive approach for uncertainty management could be useful [5], [8], [9].

In this paper, the accuracy of a high performance optical caliper is studied with reference to the measurements of dimensional parameters of turbochargers to be carried out on the production line for quality control. Depending on the chosen measurement procedure, different dimensional parameters could be measured, like turbine profile and angles, turbine/stator clearance, turbine axis alignment with a reference one, relevant plane distance to each other and so on. In order to optimize the working settings for the different parameters to be measured, a model of the measurement uncertainty is developed, depending on measurement and production considerations, in order to get a physical interpretation of the different measurement procedures.

Different models can be set, depending on the specific parameter to be measured, in order to define the best approach in any case. The methodology allows us to identify the main causes of variability together with the possible causes of systematic error, for accuracy improvement.
Three main international reference standards are taken into account:

- the reference given by the manufacturer in order to define the Maximum Permissible Error (MPE) [10], for coordinate measuring systems;
- the reference to conduct the test uncertainty for the measuring machine [11], where the main contributions to the error of indication are defined (e.g. the calibration error of the material standard of size, the error due to the misalignment, the error due to fixturing the material standard of size, ...);
- the reference for determining the uncertainty of measurement [12], where the uncertainty of the measurement procedure is evaluated through a Type A evaluation (components evaluated by statistical methods).

This approach allowed us to define a set of practical suggestions useful to improve the measurement procedure and the accuracy of data, taking into account the characteristics of the manufacturing process. The effects of on-line applying this improved methodology are also described in the paper, showing that a significant reduction of scrapes and of time devoted to control of production could be achieved [1]. More effective control procedures could then be set, allowing a significant reduction of cost of the control activities.

The paper is organised as follows. In Section 2 the methodology and the test case are described.

The framework is approached by highlighting the scenario, the requirements and the materials used for the analysis, coming from a finer identification of the boundaries for the core analysis itself, which is taken into account [5]. The preliminary description of this, is intended to define the main uncertainty causes that cannot be neglected (*2.1 Uncertainty causes identification*). Then, the uncertainty budget, the indicators for measurement uncertainty and the measurement uncertainty results are provided for two different measurement methods, named *Method 1* and *Method 2*, respectively (*2.2 Measurement uncertainty evaluation: Method 1*, and *2.3 Measurement uncertainty evaluation: Method 2*). Section 3

IOP Conf. Series: Journal of Physics: Conf. Series **882** (2017) 012007 doi:10.1088/1742-6596/882/1/012007

shows the results obtained comparing the two measurement methods, carrying out a set of measurements by means of the optical calliper on the components of interest. The results are expressed in terms of standard statistical indicators [13], which represent a common practice in the industrial scenario, which is the main field of application of this paper. Short conclusions end the paper.

2. Methodology and test case

2.1. Uncertainty causes identification.

2.1.1. The scenario. The increasing of the performances of automotive components and systems impose to fulfil stringent requirements for both design and production processes. The scenario analyzed refers to the procedure for realization and control of highly-performing turbochargers, by means of advanced and complex measuring apparatuses.

2.1.2. The Requirements. The increasing performance required for turbochargers imposes the fulfilment of more stringent requirements in terms of the mechanical design and tolerances, and consequently also in terms of the performances of the measuring system for productive process control. The reduction of the leakage of fluid under pressure together with the increase in the required power to the new models are the main goals fixed, which are translated into the following:

- a stringent tolerance on the minimum clearance between the turbine wheel and the turbine house, in the order of 0.05 mm,
- the design of a turbine with a longer axial extension, whose outer profile ends with angles between turbine wheel profile and rotation axis, which are larger with respect to the normal axis, in order to exploit the enthalpy gap as far as possible.
- The stability of the improved solution should be also fulfilled, in accord with the need of consolidating it.

2.1.3. The boundaries of the analysis. The guarantee that along the axis of rotation of the turbocharger (the x-axis, in figure 1), there is no interference between the profile of the turbine wheel and the volute of the turbine housing requires that, for each abscissa x, the clearance g(x) is positive but lower than the dimensional tolerance (Eq. 1). Therefore, the true value of g(x), which is defined as the difference between the radius $r_v(x)$ of the volute and the radius $r_g(x)$ of the impeller (Eq. 2), is dependent on the true values of $r_v(x)$ and $r_g(x)$, and this must be ensured along the entire circumference.

The adjustment of the measurement procedure to the new requirements carried out according to the six-sigma approach, led the company to provide 100% control of production, with reference to both dimension B and run-out, since the process resulted not capable.

$$\forall x \in X,\ g(x) > 0,\ g(x) < t \tag{1}$$

$$g(x) := r_v(x) - r_g(x) \tag{2}$$

2.1.4. Materials. The above-mentioned developments strain the existing measurement methods, even if based on sophisticated optical gauges [3], [6]. Basically, these systems work on the measurement of two quantities: the distance between the laser head and the measuring item (measurement of the radius) and the axial position of the laser head itself (measurement of the length). The principle of operation of the optical calliper [3] allows us to obtain directly the radius or diameter of the impeller body.

As declared by the instrument manufacturer, the system is able to achieve for both diameter and length directions of measurement, a precision in the order of the micrometre, thus resulting appropriate for the use required by the company [6]. Taking into account the previous experience and the preliminary analysis on a new measurement approach, two different possibilities are considered:

- Method 1: As far as for the method 1, it considers the measurement of the abscissa for which the axial profile of the turbine realizes a given diameter, replicating the working condition of a mechanical gauge. This method takes into account the tolerance B, given as an interval along the axial direction (figure 2);
- Method 2: As far as for the measurement method 2, it directly considers the measurement of the radius (or the diameter) of the impeller body. The tolerance acts radially.

It appears evident that the measurement method 1 is physically meaningful for low values of angle alfa, since in this case axial tolerance practically coincides with a clearance. The problem is maintaining a physical meaning of measurements of parameters to be measured also for higher values of α, linked to technological development of turbochargers.

A set of 30 turbochargers are measured, by means of the optical gauge. The results obtained by using the two measurement methods are compared in terms of their capability indices Cp, Cpk, Pp, Ppk [13].

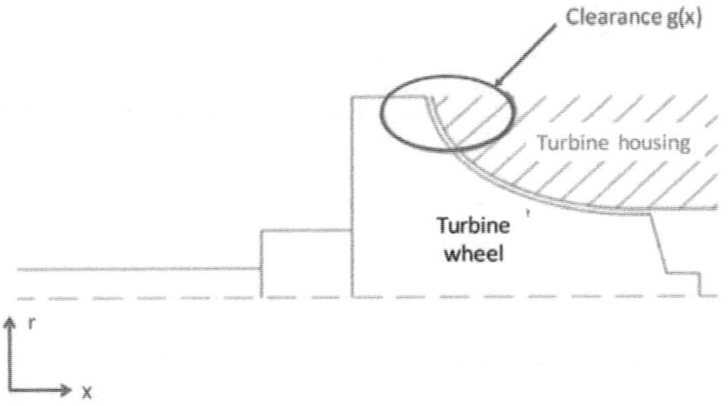

Figure 1. Tolerance on the turbocharger (schematic representation, adapted from [1])

2.2. *Measurement uncertainty evaluation: Method 1*

2.2.1. Uncertainty budget and test plan. The measurement method 1, reproduce the mechanical verification through a calibrated gauge able to measure the position B along the rotation axis, where the axial profile of the turbine achieves a diameter $\o_{gauge}$ of given value, with reference to the plane A (figure 2). Carefully analyzing the method of measurement, the following uncertainty causes are envisaged:

- a lack of significance of the tolerance B: as the angle α increases, the tolerance gradually loses physical meaning and tends to align with the axis of rotation, while the clearance has a radial direction;
- a strong influence of the slope of the turbine profile at the point where the measurement is performed, on the control of the profile itself, performed through a single point identified by a predetermined diameter.

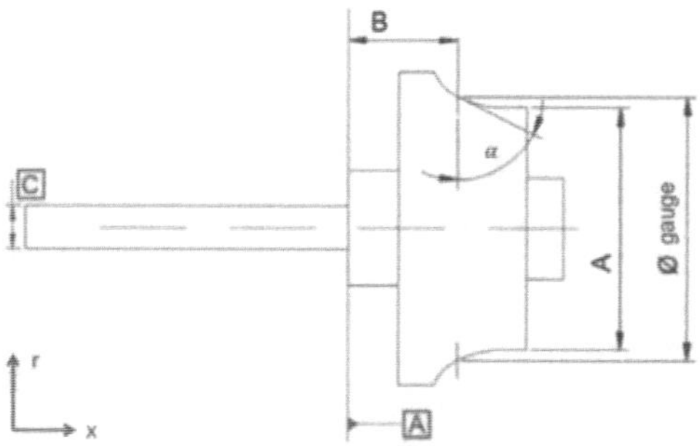

Figure 2. Measurement design for tolerance checking on the turbine wheel, measurement method 1 (schematic representation, adapted from [1])

Thus, the main uncertainty components may be summarised as follows (figure 3):

- the uncertainty of the instrument itself, when measuring an axial distance, dx';
- the variability associated to the reference plane, needed to perform the differential measurement, dx_A;
- the influence of the slope of the profile, at the point where the measurement is performed, $dx(\alpha)$;
- the misalignment between the axis of the component and the reference axis of the measuring system, $dx(\theta)$.

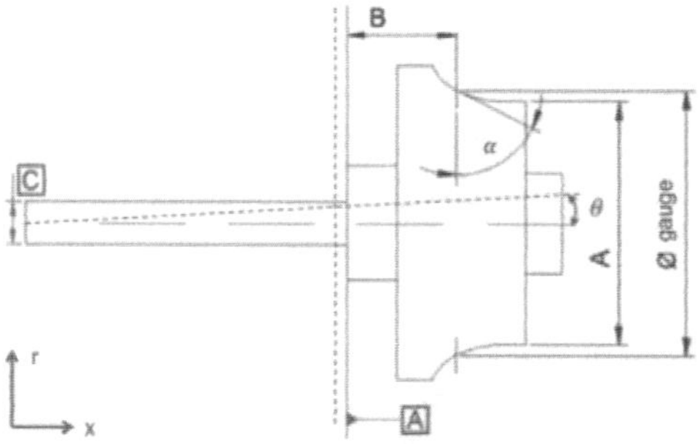

Figure 3. Schematic representation of the main uncertainty causes for measurement method 1

2.2.2. Indicators for measurement uncertainty. If we wish to use the optical instrument to obtain the abscissa for which the turbine realizes a given diameter (as required by the set tolerance, B, in Fig. 3) the instrument is required to provide a differential indication, of the type in equation (3) [1]:

$$dx = dr\ tan\alpha \tag{3}$$

where α is the angle of the profile with respect to the normal to the x-axis. It is clearly noted that the variability corresponding to dr is amplified with increasing angle α and that this makes unsatisfactory capability indices calculated with respect to the set tolerance. As far as for the measurement method 1, by applying equation (1), the measurement uncertainty dB for the abscissa B is given by equation (4):

$$dB = dr_g'(x) * \tan(\alpha) + dx_A + dx' + dr(\theta) * tg\theta \tag{4}$$

2.2.3. Measurement uncertainty results. As far as for the measurement method 1, the analysis of the uncertainty components leads to the following:

- the variability corresponding to dr is amplified with increasing angle α;
- as the angle α increases, the tolerance gradually loses physical meaning and tends to align with the axis of rotation, while the clearance has a radial direction;
- the control of the profile through a single point is strongly influenced by the slope of the profile itself at the point where the measurement is performed.

2.3. Measurement uncertainty evaluation: Method 2

2.3.1. Uncertainty budget and test plan. As far as for the measurement method 2, it directly considers the measurement of the radius (or the diameter) of the impeller body. The reference axis for the measurement system has been chosen according to the grinding process of the turbine wheels. This measurement method leads to the following uncertainty components (figure 4):

- the uncertainty of the instrument itself, when measuring a radial distance, $dr_g{}'$;
- the influence of the slope of the profile, at the point where the measurement is performed, $dr(\alpha)$;
- the misalignment between the axis of the component and the reference axis of the measuring system, $dr(\theta)$.

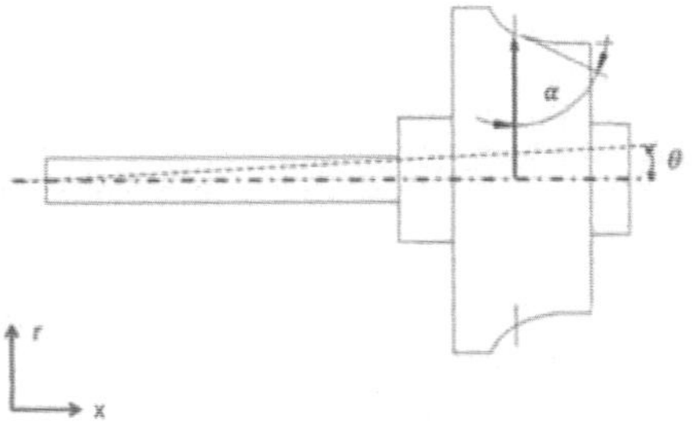

Figure 4. Schematic representation of the main uncertainty causes for measurement method 2

2.3.2. Indicators for measurement uncertainty. On the other hand, the measurement uncertainty of the radius r_g, which is the expected quantity as for the new measurement method, is given by equation (5):

$$dr_g = \frac{dx'}{\tan(\alpha)} + dr(\theta) + dr_g{}' \qquad (5)$$

2.3.3. Measurement uncertainty results. The new measurement method, on the other hand, shows that:

- high values of α reduce the whole uncertainty, since it reduces the effect of the uncertainty of $\frac{dx'}{\tan(\alpha)}$ of the axial positioning of the optical calliper;
- the misalignment between the optical axis of the calliper and the positioning axis of the shaft of the turbocharger is made negligible by the ability of the optical gauge to recreate virtually the rotation axis of the turbocharger, which is the reference axis of the grinding process; the strength of the manufacturing process becomes the strength of the control process as well, being this component of the measurement uncertainty really negligible;

IOP Conf. Series: Journal of Physics: Conf. Series **882** (2017) 012007 doi:10.1088/1742-6596/882/1/012007

- the profile is the result of an interpolation based on more measurements taken along the profile itself, thus reducing the whole uncertainty, with respect to a single measurement point.

Ultimately, the efficaciousness of the measurement method 2, with respect to the measurement method 1, is proved.

3. Results

A set of 30 turbochargers are measured, by means of the optical gauge. The results obtained by using the two measurement methods are compared in terms of their capability indices Cp, Cpk, Pp, Ppk.
Tests are carried out at different values of the angle α in order to understand the differences between Method 1 and Method 2, for different geometries of turbochargers.

The differences between the two measurement methods are evaluated by means of the indices for statistical process control. With reference to the former specification of the tolerance B (Method 1), in figure 5a and figure 6a the process capability evaluation data for a turbine profile with a small angle α are given, with a lower tolerance specification limit (LSL) of 20.1 mm and an upper tolerance specification limit (USL) of 20.2 mm. Data processing allows us also to predict, based on these performance values, the out-of-specification value, which resulted unsatisfactory.

With reference to the latter specification of tolerance and the general profile form (Method 2), the data of the process capability evaluation for a turbine profile with a low angle α are shown, with an LSL of 0 mm and a USL of 0.1 mm, are shown in figure 5b and figure 6b. The performance indices are greatly improved [20].

Similar satisfactory results are obtained with reference to the potential capability (within-capability) data and the overall capability data in terms of Cp and Cpk and in terms of Pp and Ppk, for a turbine profile with a large angle α [20].

The actions realised demonstrated that the measurement uncertainty reduction can be seen as a contribution to the knowledge of the system, also in respect to other competence areas.

The analysis of differences between the presented methods in terms of SPC indicators aims at predicting possible improvements in terms of the following productive advantages:

- a reduction in the necessary controls;
- a reduction in the dedicated manpower;
- a reduction in the observed scrapes, which are often false alarms;
- a reduction in the time taken for the operation.

4. Conclusions

In this work, with reference to the online control procedure of the critical geometrical dimensions of a high-performance turbocharger for automotive applications, the uncertainty budget and the modelling of the measurement process are carried out for two different measurement methods, both implemented on the same high-performances optical calliper.

The preventive analysis of the uncertainty budget of both methods allowed to correctly address the use of the calliper, depending on the dimensional characteristics of products (e.g. low/high values of angle α), basing the uncertainty budget on the modelling of the uncertainty causes and their effects for each measurand.

The construction of the uncertainty model requires to take into account measuring, production and operating topics. Furthermore, the following aspects have been achieved:

- concepts translation in order to interpret them from a physical perspective;
- studying the past actions for tolerances designing;
- tolerances re-designing;
- exploiting the force of the technological processes to define measurement references.

In conclusion, a structured and iterative approach supports the possibility of both addressing some improvement actions and of validating their effects. This is expected to reduce the randomness of opportunity of further actions, that is the "uncertainty" of the improvements themselves.

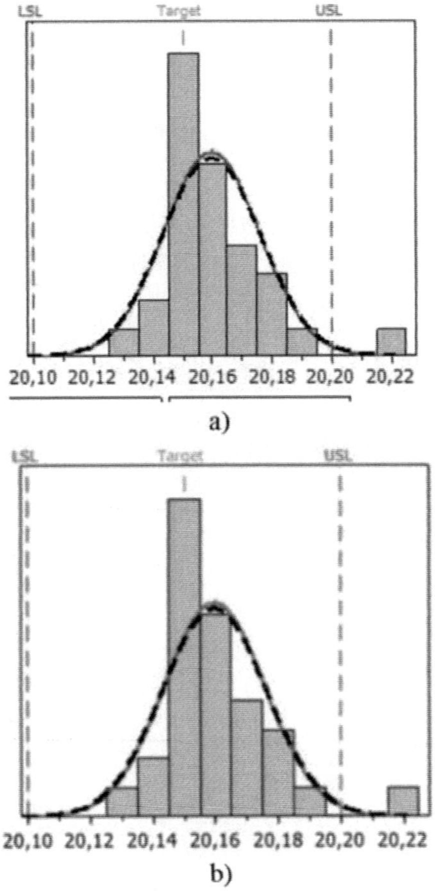

Figure 5. SPC of turbocharger – P_{pk} with a low α angle: Method 1 (a) and Method 2 (b) – redrawn from [1].

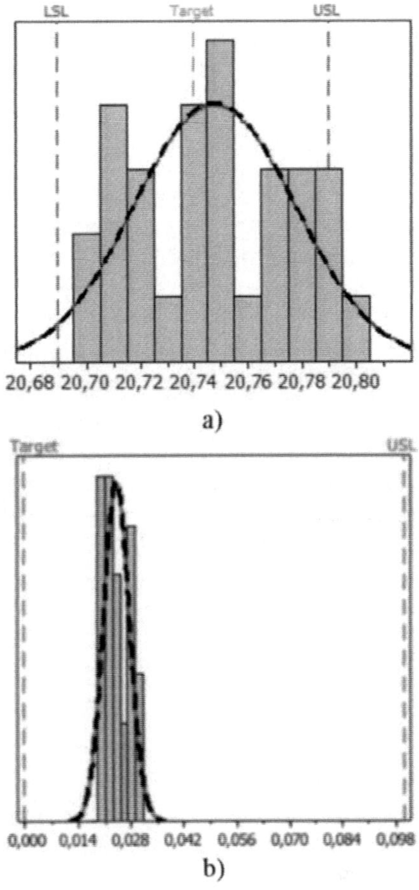

Figure 6. SPC of turbocharger – P_{pk} with a high α angle: Method 1 (a) and Method 2 (b) – redrawn from [1].

References

[1] D'Emilia G, Di Rosso G, Gaspari A and Massimo A 2015 Metrological interpretation of a six sigma action for improving on line optical measurement of turbocharger dimensions in the automotive industry *Journal of Automobile Engineering* **229**-2 pp. 261-269

[2] Kiraci E, Franciosa P, Turley G A, Olifent A, Attridge A and Williams M A 2016 Moving towards in-line metrology: evaluation of a Laser Radar system for in-line dimensional inspection for automotive assembly systems *International Journal of Advanced Manufacturing Technologies* DOI 10.1007/s00170-016-9696-8

[3] Jenoptic, HOMMEL-ETAMIC opticline C300, [Online]. Available: https://www.jenoptik.com/us-optical-shaft-metrology-hommel-etamic-opticline-c300 [Accessed 05 16 2016]

[4] Sładek J 2016 A Coordinate Metrology, Measurement Uncertainty and Requirements of Production System. *Selected Issues of Measurement Uncertainty Theory Springer Tracts in Mechanical Engineering* Springer-Verlag Berlin Heidelberg DOI 10.1007/978-3-662-48465-4_2

[5] Gaspari A 2016 Measurement-uncertainty: an innovative management tool in modern engineering systems *Ph.D. Thesis, University of L'Aquila (Italy)*

[6] D'Emilia G, Gaspari A 2016 Measurement-uncertainty as an innovative management tool in modern engineering systems *14th IMEKO TC10 Workshop on Technical Diagnostics 2016: New Perspectives in Measurements, Tools and Techniques for Systems Reliability, Maintainability and Safety* pp. 197-202

[7] Baia O, Jamshidib J, Kiracib E, Williams M A, Galettoa M, 2017 Measurement strategy impact on dimensional inspection by portable camera-based measuring systems, *Precision Engineering* **47** pp. 516–527

[8] JCGM – 100:2008 Evaluation of measurement data — Guide to the expression of uncertainty in measurement

[9] ISO 14253-2:2011 Geometrical product specifications (GPS) - Inspection by measurement of workpieces and measuring equipment - Part 2: Guidance for the estimation of uncertainty in GPS measurement, in calibration of measuring equipment and in product verification

[10] ISO 10360-8:2013 Geometrical product specifications (GPS) - Acceptance and reverification tests for coordinate measuring systems (CMS) – Part 8: CMMs with optical distance sensors

[11] ISO/TS 23165:2006 Geometrical product specifications (GPS) -- Guidelines for the evaluation of coordinate measuring machine (CMM) test uncertainty

[12] ISO 15530-3:2011 Geometrical product specifications (GPS) - Coordinate measuring machines (CMM): Technique for determining the uncertainty of measurement - Part 3: Use of calibrated workpieces or measurement standards

[13] Montgomery D C and Runger G C 2003 *Applied Statistics and Probability for Engineers*, 3rd Ed., John Wiley and Sons, Inc.

IOP Conf. Series: Journal of Physics: Conf. Series **882** (2017) 012008 doi:10.1088/1742-6596/882/1/012008

Laboratory and field tests on photo-electric probes and ultrasonic Doppler flow switch for remote control of turbidity and flowrate of a water-sand mixture flow

M Pellegrini[1] **and C Saccani**[2]

[1] Department of Industrial Engineering, University of Bologna, Via Fontanelle 40, 47100, Forlì, Italy.
[2] Department of Industrial Engineering, University of Bologna, Viale del Risorgimento 2, 40100, Bologna, Italy.

E-mail: marco.pellegrini3@unibo.it

Abstract. The paper describes the experimental apparatus and field tests carried on to remotely control through non-invasive and non-intrusive instruments turbidity and flowrate of a water-sand mixture flow conveyed by a pipeline. The mixture flow was produced by an innovative plant for seabed management. The turbidity was monitored by thru-beam infra-red photo-electric sensors, while flowrate was monitored by an ultrasonic Doppler flow switch. In a first phase, a couple of photo-electric sensors and a mechanical flow switch were preliminary tested in laboratory to verify installations concerns and measurement repeatability and precision. After preliminary test completion, photo-electric sensors and mechanical flow switch were installed in the real scale plant. Since the mechanical flow switch did not reach high reliability, an ultrasonic Doppler flow switch was identified and tested as alternative. Then, two couple of photo-electric sensors and ultrasonic Doppler flow switch were installed and tested on two pipelines of the plant. Turbidity and minimum flow signals produced by the instruments were integrated in the PLC logic for the automatic management of the plant. The paper also shows how ultrasonic Doppler flow switch measurement repeatability was negatively affected by the presence of the other ultrasonic Doppler flow switch working in a close pipeline and installed inside a steel casing.

1. Introduction

In the last decade an innovative technology for seabed plant management has been developed, designed and tested with the contribute of the University of Bologna [1,2]. The main element of the plant, called "ejector" (figure 1), is an open jet pump (i.e. without closed suction chamber and mixing throat) with a converging section instead of a diffuser. The ejector transfers momentum from a high speed primary jet flow to a secondary flow. The primary jet flow contacts the suction fluid at the nozzle exit and drags it into the sea bottom, thus starting up and sustaining the secondary flow of suction fluid from surrounding water mass. If present, solid particles are entrained in the secondary flow, where jet stream and suction fluid are further mixed, exchanging momentum and recovering pressure. The water-sediment slurry then passes through a converging and into a discharge pipe for delivery to a discharge point (or into a booster). The ejector can be used as a fixed or mobile device [3]: when used as a fixed device, one ejector works on a limited area whose diameter depends on the sediment characteristic as, for example, the angle of repose. The water for ejectors feeding is pumped from a pumping station placed near the ejectors installation area. The plant works 24 hours a day, 7 days a week, and by ejectors integration in series and in parallel it is possible to create and maintain over the time a seaway (figure 2).

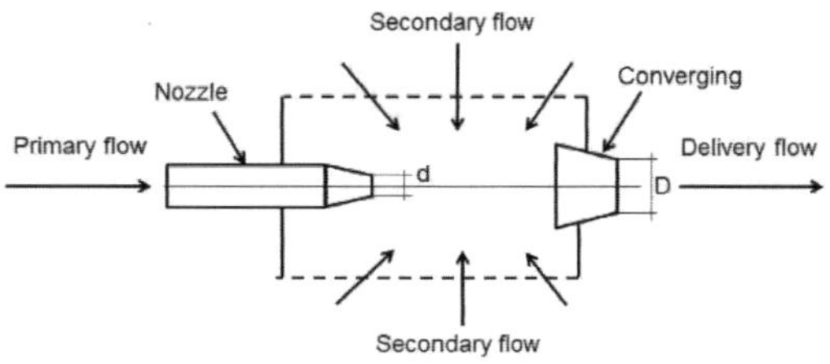

Figure 1. Schematic of the "ejector" developed by DIN.

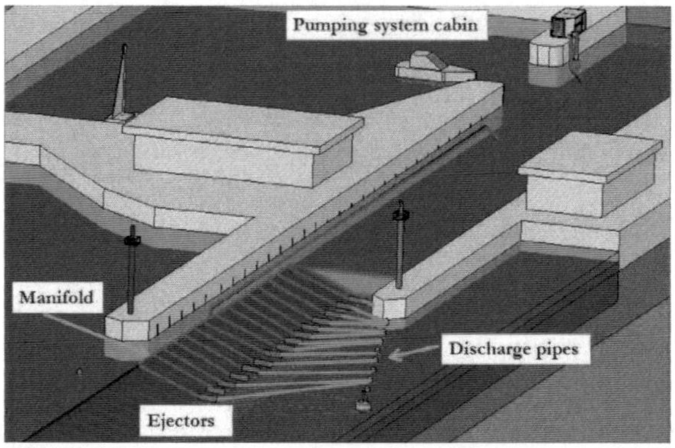

Figure 2. 3D image of seabed maintenance plant lay-out.

In 2005 [1] the first full-scale experimental plant was designed and carried out in the port of Riccione (Italy). The results of the experimental campaigns demonstrated the functionality of the system itself, the cost-effectiveness and low environmental impact if compared to the use of the dredge. In 2011 [2] a small scale industrial plant has been realized in the Portoverde Marina (Italy). This plant is characterized by better performances in automation and control higher than the first experimental plant. The monitoring of discharge pipelines was one of the critical issues to be further developed. In particular, two parameters need to be monitored: sediment content in the discharge pipelines and discharge flowrate. Both parameters can be used to regulate ejector water feeding: when discharge flowrate is too low and/or solid content is too high, the water flowrate that feeds the ejector must be increased up to a maximum level to avoid clogging risk, while if discharge flowrate and solid content are about nominal values, the ejector can work at a reduced load (i.e. less water feeding flowrate). The instruments to be applied for such measurements should have the characteristics described in table 1. After a market survey, two different instruments were identified as best suitable for the specific application: thru-beam infra-red photo-electric sensors for the measure of the solid content, ultrasonic Doppler flow meter for flowrate monitoring. Nevertheless, also a mechanical flow switch was taken into consideration to evaluate a low cost solution for flowrate monitoring, despite the fact that the instrument is both intrusive and invasive.

Table 1. Characteristics requested by the instrumentation to be used.

Characteristic	Reason
Continue monitoring	Avoid risk of clogging between two consecutive measurements.
On-off electric signal	Only need a command to bringing ejector water feeding to the maximum level from the nominal value.
High durability	Limit maintenance activities (and related costs).
High resistance	Work in aggressive environment (seawater and sand).
Low energy consumption	No access to the electric grid (use of electric batteries).
Non-intrusive and non-invasive	Avoid risk of clogging and limit pressure losses.
Good repeatability	More than precision, high repeatability is requested since the objective is not to continuously measure turbidity and flowrate, but to detect the exceeding of a critical point.

An infra-red photoelectric sensor emits a light beam from its light-emitting element. In the thru-beam configuration the light emitting element and the light receiving element are separated (figure 3). When a target passes between the emitter and the receiver it may interrupt the light transmission.

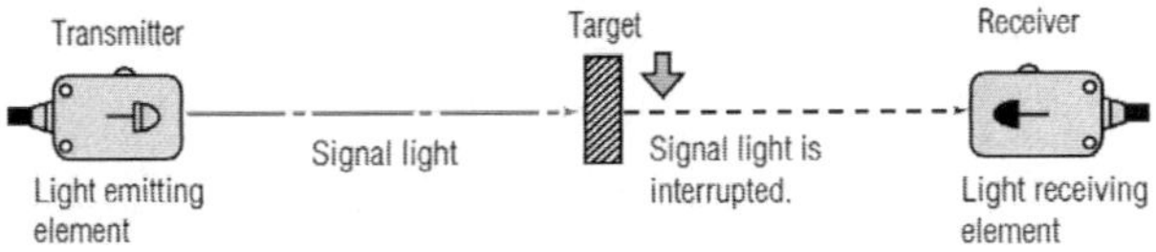

Figure 3. Example of thru-beam infra-red photo-electric sensor configuration, where signal light is interrupted by the presence of a target between light emitter and receiver.

Photo-electric sensors are widely used for different purposes in several industrial and non-industrial fields and so they represent a reliable technology [4,5,6]. The application of photo-electric sensors in solid content measurement (called "turbidity sensors") has been already applied [7,8] and many products are available, but they are quite expensive since sensors need a complex set up to result functional for the specific application. Moreover, turbidity sensors are usually quite large in size, since they are often installed for the turbidity monitoring of river, lake or sea. So, they cannot be easily adapted for a measure on a close and small duct (diameter size DN80 in the application described in the paper). On the other hand, it is possible to use commercial photo-electric sensors to measure turbidity, thus reducing cost and size, but the calibration must be realized by the user [9].
In the ultrasonic Doppler flowmeters (figure 4) an emitter sends ultrasonic waves at frequency f_1 (approximatively between 1-5 MHz) at angle α in a flowing fluid. The ultrasonic waves strike particles or air bubbles moving through the sound field at velocity v_P. The wavelength λ_1 of the emitted wave at frequency f_1 amounts to c/f_1. Due to its rate of motion v_P, the particle moving away from the emitter sees the wavelength λ_P as in equation (1):

$$\lambda_P = (c - v_P \cdot \cos\alpha) / f_1 \quad (1)$$

In turn, the receiver sees the reflected frequency f_2 (which is equal to c/λ_2) out of line because the reflecting particle is moving further away all the time, and the wavelength λ_2 changes as follows in equation (2):

$$\lambda_2 = (c - 2\, v_P \cdot \cos\alpha) / f_1 \quad (2)$$

The difference in frequency is therefore a linear measure of the rate of motion of the particles as in equation (3), where the term v_P can be neglected if compared with c.

$$f_2 - f_1 = (2\, v_P \cdot f_1 \cdot \cos\alpha) / (c - 2\, v_P \cdot f_1 \cdot \cos\alpha) \approx (2\, v_P \cdot f_1 \cdot \cos\alpha) / c \quad (3)$$

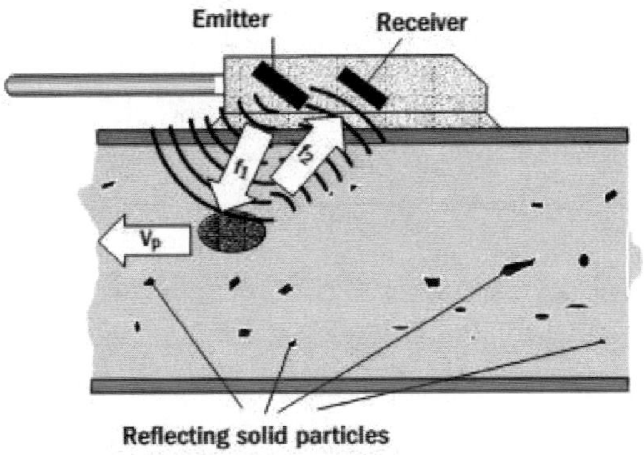

Figure 4. Principle of the Doppler flowmeter.

So, ultrasonic Doppler flowmeter needs "reflectors" like solid particles and air bubbles in the process fluid to properly work. That's why, in particular, ultrasonic Doppler flowmeters are applied in the

IOP Conf. Series: Journal of Physics: Conf. Series **882** (2017) 012008 doi:10.1088/1742-6596/882/1/012008

monitoring of flowrate discharge in wastewater plants and natural or artificial canals and channels [10,11]. Due to the complexity of the physical process, which is influenced by many factors, this kind of instrument has been perceived as un-reliable and uncertainty: in particular, wrong installation (i.e. sensor positioning) [12,13], acoustic interference [14] and variability of reflectors content [15] are the typical issues affecting the reliability of measurements made by ultrasonic Doppler sensors. So, ultrasonic flow measurement has been applied only in few and limited applications.
The paper shows the results of preliminary laboratory tests and the following test realized on field in the real scale plant on photo-electric sensors, ultrasonic Doppler flow switch and a mechanical flow switch, which was tested in the very beginning as an alternative to the ultrasonic Doppler sensor.

2. Laboratory tests

2.1 Photo-electric sensors

The thru-beam infra-red photo electric sensor Osiris XU2 produced by Schneider Electric has been identified as a potential solution for turbidity monitoring. Characteristics of the sensor are reported in table 2. A testing plant was designed and realized to verify if the degree of opacity measured by the sensor (figure 5) is compatible with sediment content level suitable for the application described in the paper. The final objective was to identify the potentiometer regulation, if any, which corresponds to a certain sediment content in the discharge pipe of the ejector.

Table 2. Schneider Electric's Osiris XU2 characteristics.

Characteristic	Value
Weight (kg)	0.155
Length (mm)	95
Diameter (mm)	M18x1
Nominal sensing distance (m)	50
Degree of protection	IP67
Rated supply voltage (V)	12-24
Current consumption (mA)	< 55
Delay time (ms)	< 15
Price, wiring included (€)	110

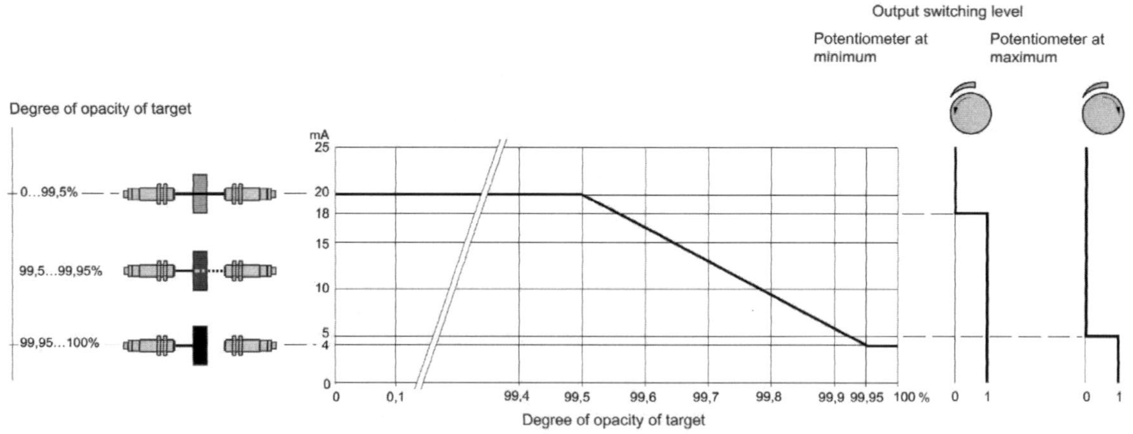

Figure 5. Degree of opacity measured by the Schneider Electric's Osiris XU2 and output signal 4-20 mA as a function of potentiometer regulation.

The testing plant was made by a centrifugal water pump, a manual regulating valve, a transparent tank wherein water and sand can be mixed, piping and a measuring pipe (figure 6). The measuring pipe is composed by a DN80 PVC duct connected with a plexiglass tube. Photo-electric sensors are installed on the plexiglass pipe. Firstly, the manual regulating valve position has been adjusted to guarantee a volumetric flowrate of about 25 m^3/h. Then, each test was characterized by the following procedure: after filling the tank up to a certain level, the pump was turned on, working only with water. Then, a certain mass of sand was introduced in the tank. The circulation guaranteed by the pump produced in few minutes a good mixing of water and sand, and so the test can start: the discharge pipe of the pump was manually connected with the aspiration duct of the pump, so the pump work in a closed ring. After observing photo-electric sensors response (green or red light, the latter being the condition of minimum opacity reached, see figure 5), the solid content in the mixture was measured by putting the discharge flow of the pump in another tank and by measuring relative water and sand level after sedimentation. The tests were carried on at different level of potentiometer regulation. It was found that the minimum opacity set with the potentiometer in the lower position corresponds to a 10% of sand in mass per kg of discharge flow (that is, more or less equal to a 5% of sand in volume per liter of discharge flow). This value is compatible with the working characteristic of ejector [2]. So, photo-electric sensors were set at the minimum value of the potentiometer and several tests were conducted with three different pairs of sensors in the same working conditions (i.e. constant increasing of sand content in the circuit). The results show a good repeatability of the instruments, since 95% of the measures follow in a variation range from the mean value of each measuring series that is ± 20 g of sand per kg of discharge flow.

Figure 6. Testing plant.

Figure 7. Measuring pipe.

2.2 Mechanical flow switch

The mechanical flow switch (figure 8) FLU25 was taken into consideration for the discharge flowrate monitoring as an economic alternative to ultrasonic Doppler sensors (FLU25 cost is about 60€). This kind of instrument is usually applied in thermal energy plant and it is used to switch off methane burner when water circulating pump suddenly stops working. The sensitive element is made by a metallic lamina that is pushed by water flow. Lamina movement is contrasted by a spring. Through spring regulation, it is possible to set the opening of the electric switch at a certain flowrate. Since fluid is characterized by the presence of sand and, more in general, solid particles, to avoid clogging the lamella was modified as in figure 9 with the aim of reducing its intrusion within the pipe. Obviously, also the spring was substituted as well. Due to the modification of the sensitive element, tests were needed to identify the correct position of the calibration screw. The mechanical flow switch was tested in the same plant shown in figure 6. After several test, it was found the new calibration screw position corresponding to a volumetric flowrate of 15 m^3/h. This value was chosen because it corresponds to a velocity in the ejector discharge duct that is lower than 1 m/s, which is the limit under which sand sedimentation in the pipeline can occur with a high grade of probability [1,2]. After being set, the mechanical flow switch was tested to verify its repeatability: several tests were carried on with the same procedure (flowrate decreasing from 25 m^3/h down to 10 m^3/h in a certain time), and the mechanical flow switch show a good repeatability, since 95% of the measures follow in a variation range from the mean value of each measuring series that is ± 1 m^3/h.

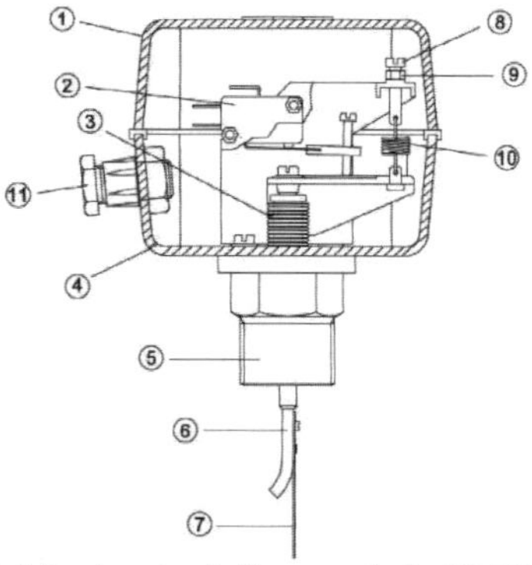

Figure 8. Mechanical flow switch FLU25, which is composed by: cover (1), micro switch (2), bronze bellow (3), body (4), threated fitting (5), actuator (6), lamina (7), calibration screw (8), locknut (9), return spring (10) and fairlead (11).

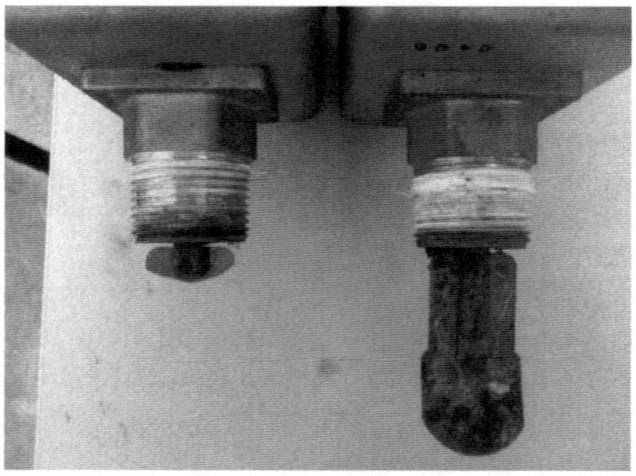

Figure 9. The lamella of the mechanical flow switch, respectively, after and before length modification made to avoid clogging risk.

2.3 Ultrasonic Doppler flow switch
The ultrasonic Doppler flow switch DFS-II produced by Greyline Instruments has been identified as a potential solution for flowrate monitoring. Characteristics of the sensor are reported in table 3. The instrument was installed and tested in the water pumping cabin of the Portoverde Marina plant (figure 10): in the cabin two flowrate meters were already present (two orifice plates plus differential pressure transmitters) to measure the water flowrate feeding each ejector.

Table 3. Greyline Instruments's DFS-II characteristics.

Characteristic	**Value**
Sensor frequency (MHz)	0.64
Degree of protection	IP67
Rated supply voltage (V)	24
Current consumption (mA)	< 150
Measuring diameter range (m)	0.0125-4.5
Measuring velocity range (m/s)	0.08-3.0
Price (€)	900

Installation and calibration are two important phases and may have strong impact on measurement precision [12-15]. Installation on vertical pipes has no particular needs, while installation in horizontal pipe is preferred at hour 3 or 9 to avoid the risk of presence of air (at hour 12) or solid (at hour 6). Generally speaking, the sensor must be installed away from devices able to disturb the flow like pump, valves, orifice plate, curves, section increasing or decreasing: a number of diameters D of straight tube before (10D) and after (6D) the sensors should be respected to guarantee the performance of the instrument. Very important is also the contact between the sensor and the duct. Before the installation the duct surface should be cleaned and degreased, by eliminating in particular paint, scale and rust. Silicon made compounds should be used to fill the gaps between sensor and duct, where contact is not possible due to pipe curvature. The sensor is fixed to the duct through a metallic clamp. DFS-II requires solids or bubbles minimum size of 100 microns with a minimum concentration of 75 ppm to ensure a precision of ±2%. So, if the installation is realized in accordance to previous suggestions and the characteristics (size and concentration) of reflectors in the fluid are over certain values, the linear calibration curve of figure 11 can be used to set the instrument.

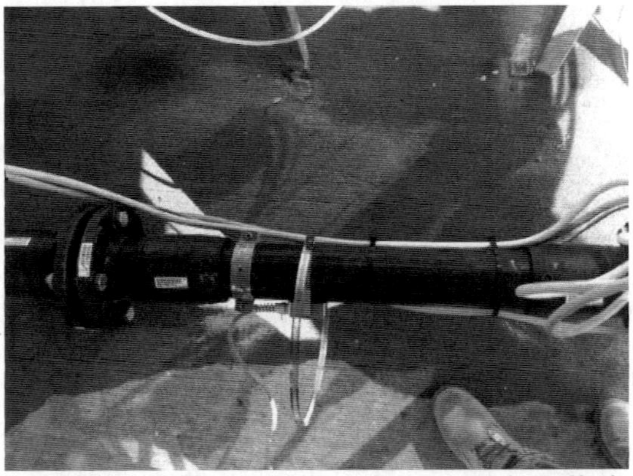

Figure 10. Ultrasonic Doppler sensor installed in the water cabin pumping in Portoverde Marina.

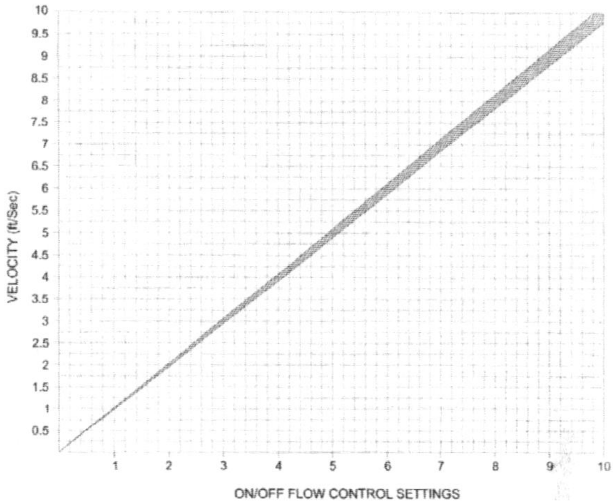

Figure 11. Calibration curve of DFS-II.

Two set-points can be defined on the instrument panel: one for the off-trip, and one for the on-trip. So, for example, if the ON set point is fixed at value 5, while the OFF set point is fixed at 3, it means that the switch is closed (position on) at a velocity of about 1,5 m/s (5 ft/s) and then is opened (position off) at a velocity of about 0,9 m/s (3 ft/s). Delay time (from 0 to 60 seconds) and sensitivity can be adjusted too.

The DFS-II was tested by simulating a cyclic increasing-decreasing of water flowrate. The pump was regulated by an inverter and the flowrate was measured by differential pressure drop on an orifice plate, working as a reference system. The test values were set at 25 m^3/h and 15 m^3/h, which means ON set point regulated at 5 (1,5 m/s of velocity) and OFF set point regulated at 3 (0,9 m/s of velocity). With these values no correspondence was found between control setting and flowrate measured by the reference measuring system. It was found that, if set with ON at 7 and OFF at 5, the DFS-II works well (switch on at 25 m^3/h, switch off at 15 m^3/h). Then, the system was tested to verify measure repeatability: several tests were carried on, also varying flowrate variation gradient, and an acceptable repeatability was observed since 95% of the measures follow in a variation range from the mean value of each measuring series that is ± 2 m^3/h. So, probably due to a not perfect installation and/or insufficient reflectors characteristics, DFS-II does not work coherently with calibration curve showed in figure 11, but, after an on-site calibration, shows satisfying performances.

3. Field tests

After preliminary test results the three instruments were installed at the Portoverde Marina seabed maintenance plant. The instruments were placed in two measuring pipes (similar to the one shown in figure 7) and closed in a metallic box (figure 12), that was installed near the dock (figure 13) and connected to the discharge lines of the ejectors. The on-off signals are communicated to the PLC of seabed maintenance plant through a wireless LAN bridge (distance about 150 meters) and are used to determine the value of the water feeding flowrate of the ejectors. In a first phase, two thru-beam infra-red photo electric sensors and two mechanical flow switches (thus, a couple for ejector discharge pipeline) were installed. Both mechanical switches and photo-electric sensors demonstrated to work well, in particular by confirming good repeatability and the absence of false positive: checks were made both through on-site measurement on discharge flowrate and solid content, as well as through rapid verification of correspondence of critical condition after the generation of high turbidity and/or low flowrate signals. On the other hand, the mechanical flow switch showed, after 2 working months, a problem related to clogging. Figure 14 and figure 15 show, respectively, a picture of the lamella inside the discharge duct and a picture of the duct clogged due to shells presence.

Figure 12. Instruments installed in the metallic box.

Figure 13. Metallic box installed near the dock and connected with discharge pipes of the ejectors.

Figure 14. Lamella profile of the mechanical flow switch within the discharge duct.

Figure 15. Discharge duct clogging produced by shells.

So, despite the good performances, the mechanical flow switch was substituted by the ultrasonic Doppler flow switch. In a first period (about eight month) there was only one ultrasonic Doppler installed. The ultrasonic flow switch works for several weeks with good results, which means absence of false positive (i.e. when discharge flowrate actually reaches critical values, the Doppler switch always signalled them). After one month, a false negative event occurs, which means that the discharge flowrate of one ejector was zero, but the instrument did not recognize it. So, the instrument was re-installed and re-calibrated on-site. In particular, the new installation focused on a more well distributed and abundant application of silicon compound. Then, the ultrasonic Doppler flow switch worked well up to about seven months. A second ultrasonic Doppler flow switch was then installed (figure 11). The calibration of the two ultrasonic Doppler flow switches was unsatisfying, since it was not possible to reach previous performance in terms of measurement repeatability. In particular, a dramatic decreasing in sensor repeatability was observed, since only 25% of measures follow in the range ± 2 m^3/h (while the percentage was over 95% in previous tests). The following were checked: right installation, number of diameters of straight tube before and after the sensor (measured), electric power supply, switch functioning, silicon compound characteristics. Number of diameters D of straight tube before (10D) and after (6D) the sensors were found to be lower than the one recommended by the manufacturer (5D and

3D, respectively); there also some little diameter variation in the pipeline (in the order of some millimetres), but the whole effects should be negligible in terms of repeatability, thus affecting only measurements precision. The others items were found to be correct. So, the sensors were re-calibrated and tested separately, and then together. It was found that the two sensors, by working separately, worked well (i.e. acceptable repeatability, 95% in the range ± 2 m^3/h). So, the hypothesis is that the two sensors influenced and disturbed each other.
The manufacturer confirmed (even if not mentioned in the instrument manual) that acoustic interference can be observed, but usually when two instruments work on the same pipeline, since the echo of the signals produced by one ultrasonic Doppler sensor can be detected also by the other one. The mutual interference of ultrasonic Doppler sensors on different pipelines needs further analysis, since the both literature and normative reference (ISO 15679:2010) do not face this kind of problem. Test will be carried on at laboratory scale to investigate the phenomenon.

4. Conclusions

Different solutions were tested both in laboratory and in real environment for the monitoring of turbidity and flowrate of a sand-water mixture inside a pipeline. The technological solutions were applied in an innovative plant for seabed maintenance. It was found that:
i. thru-beam infra-red photo electric sensors can be effectively applied for turbidity monitoring;
ii. mechanical flow switches were tested and demonstrated to work with good repeatability, but they dramatically increased the risk of pipe clogging;
iii. ultrasonic Doppler flow switch were tested and demonstrated to work with sufficient repeatability (in relation to the application), but some malfunctioning (i.e. acoustic interference) were observed when two instruments are installed on nearby pipes.
A new seabed maintenance plant is under realization in Cervia (Italy), the project being financed by the European LIFE program. So, further investigations are needed and will be carried on to verify if mutual interference is the main source of the problems revealed by ultrasonic Doppler flow switch when installed on nearby pipes.

Acknowledgment
The research activities have been financed by the *Programma Operativo Regionale – Fondo Europeo di Sviluppo Regionale* (POR-FESR) 2007-2013 of the Emilia-Romagna region. The authors also acknowledge ing. Carlo Albertazzi (*Servizio Difesa del Suolo, della Costa e Bonifica* of Emilia-Romagna region) for the support and the precious suggestions.

References

[1] Amati G and Saccani C 2005 Impianto sperimentale per il desabbiamento dei fondali nelle aree portuali *Proc. 32nd National Symp. on Engineering and Plants* (Rimini, Italy).
[2] Bianchini A, Pellegrini M and Saccani C 2014 Zero environmental impact plant for seabed maintenance *Proc. 4th Int. Symp. on Sediment Management* (Ferrara, Italy) pp 352-9.
[3] Bianchini A, Pellegrini M, Preda G, Saccani C and Vanni D 2013 Integrated technologies for sediment treatment *Environ. Eng. Manage. J.* **12** 253-6.
[4] Shahril Shah MG, Elmi AB, Nazir A and Shukri S 2012 Photoelectric sensor based intelligent track racing car *Procedia Eng.* **41** 588-92.
[5] Yijing G, Yuying S, Wei W, Jiahe Z, Lintao L and Jingdong L 2016 Field test study of a novel defrosting control method for air-source heat pumps by applying tube encircled photoelectric sensors *Int. J. Refrig.* **66** 133-44.
[6] Naaim-Bouvet F, Bellot H, Nishimura K, Genthon C, Palerme C, Guyomarc G and Vionnet V 2014 Detection of snowfall occurrence during blowing snow events using photoelectric sensors *Cold Reg. Sci. Technol.* **106-107** 11-21.
[7] Haijiang T, Daoliang L, Cong W, Qisheng D, Chengwu W and Shuangyin L 2012 Design and characterization of a smart turbidity transducer for distributed measurement system *Sensor Actuat. A-Phys.* **175** 1-8.

[8] Rai AK and Kumar A 2015 Continuous measurement of suspended sediment concentration: technological advancement and future outlook *Measurement* **76** 209-27.

[9] Ochiai S and Kashiwaya K 2010 Measurement of suspended sediment for model experiments using general-purpose optical sensors *Catena* **83** 1-6.

[10] Nord G, Gallart F, Gratiot N, Soler M, Reid I, Vachtman D, Latron J, Martín-Vide JP and Laronne JB 2014 Applicability of acoustic Doppler devices for flow velocity measurements and discharge estimation in flows with sediment transport *J. Hydrol.* **509** 504-18.

[11] Tezuka K, Mori M, Suzuki T and Kanamine T 2008 Ultrasonic pulse-Doppler flow meter application for hydraulic power plants *Flow Meas. Instrum.* **19** 155-62.

[12] Carlander C and Delsing J 2000 Installation effects on an ultrasonic flow meter with implications for self diagnostics *Flow Meas. Instrum.* **11** 109-22.

[13] Bonakdari H and Zinatizadeh AA 2011 Influence of position and type of Doppler flow meters on flow-rate measurement in sewers using computational fluid dynamic *Flow Meas. Instrum.* **22** 225-34.

[14] Hirayama N, Onodera T, Kashimura O, Suzuki S, Yao H, Ohgawara G, Yamamoto T and Takeda Y 2004 Accuracy improvement on non-invasive ultrasonic-doppler flow measurement by utilizing shear waves in metal pipe *Proc. 4th Int. Symp. on ultrasonic Doppler method for fluid mechanics and fluid engineering* (Sapporo, Japan) pp 101-6.

[15] Furuichi N 2013 Fundamental uncertainty analysis of flowrate measurement using the ultrasonic Doppler velocity profile method *Flow Meas. Instrum.* **33** 202-11.

IOP Conf. Series: Journal of Physics: Conf. Series **882** (2017) 012009 doi:10.1088/1742-6596/882/1/012009

Quantitative visualization of oil-water mixture behind sudden expansion by high speed camera

Parham Babakhani Dehkordi, Luigi P.M Colombo, Manfredo Guilizzoni, Giorgio Sotgia, Fabio Cozzi

Politecnico di Milano, Department of Energy, via Lambruschini 4, 20156 Milan, Italy

Parham.babakhani@polimi.it

Abstract: The present work describes the application of an image processing technique to study the two phase flow of high viscous oil and water through a sudden expansion. Six different operating conditions were considered, depending on input volume fraction of phases, and all of them are resulting in a flow pattern of the type oil dispersion in continuous water flow. The objective is to use an optical diagnostic method, with a high speed camera, to give detailed information about the flow field and spatial distribution, such as instantaneous velocity and in situ phase fraction. Artificial tracer particles were not used due to the fact that oil drops can be easily distinguished from the continuous water phase and thus they can act as natural tracers. The pipe has a total length of 11 meters and the abrupt sudden expansion is placed at a distance equal to 6 meters from the inlet section, to ensure that the flow is fully developed when it reaches the singularity. Upstream and downstream pipes have 30 mm and 50 mm i.d., respectively. Velocity profiles, holdup and drop size distribution after the sudden expansion were analyzed and compared with literature models and results.

1. Introduction and literature

This work presents some preliminary results of an experimental investigation about the in situ oil volume fraction (holdup), velocity profiles and drop size distribution for two phase oil-water flow behind abrupt expansion in a horizontal pipe, by means of a visualization technique.

Laser-based techniques such as Particle Tracking Velocimetry (PIV) are widely used to characterize the hydrodynamic behavior of multiphase flow, feeding the continuous phase with small tracer particles which follow the main flow. A laser is used to illuminate the tracer particles in cross-section of pipe, while a digital camera is capable of capturing images frame by frame separated by short time delay. In PIV the frame acquisition is synchronized with the laser pulses, and each frame is subdivided into small interrogation window size. Each interrogation area in an image is linked to the corresponding interrogation area in the following frame. The second interrogation window must be sufficiently large to include all particles contained in the first interrogation window; on the other hand, care must be taken to properly select the second interrogation areas because, enlarging interrogation window would considerably reduce pixel resolution in the first window and increase computation time, as emphasized by Keane and Adrian [1]. After filtering and thresholding, pixel areas representing dispersed phase particles can be identified. The particles would move and shift from one image to another. Hence, the cross-correlation function is able to statistically measure the level of match between two frames. The result of above procedure is a correlation plane, whose highest value can be used as a direct appraisal of the particle image displacement. The velocity of particles is then estimated by dividing the measured displacement and the time delay between consecutive frames. In the past decades, many researchers used PIV to characterize two-phase flow, mainly gas-liquid, in different pipe configurations. Delnoij [2] developed a new ensemble correlation PIV technique for a bubble column (so in bubbly flow regime), where movement of both the liquid phase (water) and the bubbles is recorded by a single CCD camera and using bubble-tracer patterns. The instantaneous velocity field and bubble concentration were detected using cross-correlation technique. The flow regime under investigation was bubbly flow. Polystyrene particles with sauter mean diameter of 250 μm, a density of 1070 $kg \cdot m^{-3}$ were used as tracer particle. The results showed that velocity difference between water and air must be sufficiently large to avoid overlap between the two displacement correlation peaks. Zhou et al. [3] used PIV to perform liquid phase turbulence measurement of liquid phase in an air-water two phase flow in a 50 mm i.d. horizontal pipe using PIV. Moreover, an optical phase separation approach/planar laser-induced fluorescent (PLIF) method was also used to differentiate bubbles from tracer particles in domain, and the local void fraction was measured using a four sensor conductivity probes. They compared the results from PIV measurements with literature ones using thermal anemometry for the same condition. Then, superficial liquid velocity was computed from local velocity and void fraction measurements. Difference in superficial liquid velocity between PIV and hot film anemometry (HFA) was found to be ±7% for void fraction less than 20%. Application of the PIV technique for two-phase gas-liquid for different configurations can be found also in Hassan et al. [4] and Gui et al. [5].

Although many works have been accomplished about gas-liquid flow, a few studies can be found regarding the application of particle velocimetry to liquid-liquid two phase flow. Augier et al. [6] performed tests about a dense liquid-liquid homogeneous flow based on PIV, considering a refractive index matching technique, in a vertical column. The objective was to measure local velocity field and phase fraction. The liquid-liquid system was composed of n-Heptane (dispersed), with viscosity of 4.5×10^{-4} $Pa \cdot s$ and density 683.7 $kg \cdot m^{-3}$, as well as mixture of 50% Glycerol-water solution as continuous phase. Sintered glass particles with diameter of 10 μm were selected as tracer particles because they have hydrophilic features, thus they would not penetrate to the organic dispersed phase. The laser sheet with 0.2×10^{-3} m thickness is responsible for illuminating the flow generated by a 20 mJ double pulsed Nd:Yag laser. The frequency of laser is ranged from 1 to 15 Hz with wavelength of 532 nm. The time delay between two flashes was between 5 and 10^4 μs, and a CCD camera with a 1008×1018 pixels resolution was used to record the flow. A sophisticated post processing was carried out, considering a 64×64 pixels interrogation area. The oil phase fraction was varied between 0 and 0.4 and local drag coefficients were measured from tests. The comparison of local drag coefficients (as a function of phase fraction) and models in literature was made, showing that classical drag coefficient laws underestimated the influence of phase fraction on drag coefficient.

The work by Pulvirenti and Sotgia [7] may also be cited, considering the application of PIV to stratified wavy oil-water flow. Mineral oil (μ_o=0.95 $Pa \cdot s$ and ρ_o=890 $kg \cdot m^{-3}$) and tap water were used as test fluids. The experimental tests were conducted in a horizontal 40 mm i.d. pipe to investigate the wavy structure of stratified flow. A mixture velocity of 0.7 $m \cdot s^{-1}$, and input water volume fraction of 0.37 were considered for analysis. Silver coated hollow glass spheres were selected as particle tracer for seeding water. Instantaneous water velocity fields were detected using 2D PIV technique and water velocity maps developed. Furthermore, the position of the oil-water interface was detected and circulation of water in the region between each couple of consecutive crests of oil core waves was recognized.

Morgan et al. [8] performed experimental tests about liquid-liquid flow in a horizontal circular tube. An aliphatic hydrocarbon oil (Exxsol D80, μ_o=2.3x10-3 $Pa \cdot s$ and ρ_o=796 $kg \cdot m^{-3}$) and glycerol-water solution (μ_{w-g}=47×10^{-3} $Pa \cdot s$ and ρ_{w-g}=1,205 $kg \cdot m^{-3}$) were used as test fluids. The concentration of glycerol must be regulated in order to match refractive index between oil and glycerol-water mixture (RI=1.44), the authors also emphasized the importance of refractive index between fluids and pipe wall. Hence, a

IOP Conf. Series: Journal of Physics: Conf. Series **882** (2017) 012009 doi:10.1088/1742-6596/882/1/012009

proper choice of pipe wall materials is crucially important for image processing technique. In their work, light distortion due to curvature of pipe cross-section was corrected by using a graticule technique. They focused on the flow regime transition from stratified to dual continuous flow, used high speed PLIF, PIV, and Particle Tracking Velocimetry (PTV). They realized that the mixture of glycerol-water solution (heavier and more viscous phase) moved in laminar, while Exxsol oil was in a turbulent flow.

As evident from the cited literature, PIV characterization of liquid-liquid flow in straight pipe has been thoroughly performed. On the contrary, information regarding oil-water flow through singularities is still lacking. Concerning flow of oil-water through abrupt change of pipe cross-section, the authors were able to find four contributions in literature, Hwang and Pal [9], Balakhrisna et al. [10], Kaushik et al. [11], and Colombo et al. [12], all of them unrelated to PIV or similar techniques. The present study is therefore aimed at extending the application of image processing techniques, with a high speed camera, to two-phase oil-water flows behind a sudden expansion, in order to quantitatively explore the effect of such singularity on the flow field and holdup. Instantaneous velocity fields for oil drops, in-situ volume fractions, and oil drop size distribution were determined and the results were compared to literature models and correlations.

2. Experimental setup

The oil-water facility is available at Multiphase Thermo-Fluid laboratory, Department of Energy, Polytecnico di Milano. The schematic of test facility is shown in Fig. 1. The test pipe, having a total length of 11 m, is made Plexiglass® to allow the visual inspection of flow, that was performed by a high speed camera. Oil and water injectors were originally designed to ensure the onset of a stable core-annular flow, with water touching the duct wall and an oil core. However, changing the input flow rate of the phases it is also possible to obtain a dispersed flow, (for further information see e.g. Sotgia et al. [13]). For the present experimental campaign, the test section includes a sudden expansion at a distance of 7 m from the inlet mixing section, with upstream and downstream internal pipe diameters of 30 mm and 50 mm, respectively. A total of 12 pressure taps are mounted, almost uniformly distributed along the pipe, and a pressure transducer (Setra® model 230, full scale: 1 psi=6895 Pa) is installed to measure the pressure drop. The rheological properties of oil and water are reported in Tab. 1. Oil and water interfacial energy was measured at the Chemistry, Material, and Chemical Engineering Department, Politecnico di Milano, by means of a LAUDA tensiometer, while the oil viscosity by the laboratory "INNOVHUB-Stazioni Sperimentali per l'Industria- Divisione Stazione Sperimentale per le Industrie degli Oli e Grassi (SSOG)" and its temperature dependence may be expressed as:

$$\mu_o = 3.44\exp(-0.71\,T) \quad (1)$$

Table 1. Rheological properties of mineral oil and tap water for experimental tests

Rheological properties at $20\,^{\circ}C$	Tap water	Mineral oil
ρ [kg·m^{-3}]	998	910
μ [Pa·s]	1.026×10^{-3}	0.838
$\sigma_{o\text{-}w}$ [J·m^{-2}]	0.02	

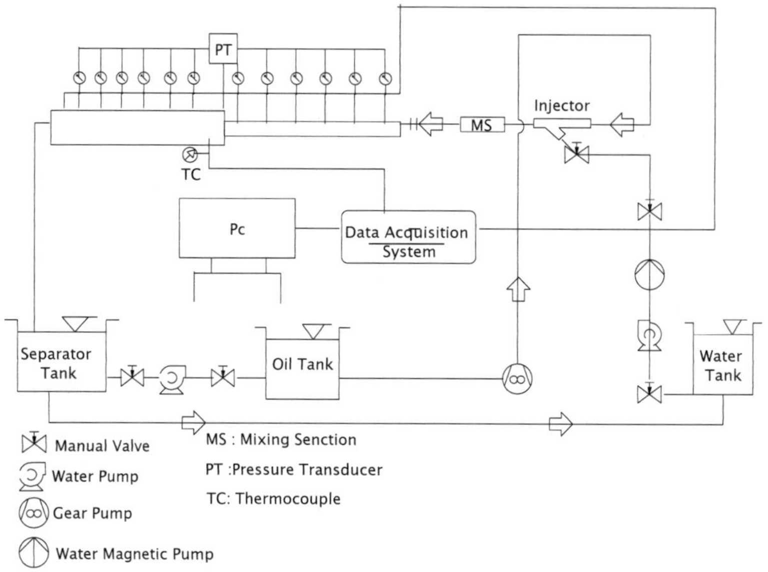

Figure 1. Schematic of the test facility

The water flow rate was measured by a magnetic flow meter (accuracy ±0.5% of reading), while the oil volume flow rate was measured by a calibrated gear metering pump. Figure 2 shows the data acquisition system, including the high speed camera, and the test section, within which the mixture is flowing from right to left. A white sheet was placed at the back of pipe section to avoid disturbances in the pictures from the back wall. The high speed camera was placed perpendicular to the pipe section and at the same height, while two lamps (halogen, 800 W each) were used to provide the necessary illumination. As a side view and not a cross-section of the duct is acquired, what is in fact obtained is a 2D projection of the flow, including perspective effect and optical distortions due to the different refraction indices of water, Plexiglass® and air. Consequently, there is no way to identify coplanar drops and to discriminate three-dimensional components of the drop velocity. As this is preliminary analysis, these issues were not taken into account. For this study, tracer particles were not used for seeding as done for PIV application because objective of the work is to measure velocity profile of oil drops, which can be easily distinguished from water continuous phase. Tracer particles are usually used to characterize the flow of continuous phase. Furthermore, the use of tracer particles to measure also velocity of continuous phase requires two cameras, one for dispersed oil drops and another for tracer particles. Post-processing of the acquired images was done using Dantec Dynamic Studio v3.12 [14] software. Table 2 reports the camera settings for the experimental tests.

Table 2. Camera setting for experimental tests

Pixel resolution [pixels]	1024×400
Time delay between frames [μs]	1666.50
Frame rate [fps]	600
Field of view [m^2]	0.37×0.05
Interrogation window[pixels]	64×64

Figure 2. Data acquisition system and test section

3. Governing parameters and experimental conditions

In the following section, the most important governing parameters for oil-water mixtures will be summarized. The superficial velocity or volumetric flux J (ms^{-1}) is defined as the ratio between the volumetric flow rate of each phased and the cross-sectional area of the pipe. The sum of the two superficial velocities of the phases is the total mixture velocity:

$$J_{o-w} = J_o + J_w \quad (2)$$

The oil volume fraction, ε_o is defined as the ratio between superficial velocity of oil and the superficial velocity of the mixture:

$$\varepsilon_o = \frac{J_o}{J_{o-w}} \quad (3)$$

The water volume fraction can be defined in a completely analogous way, or simply obtained by $\varepsilon_w = 1 - \varepsilon_o$. The phase holdup is calculated by dividing the superficial velocity of each phase and the actual (effective) velocity as:

$$H_o = \frac{J_o}{U_o} \quad , \quad H_w = \frac{J_w}{U_w} \quad (4)$$

The slip ratio is the ratio between the effective oil velocity and the effective water velocity:

$$S = \frac{U_o}{U_w} \quad (5)$$

Obviously, the phase volume fraction is in general different from the phase holdup and the slip ratio is different from 1; they both depend on flow regime under investigation. For dispersed flows where the dispersed phase moves at the same velocity of the continuous carrier phase, the phase volume fraction and the holdup becomes equal and the slip ratio becomes unitary ("homogeneous flow"). The investigated experimental conditions are tabulated in Tab. 3 (with superficial velocities evaluated downstream of singularity, thus in the larger pipe).

Table 3. Summary of experimental test conditions

Case test	J_o [m s^{-1}]	J_w [m s^{-1}]	ε_o [-]	Re_{os} [-]	Re_{ws} [-]
1	0.29	0.70	0.29	16	35,297
2	0.29	0.84	0.25	16	42,356
3	0.44	0.70	0.38	24	35,297
4	0.44	0.56	0.46	24	28,237
5	0.59	0.70	0.45	32	35,297
6	0.59	0.84	0.41	32	42,356

Values of Reynolds number for Phases (Re_{os} and Re_{ws}) are calculated with the superficial velocities. In terms of flow pattern, all the conditions under investigations are related to the dispersion of oil drops in water.

4. Result and Discussion

Measurements were performed in the region behind the sudden expansion to evaluate the effect of the latter on in-situ velocity profiles, holdup and drop size distribution, in the flow re-development region region. Figure 3 shows a typical example of dispersed flow of oil drops in water continuous flow for J_o=0.29 m s^{-1} and J_w=0.84 m s^{-1}. Two frames, A and B, are required to create a correlation plane. The images A and B are recorded at time instant t and t+Δt, respectively, with Δt=1666.5 μs. It is worth noting that larger values of the time interval cause as higher probability for oil particles to leave the light plane. The images are converted in grey scale (Fig. 3-b) and then they must be binarized by a threshold technique to distinguish between oil drop particles and water continuous flow, see Fig 3-c. The particle shift between image A and B allows measuring the local velocity profile over the entire domain, from which spatially-averaged values can be determined (within the already cited limits related to the use of side views). This can be done over horizontal lines, to get vertical (so radial for the duct) velocity profiles, or over the whole domain. The calculations are then repeated over a large number of image couples, to obtain statistically significant values of the time-average of such velocities. As an example, Fig. 4 shows the evolution of the time-averaged velocity of the oil particles over the whole domain as a function of number of image couples used for the calculation, for test case 2.

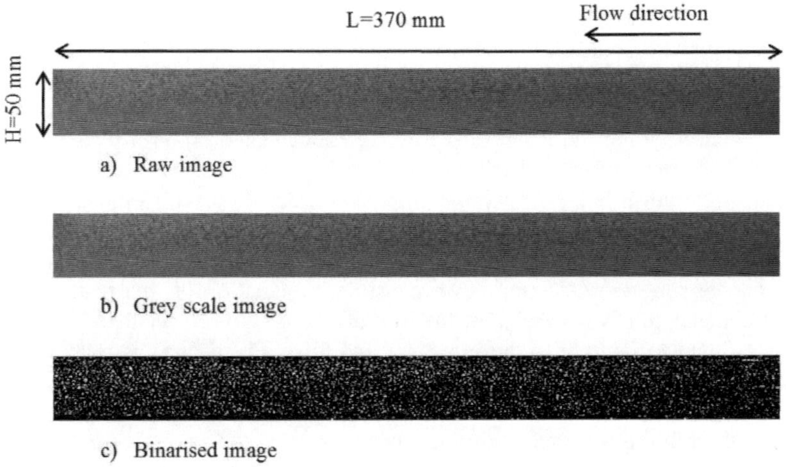

Figure 3. Typical example of high viscous-oil dispersion in water continuous flow: a) raw image, b) grey-scale image, c) binarized image

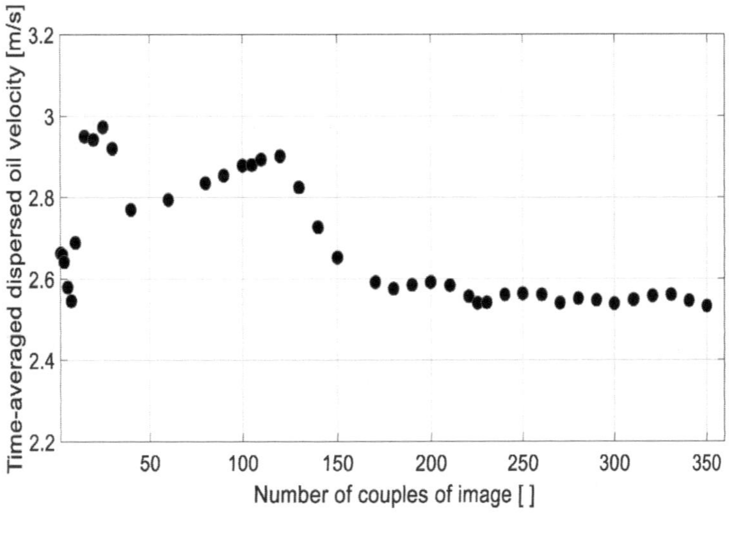

Figure 4. Cross-sectional time-averaged oil particle velocity over the whole investigated domain as a function of the number of the used image couples for J_o=0.29 m·s^{-1} and J_w=0.84 m·s^{-1}

As it is evident, the average begins to stabilize after 150 images. Therefore, in the current study, a total of 400 frames are used to analyze in-situ oil particle velocity and holdup. The measurement tests were acquired at 4 axial positions downstream the sudden expansion: L/D=1, 2.5, 4, 5.5. Each side view covers a horizontal length of 370 mm and in vertical the full duct diameter of 50 mm.

4.1 Velocity profiles

Figure 5 illustrates the velocity profile for the six cases reported in Tab. 1. For each test, the in-situ time-averaged velocity of oil particles is normalized by the mixture superficial velocity. It is evident from Fig.5 that the velocity profile has a 'top-hat' shape for the farthest axial positions from sudden expansion, while it has a Gaussian-like shape very close to the singularity. The flat velocity profile for positions far from singularity would seem to show that the oil-water flow is becoming fully developed. This seems also consistent with the fact that velocity distributions for cases 3-6 at axial distances of L/D=4 and L/D=5.5 are perfectly matched. Nevertheless, for lower input oil fraction the difference between the profiles at L/D=4 and L/D=5.5 gives an opposite indication of still undergoing development. For velocity profiles at axial distance of L/D=5.5, it can also be observed that the time-averaged velocity of oil particles becomes zero at bottom part of pipe cross-section. Two possible explanation of this result may be given. One is that after the abrupt change of cross section, the mixture velocity is lower than in the upstream pipe, and dispersed oil drops tend to migrate to the upper part of tube. The second is related to the fouling of the internal surface of the pipe, as Plexiglass® is an oleophilic material. The analysis of the velocity profiles at the axial distance L/D=1 shows that time-averaged velocity of oil particles is much higher at the centerline than in the regions very close to the pipe wall, as it can be reasonable given the fact that the mixture has just entered the larger downstream pipe, so the central part behaves partially as a "jet". The results of measurements from image processing reported in Fig. 5 proved that development length after this kind of singularity for high viscous oil-water is dependent on the oil input volume fraction. For higher values of oil volume fraction, the flow becomes fully developed at axial distance of L/D=4.

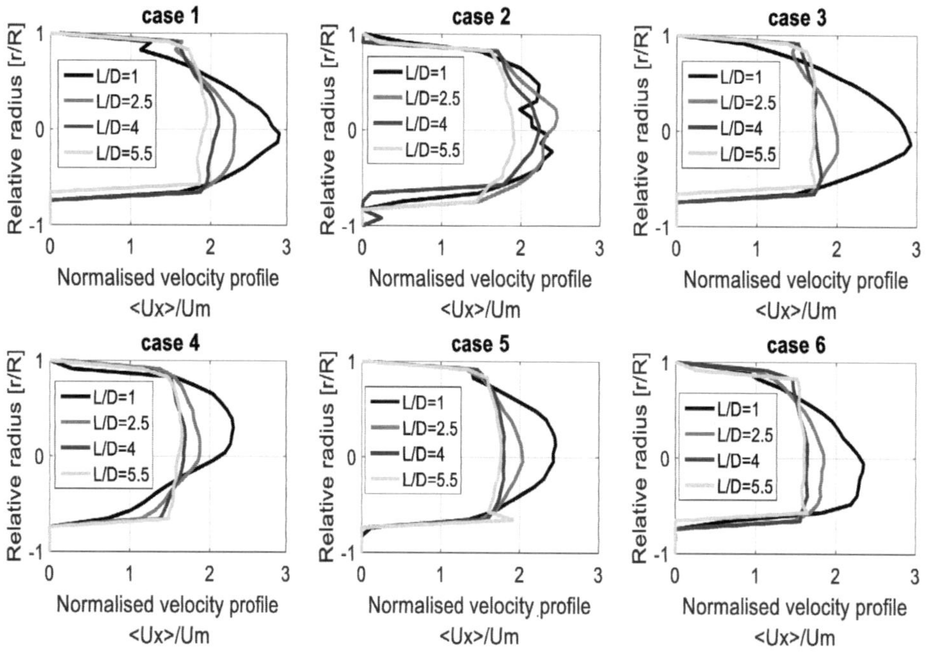

Figure 5. Time-averaged velocity profile at different axial position of L/D=1, L/D=2.5, L/D=4 and L/D=5.5 for the six cases reported in Table 1

IOP Conf. Series: Journal of Physics: Conf. Series **882** (2017) 012009 doi:10.1088/1742-6596/882/1/012009

4.2 Time-averaged oil holdup

To calculate the cross-sectional average of the oil holdup, it is necessary to calculate the cross-sectional average velocity profile for each operating condition as:

$$< U_o >= \frac{1}{A}\int u_o dA \qquad (6)$$

The time-averaged velocity in domain is automatically extracted by Dantec's Dynamic Studio 3.12 software. Thus, in each point in radial direction, time-average velocity is obtained. The oil holdup can be estimated as the ratio between the oil superficial velocity and the cross-sectional average velocity:

$$< H_o >= \frac{J_o}{<U_o>} \qquad (7)$$

The time-averaged cross-sectional water holdup can be simply calculated as; $<H_w>=1- < H_o >$.

Arney et al. [15] developed an empirical correlation based experimental data using quick closing valve for core-annular regime:

$$H_o = 1 - \varepsilon_w[1+0.35(1-\varepsilon_w)] \qquad (8)$$

Colombo et al. [12] measured liquid holdup by quick closing valve method over some hundreds experimental points for oil-water flows in pipes having constant diameter and including two different sudden contractions, from 50 mm to 40 mm i.d. (contraction ratio, ζ=0.64) and from 50 mm to 30 mm i.d. (ζ=0.36). Oil was the same than in the current study, while the flow regime under investigation were both dispersion of oil in water and core-annular flow. The experimental results about oil holdup were compared to Arney correlation, showing a very good agreement (average absolute error 5.15%, and 5.88% for ζ=0.64 and ζ=0.36, respectively). This proved that Arney correlation works well for dispersed flow regime. Thus, comparison of cross-sectional time-average oil holdup estimated by image processing technique to Arney correlation was performed in this study. Figure 6-a shows oil holdup versus input volume fraction for the axial distance L/D=5.5, which should be the one less influenced by sudden expansion. All data falls below bisector, showing that the oil-water mixture cannot be treated as a homogeneous flow, even if it is an oil dispersion in water. Furthermore, as the oil holdup is lower than the oil input volume fraction, the oil velocity is higher than the water velocity and the oil-to-water slip ratio is always significantly larger than unity, as it will be further discussed in the following. These findings may appear surprising for a dispersed flow regime, that is often modeled as if mixture behaved homogeneously; but it must be considered that it is evident from visual inspection that all the oil particles move to the upper part of tube due to presence of the larger pipe downstream of the singularity, thus the distribution of oil particles becomes inhomogeneous.

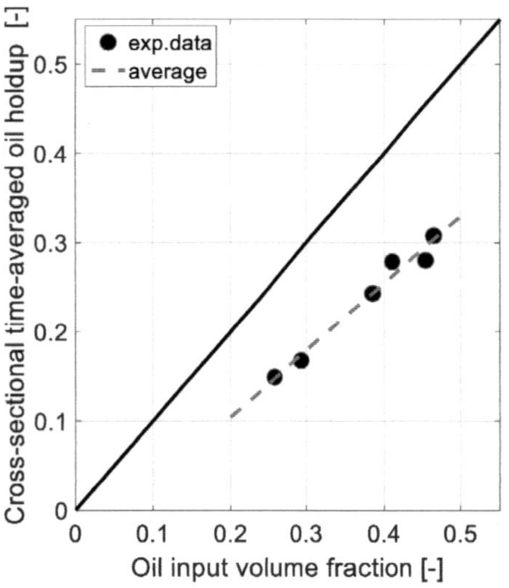

Figure 6. a) cross-sectional time-averaged oil holdup against oil input volume fraction

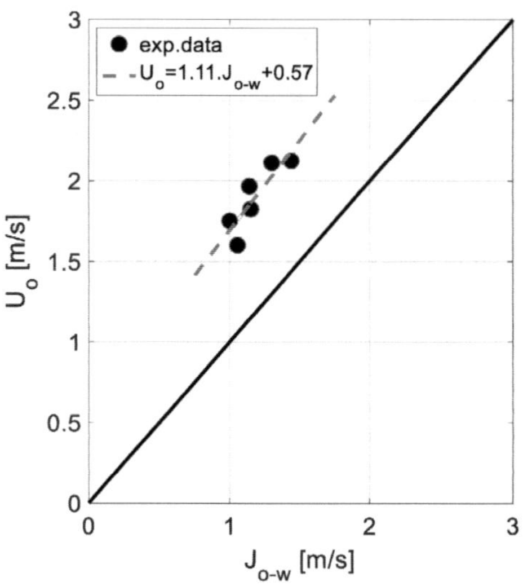

Figure 6. b) cross-sectional time-averaged oil velocity against mixture superficial velocity

The fact that oil is moving faster is consistent with the results obtained by Morgan et al. [8] and Colombo et al. [12]: the quantitative values on the contrary are different from the results for uniform-diameter pipes and sudden contractions, very likely due to the significant effect of the expansion in the first diameters of the downstream pipe. The results of oil holdup in Fig 6-a are consistent with the work of Colombo et al. [12] and Morgan et al. [8].

Brauner [16] emphasized that for low mixture superficial velocity where water is the continuous phase (conditions that applies to the present study), relatively large oil drops can be formed which considerably affect slippage between oil and water continuous phase. She suggested the use of Zuber-Findlay [17] drift flux model to characterize two phase oil-water dispersion flow. In the framework of the drift-flux model, the oil velocity can be modeled as:

$$U_o = C_o J_{o-w} + u_d \tag{9}$$

where C_o is a distribution parameter, which is in relation to the phase velocities and distribution within the duct, while u_d is the drift velocity. Figure 6-b shows average oil drop velocity as a function of mixture superficial velocity, where linear fitting of experimental data by image processing evidences that $C_o>1$, meaning that drop concentration is higher at the center of pipe, as supported by the work of Colombo et al. [12] and Brauner [16]. The amount of available data from the present campaign is regrettably too limited for a really significant validation of the Zuber-Findlay drift flux model after singularity, and this can be topic for further investigation. Figure 7-a. reports the values of the slip ratio between oil particles and water, defined by Eq. 5, as a function of the oil input volume fraction. The homogeneous flow model is represented by the solid line at s=1. The results are consistent with those about holdup, evidencing how the homogeneous model largely under-predict the measurements from image processing. This under-prediction is not varying significantly with the oil input volume fraction, it just oscillates in a range between 1.75 and 2.25. These results are contrary to what Morgan et al. [8] observed for low viscosity oil-water flow in straight pipe, where slip ratio under-prediction increases at increasing oil input volume fraction. Finally, the results of comparison between oil holdup estimated using Arney correlation and oil holdup by image processing are reported in Fig. 7-b. Data align themselves around the +25% line with respect to the predictions of the correlation, in contrast with the very good agreement found for uniform-diameter pipes and contractions (Colombo et al. [12]). This is a further evidence that the influence of the sudden expansion at L/D=5.5 is still considerable, even if the velocity profile appears fully developed.

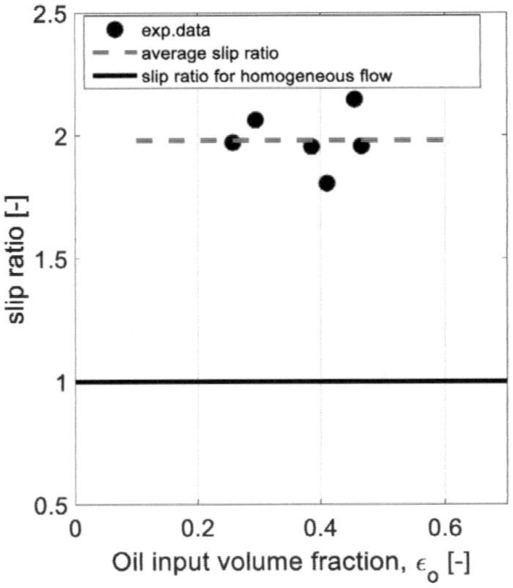

Figure 7. a) slip ratio versus oil input volume fraction

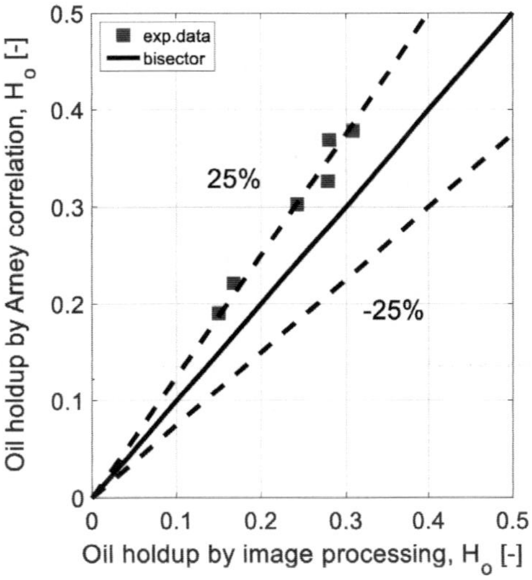

Figure 7. b) measured oil holdup by image processing against oil holdup predicted by Arney correlation [15]

4.3 Drop size distribution

The identification of oil drops in the processed images allowed also performing an analysis of the drop size distribution. After image binarization, white blobs corresponding to oil drops can be identified and their dimensions calculated. Once again neglecting optical distortions, and assuming a spherical shape for the drops, an effective drop diameter is calculated as:

$$d_{eff} = 2\sqrt{\frac{A_d}{\pi}} \quad (10)$$

From which the average drop diameter can be then calculated:

$$\mu_d = \frac{1}{N}\sum_{i=1}^{N} d_{eff} \quad (11)$$

A major problem in measuring drop size is that depending on the thresholding/filtering steps and on the parameters used during the image processing, blobs representing drops may merge. A a consequence, different threshold values result is different drop size distribution and much care must be taken to select a proper threshold value. Lower threshold values may result in loosing

IOP Conf. Series: Journal of Physics: Conf. Series **882** (2017) 012009 doi:10.1088/1742-6596/882/1/012009

drops, while increasing threshold values may cause fictitious merging of them. Automated selection of threshold values can prove unreliable, so direct visual comparison with the original images was used here to select suitable threshold values. Figure 8 shows a typical example of drop size distribution for case 2 (J_o=0.29 m·s^{-1}, J_w=0.84 m·s^{-1}), while for comparison Fig. 9 illustrates some instantaneous images from cases 1-4, evidencing how larger structures, more similar to "flakes", coexist with the population of oil drops. This reflects in the presence of large structures in the histogram, which in fact influence significantly the moments of the drop size distribution, despite their very limited numerical amount: the histogram is obviously right skewed (as negative values cannot be present), with mean at 1.60 mm and median at 0.93 (standard deviation 1.90 mm). Given the described issue, the median - which is less sensitive to extreme values than the mean is probably the most reliable indicator. The average drop size for this case is lower that what Morgan et al. [8] observed for low oil viscosity-water flow. Even if increasing viscosity of the dispersed phase (at the same superficial velocity) can postpone the process of drop deformation and breakup, and breakage is higher in regions near to pipe wall than in the center of pipe (see Lovick and Angeli [18]), it may be that the effect of the sudden expansion in breaking the oil particles (drops and flakes) overcomes the influence of oil viscosity.

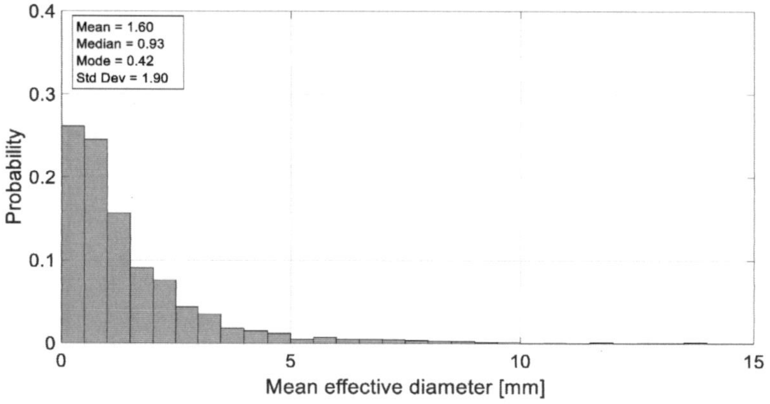

Figure 8. Drop size distribution histogram for case 2 (J_o=0.29 m·s^{-1}, J_w=0.84 m·s^{-1})

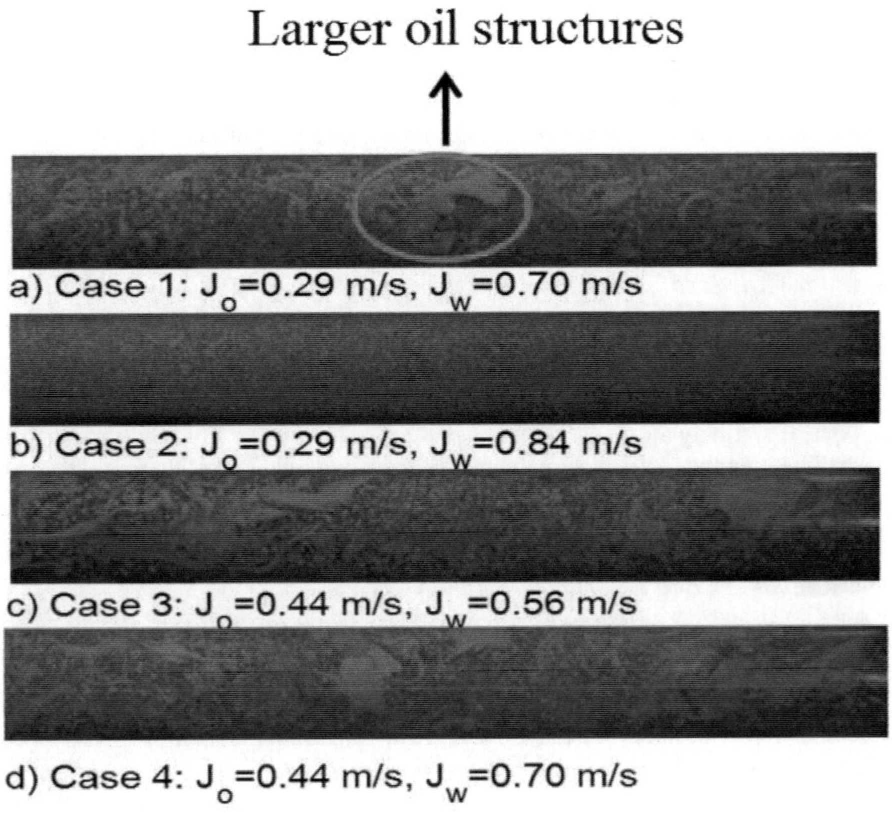

Figure 9. Flow visualization by high speed camera for cases 1-4

5. Conclusions

An oil dispersed in water continuous flow was experimentally investigated using a non-intrusive optical technique, in a 11 m long horizontal pipe with a sudden section expansion from 30 mm to 50 mm i.d. Very viscous oil and tap water were considered as test fluids. Instantaneous 2D velocity fields for oil drops were measured by a cross-correlation technique, and velocity profiles for six cases were obtained. Oil liquid holdup was then calculated, taking into account the cross-sectional time-averaged oil velocity. Based on this quantity, the cross-sectional time-averaged values of oil and water velocities, and the slip ratio between the two phases were also calculated. The results showed that the effect of the sudden expansion on the velocity profiles in the downstream pipe is significant for axial positions close the singularity, than they seem to re-develop with a "top-hat" shape. Concerning cross-sectional time-averaged oil holdup, the results show that the oil-water mixture cannot be considered as a homogeneous flow because oil always move faster than continuous water, even though flow regime under investigation is dispersed flow. This is also supported by the fact that the value of slip ratio is always higher than unity and practically independent from oil input volume fraction. The holdup and slip results, together with the results of a comparison with the drift-flux model, also evidence that despite the shape of the velocity profiles, the flow after 5 diameters from the expansion seems still not developed.

IOP Conf. Series: Journal of Physics: Conf. Series **882** (2017) 012009 doi:10.1088/1742-6596/882/1/012009

References

[1] Keane R D and Adrian R J 1992 Theory ofcross-correlation analysis of PIV images *Appl. Sci. Res.* **49** 191–215

[2] Delnoij E 1999 Fluid dynamics of gas-liquid bubble columns - A theoretical and experimental study, PhD thesis, University of Twente

[3] Zhou X and Doup B 2013 Measurements of liquid-phase turbulence in gas-liquid two-phase flows using particle image velocimetry *Meas. Sci. Technol.* **24** 125303

[4] Hassan Y A, Blanchat T K and Seeley Jr 1992 Simultaneous velocity measurements of both components of a two-phase flow using particle image velocimetry drop *Int. J. Multiphas. Flow* **18** 371–395

[5] Gui L, Lindken R and Merzkirch W 1997 Phase-separated PIV measurements of the flow around systems of bubbles rising in water Proceedings of the ASME (American Society of Mechanical Engineers) Fluids Engineering Summer Meeting, 97–3103

[6] Augier F, Masbernat O and Guiraud P 2003 Slip velocity and drag law in a liquid-liquid homogeneous dispersed flow *AIChE J* **49**(9) 2300–2316

[7] Pulvirenti B and Sotgia G 2004 Experimental analysis of water fluid dynamics in oil-water wavy-stratified regime, 3rd International Symposium on Two-Phase Flow Modelling and Experimentation, Pisa, Italy

[8] Morgan R G, Markides C N, Zadrazil I, Hewitt G F 2013 Characteristics of horizontal liquid-liquid flows in a circular pipe using simultaneous high-speed laser-induced fluorescence and particle vcelocimetry drop *Int. J. Multiphas. Flow* **49** 99–118

[9] Hwang C-Y J and Pal R 1997 Flow of two-phase oil/water mixtures through sudden expansions and contractions *Chem. Eng. J.* **68**(2-3) 157–163.

[10] Balakhrisna T G 2010 Oil-water flows through sudden contraction and expansion in a horizontal pipe – Phase distribution and pressure drop drop *Int. J. Multiphas. Flow* **36**(1) 13–24

[11] Kaushik V V R, Ghosh S, Das G and Das P K 2012 CFD simulation of core annular flow through sudden contraction and expansion *Petrol. Sci. Techol.* **86-87** 153–164

[12] Colombo L P M, Guilizzoni M, Sotgia G M and Marzorati D 2015 Influence of sudden contractions on in situ volume fractions for oil-water flows in horizontal pipes drop *Int. J. Multiphas. Flow* **53** 91–97

[13] Sotgia G, Tartarini P and Stalio E 2008 Experimental analysis of flow regimes and pressure drop reduction in oil-water drop *Int. J. Multiphas. Flow* **34**(12) 1161–1174

[14] Dantec dynamic studio user guide v3.12 2010

[15] Arney M S, Bai R, Guevara E, Joseph D D and Liu K 1993 Friction factor and holdup studies for lubricated pipelining – I. experiments and drop *Int. J. Multiphas. Flow* **19**(6) 1061–1076

[16] Brauner N 2001 The prediction of dispersed flows boundaries in liquid-liquid and gas-liquid systems drop *Int. J. Multiphas. Flow* **27**(5) 885–910

[17] Zuber N and Findlay J A 1965 Average volumetric concentration in two-phase flow systems *J. Heat Transf.* **87** 453–468.

[18] Lovick J and Angeli P 2004 Droplet size and velocity profiles in liquid-liquid horizontal flows *Chem. Eng. Sci.* **59** 3105–3115

IOP Conf. Series: Journal of Physics: Conf. Series **882** (2017) 012010 doi:10.1088/1742-6596/882/1/012010

VELOCITY MEASUREMENTS IN NASAL CAVITIES BY MEANS OF STEREOSCOPIC PIV - PRELIMINARY TESTS

Fabio Cozzi[1,4], Giovanni Felisati[2], Maurizio Quadrio[3]

[1] Department of Energy, Politenico di Milano, Via Lambruschini 4, 20156, Milan, Italy

[2] Otorhinolaryngology Unit, Head and Neck Department, San Paolo Hospital, Università degli Studi di Milano, Via di Rudinì 8, 20142 Milan, Italy

[3] Department of Aerospace Sciences and Technologies, Politecnico di Milano, Via La Masa 34, 20156 Milan, Italy

[4]Corresponding author: fabio.cozzi@polimi.it

Abstract. The prediction of detailed flow patterns in human nasal cavities using computational fluid dynamics (CFD) can provide essential information on the potential relationship between patient-specific geometrical characteristics of the nasal anatomy and health problems, and ultimately led to improved surgery. The complex flow structure and the intricate geometry of the nasal cavities make achieving such goals a challenge for CFD specialists. The need for experimental data to validate and improve the numerical simulations is particularly crucial. To this aim an experimental set-up based on Stereo PIV and a silicon phantom of nasal cavities have been designed and realized at Politecnico di Milano. This work describes the main features and challenges of the set-up along with some preliminary results.

1. Introduction

The human nasal cavity is devoted to several important physiological functions such as: filtering, warming and humidifying of the inspired air before it reaches the lungs and to facilitate the sense of smell. The nasal cavity is also important for taste. Any dysfunction of this organ causes a reduction in the personal comfort, and often surgery is employed to eliminate or just to alleviate such problems. Nasal breathing difficulties (NBD) represent one of the most common medical conditions, afflicting tens of million people and accounting for billions of Euros in health-care costs annually [1,2]. The surgery approach is hindered to a various extent by the incomplete knowledge of the behavior of the air-flow along the nasal cavity; any improvement in this respect would lead to better surgery planning and greater effectiveness: i.e. benefits for the patient and significant costs reductions for the healthcare system. From this point of view, computational fluid dynamics (CFD) in principle enables detailed prediction of flow patterns in nasal cavities and provides essential information on the potential relationship between patient-specific geometrical characteristics and health problems. Obviously, the daily clinical practice requires the numerical model to be not only accurate but also able to provide useful results in a reasonably short time. The complex flow structure along with the intricate geometry of the nasal anatomy pose significant challenges if one sets out to achieve such goals.

IOP Conf. Series: Journal of Physics: Conf. Series **882** (2017) 012010 doi:10.1088/1742-6596/882/1/012010

Experimental data to validate and improve the numerical simulations are particularly needed in this context, and at the same time particularly difficult to obtain, as *in-vivo* measurements are nearly impossible. Hence, an experimental set-up based on Stereo PIV (SPIV) and a silicon phantom of nasal cavities has been designed and realized at Politecnico di Milano. To the authors' knowledge, the very few literature works concerned with *in-vitro* experiments analyzing the airflow inside the upper airways have so far considered only the main passageways, excluding the large volumes of the paranasal sinuses, see for example [3,4,5]. Instead we decided to build an anatomically consistent model which retains the entire air volumes, i.e. the full anatomy of the human nose, including the whole region of the paranasal sinuses, which are obviously very important in treating NBD in general, and in planning a FESS (Functional Endoscopic Sinus Surgery) in particular. The present paper describes the approach followed to develop an experimental set-up, based on S-PIV and refractive index matching technique, to analyze the airflow inside a complete silicon model of the human nose. Preliminary results and a comparison with a Large Eddy Simulation on the same anatomy are presented and briefly analyzed.

2. Experimental set-up and Techniques

2.1. The Refractive Index Matching technique

A perfectly transparent model is needed if Particle Image Velocimetry (PIV) technique has to be used. Nevertheless nasal cavities have a quite complex geometry with irregular shapes and curved wall, so even in a perfectly transparent model the mismatch between the refractive index of themodel and that of the fluid (air) generates strong optical distortion preventing the use of quantitative image-based techniques. Budwing presented a robust method for eliminating refraction of light passing through model and/or test section walls [6]. The method consists in replacing the actual fluid with one that matches the refractive index of the solid transparent object (i.e. the model or the walls) and it is known as refractive index matching (RIM) technique. Organic liquids, liquid mixtures or aqueous solutions can be used as matching fluid [6,7,8]. The RIM makes the refractive index field spatially homogeneous eliminating any optical distortion. In our case the transparent model is totally submerged in a large box containing the matching fluid. By this way the model becomes optically invisible inside the box and one can shoot the laser through the model, or look through it, without optical artifacts. Being the model disappeared in the fluid, it is called "Phantom model". To simulate the breathing cycle a volumetric pump is connected to the model in such a way to allow the fluid to be inhaled/exhaled through the nostrils of the model.

2.2. The Phantom Head.

The actual size of the model was determined after an accurate analysis of the many factors involved in such a decision. The final size (in a 2:1 scale, i.e. the model is exactly twice the real dimensions) is a weighted compromise among several factors: a larger model reduces flow velocities, increases the gaps, and improves PIV resolution, but comes at various costs, including increased size, weight and money.

The geometry of the nasal cavity is reconstructed from the head CT scan of a 67-year-old male patient who showed a normal sinonasal anatomy. The patient underwent a plain head CT scan with a 64-row multi-detector CT. The scanner has a 512 × 512 matrix, accounting for a 0.49 mm × 0.49 mm spatial resolution in the sagittal–coronal plane and a 0.625 mm gap between consecutive axial slices. The images from the CT scan are first manipulated with the open-source software 3D-Slicer [9]. 3D-Slicer is used to convert the CT images into an accurate geometrical description of the boundary of the volume of interest, through image segmentation and volume reconstruction. After selecting the volume of interest and segmenting the image, the reconstructed three-dimensional geometry is exported and saved into a STereoLithography (STL) file. A limited amount of smoothing is applied to the reconstructed surface before exporting to STL, to avoid small-scale irregularities. The chosen anatomy, and the employed STL file, can then be used in the numerical simulations. This is a key step for enabling in the future a quantitative one-to-one comparison between numerical and experimental results.

At first the shape of the air volume (in analogy with the lost-wax casting technique) is produced and then the model is eventually casted into transparent silicone. The used 3D-printing technology allowed a spatial accuracy much better than the slice thickness of the CT scan while using as a printing material a water-dissolvable substance. Among the other technical characteristics, the printer has a resolution of 600x540 dpi, and prints details down to 100 micron size, with a building speed of 28mm/hour and a 254x381x203 working volume, that is just enough for our model. The printing material is demihydrathed calcium solphate, that is easily dissolved in water. Once printed, the model is carefully coated with a blue-colored, diluted vinil-based glue in preparation to the silicone casting. For the silicone we have adopted the very stable Sylgard 184 silicone elastomer, according to the manufacturer it has a refractive index of n = 1.4118 @ 589 nm [10]. A Plexiglas-made container box has been realized for the casting process. When ready for casting, the two components of the silicone are well mixed and degassed before pouring into the box. The curing process lasts a few days at room temperature. The blue-coated model within the casting box filled by silicone is shown in figures 1(a)-(b). Figure 2(a) shows the model once extracted from the casting box, before the start of the long and delicate process of removal of the internal material, and figure 2(b) shows the fully cleaned model.

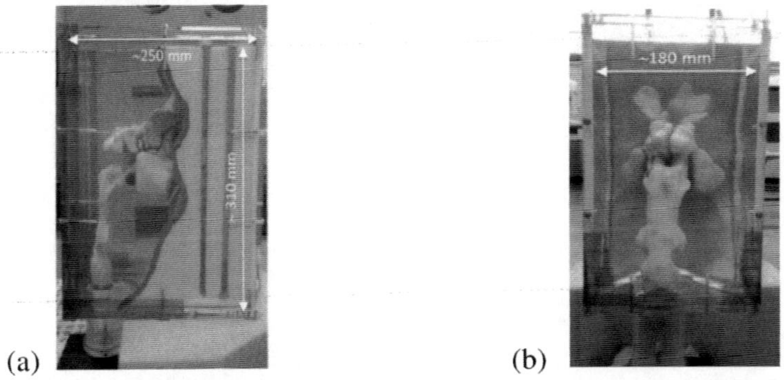

Figure 1. The silicone model inside the casting box. The nasal anatomy is evidenced by the blue coated printed material. (a) Side view, (b) front view.

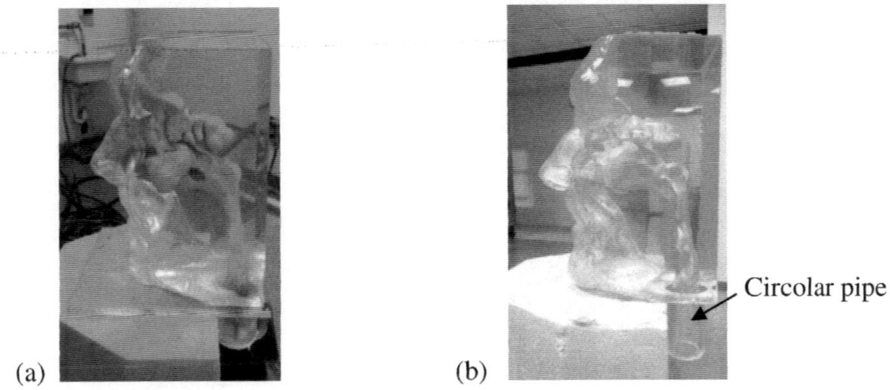

Figure 2. The silicone model once removed from the casting box. (a) Before removal of the internal material, (b) the cleaned model.

The printing material was hard to remove using only water, and a 5% Acetic Acid in water solution worked much better. Unfortunately the model suffered of few small damages due to a first attempt to mechanically remove the printing material and lastly, once cleaned, the model yellowed, figure 2(b). Anyhow these issues didn't compromise the optical clarity of the model allowing the experimental activity to be carried out. As shown in figure 2(b), the laryngopharynx of the cast model is extended in a circular pipe of about 50 mm in diameter and 100 mm in length to allow the pumping system to be connected to the model.

2.3. Fluid choice and properties

2.3.1. Refractive Index. As a first requirements the working fluid has to match the refractive index of the model. Following the work of Spence et al. [3] we chose as a matching fluid a mixture of water and glycerol. The latter is a non-toxic, clear liquid with a high refractive index and viscosity [11]. Being commonly used in the food industry glycerin is inexpensive and it is also compatible with materials commonly used in the phantom, flow circuit and pump [11]. The refractive index of the mixture, n_{mix}, can be quite easily controlled by adjusting water fraction and temperature. For their phantom model Spence et al. used a matching mixture composed by 39% water and 61% glycerin by volume [3], the same matching mixture is reported by Geoghegan et al. for phantom made of Sylgard 184 [11].

The refractive index of a binary mixture, n_{mix}, can be estimated through the use of mixture rules [7]. For water-glycerin mixture the simplest one is based on the Gladstone-Dale relationship and it gives n_{mix} as per equation (1):

$$n_{mix}-1=\left(n_{H_2O}-1\right)X_{H_2O}+\left(n_{Glyc}-1\right)X_{Glyc} \tag{1}$$

where n_{H_2O} is the refractive index of water, n_{Glyc} is the refractive index of glycerin, X_{H_2O} is the volume fraction of water, and X_{Glyc} is the volume fraction of glycerin. Another commonly used mixture rule is based on the Lorentz-Lorentz equation and shown in equation (2).

$$\left(\frac{n_{mix}^2-1}{n_{mix}^2+2}\right)=\left(\frac{n_{H_2O}^2-1}{n_{H_2O}^2+2}\right)X_{H_2O}+\left(\frac{n_{Glyc}^2-1}{n_{Glyc}^2+2}\right)X_{Glyc} \tag{2}$$

To check their accuracy, the refractive index of the water-glyceryn estimated using both equations (1) and (2) was compared to the value measured at 298.15 K and 589.3 nm by Vural et al [12]. The values $n_{H_2O}=1.3334$ and $n_{Glyc}=1.4729$ taken from [12] are used in both equations, while volume fractions are estimated from mass fractions and pure fluids density.

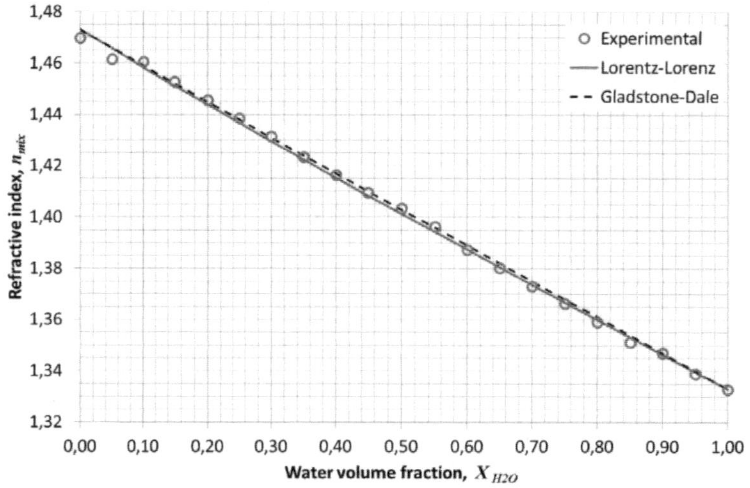

Figure 3. Comparison between the experimental values of n_{mix} (empty symbol) [12] and those predicted by equation (1) (dashed line) and equation (2) (solid line).

Results shown in figure 3 evidenced that both mixture rules work fine in predicting the measured n_{mix}. Equation (1) can be preferable being more simple, anyhow equation (2) shows a slightly better agreement with the experimental data. According to equation (2) the fluid used by Spencer et al. and by Geoghegan et al. has n_{mix} =1.417. On the other side the manufacturer quotes a refractive index of n_{syl} = 1.4118 @ 589 nm for the Sylgard 184 [10], while Geoghegan et al. reports n_{syl} = 1.43 for the same material without specifying at which wavelength [11]. The refractive index of the cured material depends on the mixing and curing processes, so the water-glycerin mixture has to be matched to the specific model used in the experiment. To this aim the technique suggested by Hopkin et al [5] is used. A mixture slightly richer in glycerin than Geoghegan et al.'s one is prepared, then the model is immersed in the fluid mixture by carefully allowing all of its cavities to be filled by the fluid. A grid of lines is placed behind the model, and while observing the grid distortion water is gradually added to the mixture until the distortion disappears completely. The mixture is thoroughly stirred and pumped through the model each time water is added so that it is homogeneous and homogenously distributed at its inside. Figures 4(a) and 4(b) show the phantom head immersed in a 39% water and 61% glycerin mixture (by volume) and in a 46% water and 54% glycerol mixture (by volume) respectively. The latter is the matching fluid mixture as evidenced by the undistorted rectangular grid visible behind the phantom, figures 4(b). According to equation (2) the matching mixture has n_{mix} =1.407. The mixture density computed from the pure fluids density and the water volume fraction is 1.14 g/cm^3.

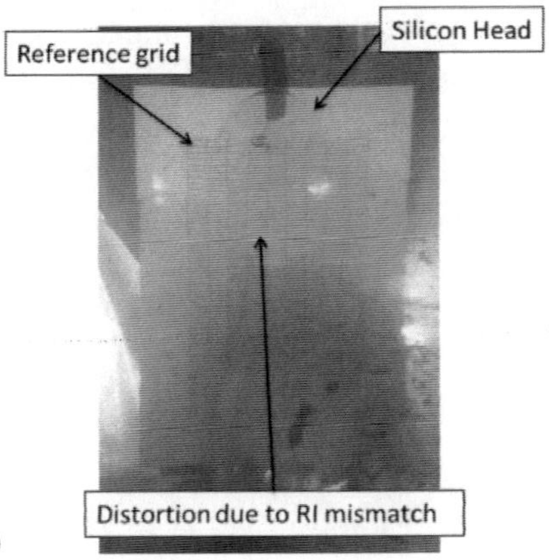

(a)

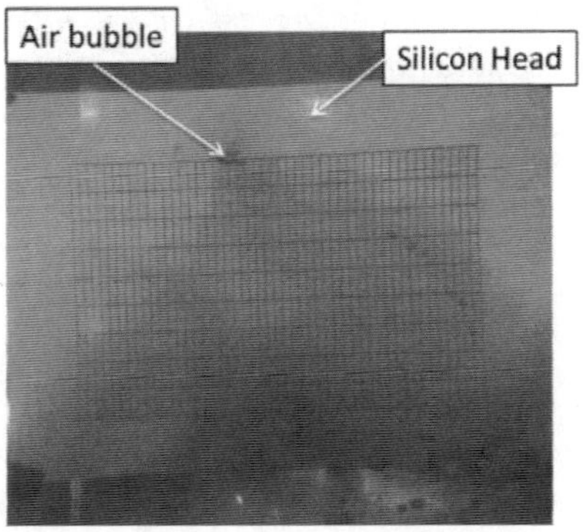

(b)

Figure 4. The phantom head immersed in the water/glycerol mixture. (a) Non-matching fluid mixture, (b) matching fluid mixture (46% water and 54% glycerol by volume).

2.3.2. Fluid-dynamic similarity. Being the working fluid not air, there are several differences in the physical parameters between the model and the real phenomenon of human breathing; these differences can be accounted for by obeying dynamical similarity: provided the fluid moves within the model at a suitably scaled velocity, the overall physics is the same and the results can be exactly related to the *in-vivo* values. To this aim the Reynolds number *Re*, equation (3), of the flow phantom is set equal to the real-world flow.

$$Re = \frac{QL}{A\nu} \tag{3}$$

where Q is the volumetric flow rate, A is the area of a reference section, L is a reference length and ν is the kinematic viscosity of the fluid. Considering equation (3) and a geometrical scale factor $M = L_{in-vitro}/L_{in-vivo} = \sqrt{A_{in-vitro}/A_{in-vivo}}$ between the model (*in-vitro*) and the real-world (*in-vivo*) anantomy, the dynamic similarity requires:

$$Q_{in-vitro} = Q_{in-vivo} M \frac{\nu_{in-vitro}}{\nu_{in-vivo}} \tag{4}$$

IOP Conf. Series: Journal of Physics: Conf. Series **882** (2017) 012010 doi:10.1088/1742-6596/882/1/012010

The dynamic similarity for the oscillatory respiration flow requires the Womersley number α to be taken into account [11]:

$$\alpha = L\sqrt{\frac{\omega}{\nu}} \tag{5}$$

where ω is the angular frequency. To achieve the same Womersley number for the *in-vivo* and the *in-vitro* oscillatory flow conditions, the corresponding time periods $T_{in-vivo}$ and $T_{in-vitro}$ should scale as per Equation (6) [3].

$$= M^2 \frac{\nu_{in-vivo}}{\nu_{in-vitro}} \tag{6}$$

. However in this preliminary work only steady flow conditions are analyzed, thus the Womersley number needs not to be taken into account. The geometrical scale factor of our model is *M=2*. The kinematic viscosity of the matching fluid (46% water and 54% glycerol) at 25 °C is $\nu_{in-vitro}$=5.51×10^{-6} m^2/s, as computed by interpolating the experimental data of Shankar and Kumar [13], while the kinematic viscosity of air at 25 °C and 1 bar is $\nu_{in-vivo}$= 15.58×10^{-6} m^2/s [14]. By substituting those values in equation (4) we get $Q_{in-vitro} = 0.7\ Q_{in-vivo}$.

2.4. *Aquarium and flow system.*

The silicone model is placed in a large container, called the aquarium, having a volume of about 0.33 m^3. Walls providing optical access for the SPIV cameras and the laser sheet are made of optical-grade glass, while the others are made of plexiglass. The glass is used because it has better optical quality and resistance to scratch than plexiglass. The shape of the aquarium is carefully designed to have the viewing angle between the two SPIV camera to form an angle of 90° and the optical axis of each camera to be perpendicular to the glass wall. These angles allow an optimal balance between in-plane and out-of-plane errors in the reconstructed velocity vectors while minimizing the optical distortion [15]. A sketch of the aquarium with the two SPIV cameras is shown in figure 5.

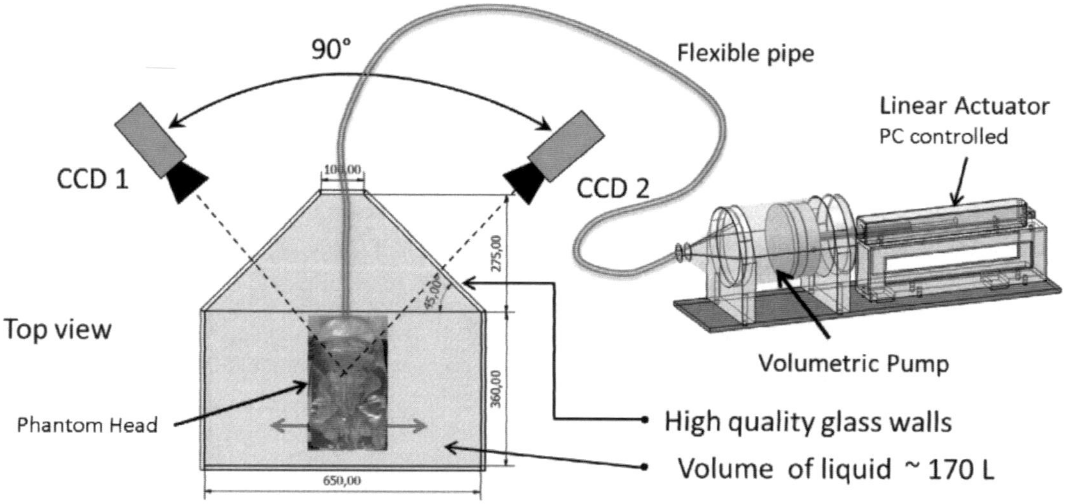

Figure 5. Sketch of the experimental set-up (top view of the aquarium and SPIV camera).

The aquarium is filled with about 170 liters of the matching water/glycerin mixture (46% water and 54% glycerol). The phantom is screwed to an aluminum plate which is placed on a rail system and connected to a linear actuator. This arrangement allows to displace the phantom head respect to the laser sheet and to perform SPIV measurement at different and parallel sagittal planes, figure 6 and figure 7.

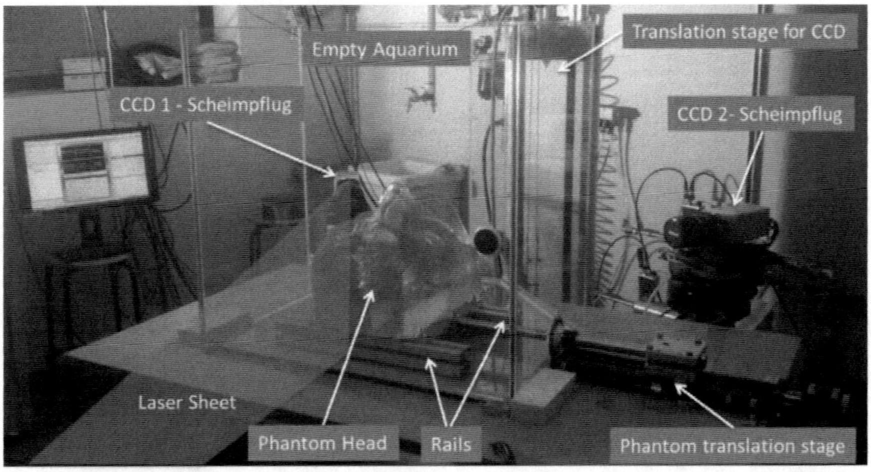

Figure 6. Photograph of the experimental set-up showing the phantom head in the empty aquarium.

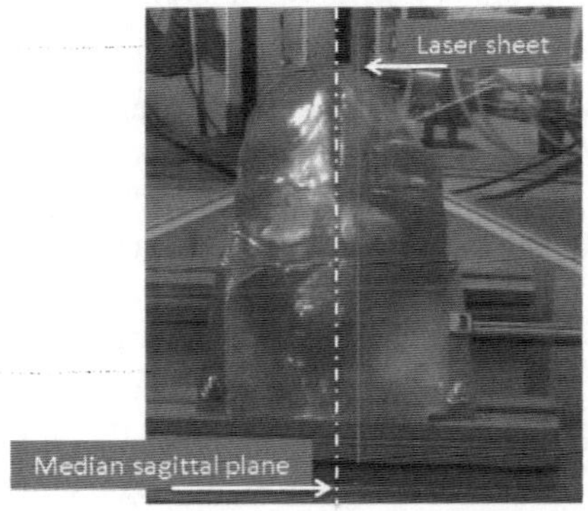

Figure 7. Photograph of the phantom head in the empty aquarium showing the location of the laser sheet during Stereo-PIV measurements.

To implement the scaled version of the human breathing cycle, a pumping system is designed and built on-purpose and connected to the model through a flexible hose. The pump is of syringe type, the piston has a diameter of about 210 mm and a stroke of 100 mm. A linear actuator is connected to the piston and controlled through a PC by using a Labview program. The voltage signal generated by the PC controls speed and acceleration of the piston, thus allowing the scaled periodic breathing cycle of a human being to be accurately reproduced. The pump is connected to the model through a flexible pipe, during the expiratory phase the fluid from the pump is injected through the model and it exhausts in the aquarium through the nose, while the reverse path occurs during the inspiratory phase.

2.5. *Stereo PIV system and measurements.*

A Stereo-PIV system is employed in order to measure the spatial distribution of the three velocity components inside the model. The Stereo-PIV system consisted of a double-pulsed Nd:YAG laser operating at $\lambda = 532$ nm with a maximum pulse energy of 200 mJ/pulse and a pair of CCD cameras with a resolution of 1344 * 1024 pixels each, equipped with Nikon lenses of 60 mm focal length, lens aperture was set to F#=8 on both cameras. The cameras were mounted on a traversing system in a Scheimpflug configuration, at opposite sides of the laser sheet. They were oriented with an angle of 45° to the forward scattering direction. The traversing system allowed to move the camera with a resolution of 0.1 mm. The laser sheet was about 1 mm thick and aligned with the sagittal plane of the phantom model. During this preliminary tests, measurements were carried out only in the sagittal plane located about 18 mm away from the median one, see figure 7. Three different fields of view were considered, they were of the same size but displaced each other of 50 mm in the vertical direction, see figure 8. In the object space each field of view has a size of about 96x54 mm^2 (horizontal x vertical). The velocity maps from those three regions were then recombined together to get the final velocity map.

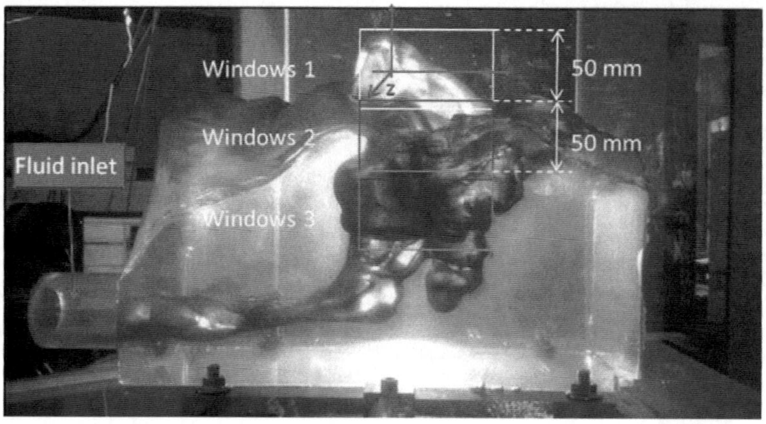

Figure 8. Photo of the phantom model in the empty aquarium showing the reference system and the regions where SPIV measurements have been carried out (the 3 rectangles).

Double images with an inter frame time of 500 µs are acquired at a rate of 5 Hz. Dantec's Dynamic Studio software is used to acquire and to process the images. Before the cross-correlation analysis PIV images are pre-processed. An image balancing filter with a smooth cell size of 5 x 5 is applied to correct the non-uniformities of laser light sheet, then the mean image is computed and subtracted from the instantaneous ones to remove background lighting and reflections.

The seeding particles are polyamide spheres, about 20 µm in diameter, thoroughly dispersed in the liquid mixture. The particle density is about ρ_P= 1.03 g/cm^3 thus slightly buoyant in the fluid mixture (ρ_{mix} = 1.14 g/cm^3). The refractive index of particles is $n_p = 1.5$, not so far from the computed refractive index of the mixture n_{mix} =1.407. The scattered light intensity decreases as the ratio between the refractive index of particles and that of the fluid come close to one, in our case particles were scarcely visible in the SPIV images, this resulted in a quite low image particle density. By increasing the seeding particles dispersed in the fluid the diffused laser light and the liquid turbidity increased deteriorating the SPIV images. Other seeding materials that could improve image quality will be tested in a future work (i.e. metallic coated hollow-glass spheres or fluorescent particles).

During this preliminary tests only a steady flow rate was considered and SPIV images were acquired only during the suction phase of the pump. During such phase the piston moved at a constant velocity of 9 mm/s, the corresponding flow rate of the fluid mixture was about 0.32 l/s. According to equation (4) the latter was dynamically similar to an *in-vivo* air flow rate of about 0.45 l/s. This type of test was planned to verify proper operation of the experimental set-up, it was not intended to simulate an actual breathing cycle which is periodical and composed of inhalation and exhalation phases. The acquisition of PIV images were synchronized with the suction phase of the pump in such a way that measurements were taken only 1.5 seconds after the piston starting and ended about 1.5 seconds before the piston stops. Only about 24 couples of images can be acquired during a single suction phase, thus at each camera position a total of 600 images were collected by replicating the run 25 times.

The standard cross correlation algorithm performed poorly on such images because of the low particle image density and the low brightness of the particle images. Being the flow steady the correlation averaging method [16] implemented in the Dantec's software have been applied to each data set composed by the 600 images. The correlation averaging method is mainly used in micro PIV and it works by averaging the cross-correlation planes of corresponding interrogation areas from different image couples. By this way the uncorrelated noise is rejected while the displacement-correlation peak is strengthened making the latter visible over the cross-correlation noise floor. The size of the interrogation area (IA) was 32x32 pixels, with an overlap of 75%. The resulting vector spacing was about 0.7 × 0.5 mm^2 ($\Delta x * \Delta y$). The target used for the stereo calibration was an array of dots printed on paper and sandwiched between two thin plexiglass plates. The calibration was performed submerging the target in the aquarium and taking its picture at nine equally spaced positions along the out of plane direction, z ($-2\ mm \leq z \leq 2\ mm$). The implemented camera model was a third order polynomial. The third velocity component was reconstructed from the two 2D vectors fields by using the method proposed by Soloff et al. [17] and implemented in the software.

3. Experimental results

A de-warped mean PIV image corresponding to the field of view indicated as Window 2 in figure 8 is shown in figure 9. The image brightness has been adjusted to enhance the visibility of details. Bright lines in figure 9 correspond to the solid-fluid interface at the laser sheet plane. Such interface is visible due to: a) light scattered by particles attached to the walls, b) a non-perfect index matching between the fluid and the model. The bright silhouette appearing in the mean image was used as a reference to manually draw a mask. The latter was subsequently applied to the velocity maps to highlight the non-fluid region. Laser light sheet non-uniformity (shown as black stripes) and light diffused by model defects or damages are also visible in figure 9. The image balancing and background removal procedures describe in section 2.5 allowed to partially compensate for such image defects.

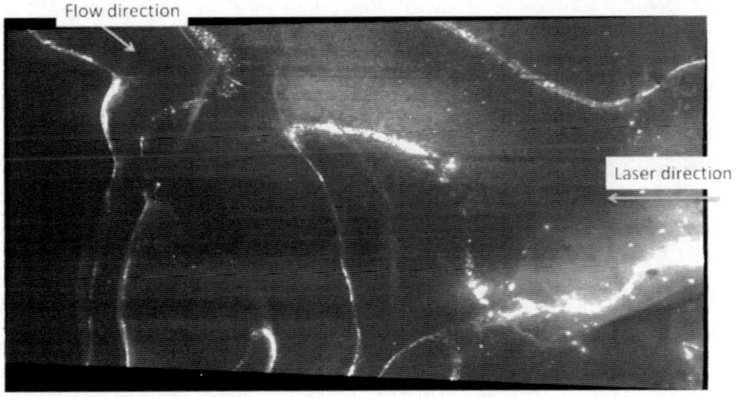

Figure 9. De-warped mean PIV image, the field of view corresponds to windows 2 in figure 8.

The velocity map (modulus of the velocity vector) obtained from the Stereo-PIV measurements is shown in figure 10(a), and compared to numerical results shown in figure 10(b).

The numerical data have been obtained via a computational procedure developed during the last few years at Politecnico di Milano, which is entirely based on open-source software. The data are taken from one case among the several considered in previous work (more details can be found in [18] and [19]), is based on the same anatomy, and portraits the same mean velocity field as computed from a Large-Eddy Simulation. The simulation has been carried out on the real model (i.e. 1:1 scale, working fluid is air), and a correction factor has been subsequently applied to the velocity and length scales in order to produce figure 10, accounting for the dynamical similarity between the two flows. The data presented correspond to a mean flow field computed via time-averaging of the unsteady and 3-dimensional LES results, for a steady inspiration, hence directly comparable to the experiment. The numerical calculations covered 0.2 seconds of physical time, and the computing time required for such a case was of approximately 100 hours by using 16 computing cores. The mesh size, of more than 5M elements, makes this LES a very well resolved one, so that the contribution from the turbulence model (the classical, non-dynamic Smagorinsky model) is almost negligible everywhere. All the terms of the incompressible Navier-Stokes equations have been discretized via second-order-accurate operators.

IOP Conf. Series: Journal of Physics: Conf. Series **882** (2017) 012010 doi:10.1088/1742-6596/882/1/012010

The mean flow field plotted in figure 10(b) is the time-average of the time-varying LES solution over the entire simulation time. The experimental velocity maps from the three fields of view of figure 8 are plotted in figure 10(a) as a single map by taking into account the vertical displacement of 50 mm existing between each of them; on the same figure, the dashed area highlight the non-fluid region. Despite the use of the correlation averaging method, regions exist where clearly the fluid velocity cannot be recovered. These regions appear as stripes in the velocity maps, and they seem to overlap with the black stripes observed in the laser sheet, see figure 9. The latter indicate regions less lighted by the laser sheet, where particles images are more faint, thus exacerbating the issue related to the low image particle density and decreasing the number of validated velocity vectors. From the figure, one can easily appreciate the complexity of the flow patterns that develop in such small-scale domain. In figure 10(b), the same plot as derived from CFD shows a similar picture, at least at a qualitative level. Local velocity maxima in correspondence of the 3 meata, where the cross-section of the nasal airways shrinks owing to the protrusion of the turbinates, can be clearly observed.

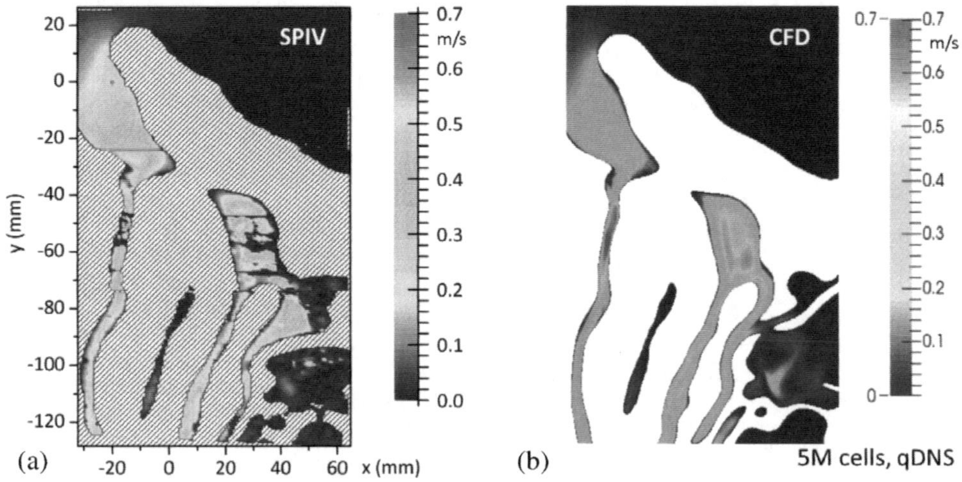

Figure 10. Modulus of the velocity vector in the sagittal plane z=18 mm. (a) SPIV measurements, (b) LES numerical simulation (mean flow).

4. Conclusion

This paper describes an experimental set-up based on Stereo-PIV and refractive index matching techniques used to analyze the flow inside a complete patient-specific silicon model of the human nose. The geometry of the model reproduces that obtained through a CT scan and then used in the numerical simulations. The model has been built in a 2:1 scale and using 3D printing technology. To the authors' knowledge, this is the first attempt to experimentally investigate the flow inside a complete human nose including all the paranasal sinuses. The making of the phantom model and the choice and tuning of the matching fluid mixture are thoroughly described in the paper. The fluid dynamic similitude is also briefly presented and discussed. An aquarium of about 0.33 m^3 having suitable optical accesses for laser and camera has been built and filled with about 170 liters of matching fluid mixture. The phantom model is submerged in the aquarium and it is connected to a piston pump through a flexible pipe that enters in the model upstream of the laryngopharynx: this allows the fluid mixture to be inhaled/exhaled through the nose to simulate the breathing cycle. Few preliminary test were performed using a constant flow rate to verify proper operation of the experimental set-up. Those tests were not intended to simulate an actual breathing cycle which is periodical and composed of inhalation and exhalation phases. A stereo-PIV system allowed to measure the three velocity components at three different location in a sagittal plane of the phantom model. Experimental results evidenced that for this quite large model the refractive index matching worked properly and that the set-up was suitable for the intended purposes. They also evidenced that seeding particles having higher light scattering capability should be sought, alternatively fluorescent seeding could also be used. Finally, despite the preliminary stage of the research, the measured flow field resulted in fair agreement with that obtained by a LES simulation on the same anatomy.

Acknowledgments
The authors wish to thank the collaborators who contributed to the present work, namely MD Carlotta Pipolo and Alberto Saibene, and Mr. Simone Raina and Mr. Martino Ermacora.
Financial support from Regione Lombardia and Italian Ministry of Health through the project CAPNAF-FD is gratefully acknowledged.

References

[1] Stewart M, Ferguson B J and Fromer L 2010 Epidemiology and burden of nasal congestion Int. J. Gen. Med. 3 37–45
[2] Smith K A, Orlandi R R and Rudmik L 2015 Cost of adult chronic rhinosinusitis: A systematic review Laryngoscope 125 1547–56
[3] Spence C J T, Buchmann N A, Jermy M C and Moore S M 2011 Stereoscopic PIV measurements of flow in the nasal cavity with high flow therapy Exp. Fluids 50 1005–17
[4] Im S, Heo G E, Jeon Y J, Sung H J and Kim S K 2014 Tomographic PIV measurements of flow patterns in a nasal cavity with geometry acquisition Exp. Fluids 55
[5] Hopkins L M, Kelly J T, Wexler a. S and Prasad a. K 2000 Particle image velocimetry measurements in complex geometries Exp. Fluids 29 91–5
[6] Budwig R 1994 Refractive index matching methods for liquid flow investigations Exp. Fluids 17 350–5
[7] Wiederseiner S, Andreini N, Epely-Chauvin G and Ancey C 2011 Refractive-index and density matching in concentrated particle suspensions: a review Exp. Fluids 50 1183–206
[8] Amini N and Hassan Y A 2012 An investigation of matched index of refraction technique and its application in optical measurements of fluid flow Exp. Fluids 53 2011–20
[9] Pieper S, Halle M and Kikinis R 2004 3D Slicer 2004 2nd IEEE Int. Symp. Biomed. Imaging Nano to Macro (IEEE Cat No. 04EX821) 2 632–5
[10] Dow Corning 2014 Product Information of Sylgard® 184 Silicone Elastomer 4
[11] Geoghegan P H, Buchmann N A, Spence C J T, Moore S and Jermy M 2012 Fabrication of rigid and flexible refractive-index-matched flow phantoms for flow visualisation and optical flow measurements Exp. Fluids 52 1331–47
[12] Vural U S, Muradoglu V and Vural S 2011 Excess molar volumes, and refractive index of binary mixtures of glycerol + methanol and glycerol + water at 298.15 k and 303.15 K Bull. Chem. Soc. Ethiop. 25 111–8
[13] Shankar P N and Kumar M 1994 Experimental Determination of the Kinematic Viscosity of Glycerol-Water Mixtures Proc. R. Soc. London A Math. Phys. Eng. Sci. 444
[14] Spurk J H and Aksel N 2012 Fluid mechanics
[15] Prasad a. K 2000 Stereoscopic particle image velocimetry Exp. Fluids 29 103–16
[16] Westerweel J, Elsinga G E and Adrian R J 2013 Particle Image Velocimetry for Complex and Turbulent Flows Annu. Rev. Fluid Mech. 45 409–36
[17] Soloff S M, Adrian R J and Liu Z-C 1997 Distortion compensation for generalized stereoscopic particle image velocimetry Meas. Sci. Technol. 8 1441–54
[18] Manara F and Lamberti G 2015 Simulazioni RANS and LES delle cavità nasali Master's Thesis in Aeronautical Engineering (Politecnico di Milano)
[19] Lamberti G, Manara F and Quadrio M 2015 RANS/LES/DNS simulations of the airflow in nasal cavities APS Division of Fluid Dynamics (Fall) 2015

IOP Conf. Series: Journal of Physics: Conf. Series **882** (2017) 012011 doi:10.1088/1742-6596/882/1/012011

Optical measures of dust velocities and direction during loss of vacuum accidents in confined environment and correlation between dust positions and properties with the resuspension degrees and the velocity modules

Malizia Andrea[1,*] with Riccardo Rossi[2] and Pasquale Gaudio[2]

1. Department of Biomedicine and Prevention, University of Rome Tor Vergata, Via del Politecnico 1, 00133, Rome, Italy.
2. Department of Industrial Engineering, University of Rome Tor Vergata, Via del Politecnico 1, 00133, Rome, Italy.

*Corresponding author: malizia@ing.uniroma2.it

Abstract. Dust explosions are dangerous events that still today represent a risk to all the industries that produce and/or handle combustible dust like the agro-alimentary, pharmaceutical and energy ones. When a dust cloud is dispersed in an oxidant gas, like air, it may reach the explosive concentration range. A model to predict the dust critical conditions, that can cause explosions, is a key factor for safety of operators and the security of the plants. The key point to predict this dust resuspension is to measure the velocity vectors of dust under the accidental conditions. In order to achieve this goal the authors have developed an experimental facility, STARDUST-U, which allow to obtain different conditions of temperature and pressurization rates characteristic of accidents in confined environment. The authors have developed also optical methods and software to analyse different dust resuspension phenomena under different conditions in confined environment. In this paper, the author will present how they measure the dust velocity vectors in different experimental conditions (and for different type of dusts) and how they have related the dust characteristics and positions inside STARDUST-U with the resuspension degree and the velocity values.

1. Introduction

Despite the phenomenon of dust explosion has been known for more than 100 years, still today it is a dangerous hazard in process, medical, pharmaceutical, food, energetic, agribusiness industries and in any industrial plant that handles, produces and/or stores combustible dust [1, 2, 3]. Between the most recent and notable accidents there are the Imperial Sugar explosion [4], the Kunshan explosion and the Formosa Fun Coast explosion [3]. Table 1 shows the dust that led to dust explosion and the number of injuries and deaths [1, 4, 5].

Table 1. The most recent and notable accidents of dust explosions

Name of Event	*Source dust*	*Deaths*	*Injuries*
Imperial Sugar explosion (2008)	Sugar dust	14	42
Kunshan explosion (2014)	Metal powder	146	114
Formosa Fun Coast explosion (2015)	Coloured starch powder	15	498

It is well know that a dust explosion is a rapid combustion of dust finely dispersed in an oxidant gas. Several tests have been conducted to understand precisely the conditions that lead to dust explosions. These conditions are usually summarized through the dust explosion pentagon [2, 5-7]. The conditions that lead to dust explosions are:

1. ***Combustible dust***: Since combustion is an oxidation of the combustible, dust have to be oxidizable;
2. ***Oxidant***: The dust must be mixed inside an oxidant (usually air);
3. ***High concentration and fine dispersion:*** The rapid combustion is strongly correlated to the concentration level and how fine the dust is dispersed inside the oxidant [1, 8, 9];
4. ***Ignition source:*** It is needed to trigger the combustion [1];
5. ***Confinement:*** In some cases, confinement is not needed but usually strong improve the damages and the severity of the explosion [1].

Methods and techniques able to avoid that all these requests are reachable must be developed in order to make prevention of dust explosions. Prevention is the most important[10, 11].
The Quantum Electronics and Plasma Physics (QEP) research group develop and experimental apparatus called STARDUST-Upgrade [13]. It allows to replicate some thermos-fluid dynamic conditions that can provoke resuspension of dust in case of an under expanded-jet flowing inside a vacuum vessel that is typical of the accidents previously mentioned and also of accidents in energetic facilities [14,15].

The authors develop two optical experimental set-ups (similar to those of well known optical techniques like PIV and SHADOWGRAPH) to get images of mobilized dust under different thermos-fluid dynamic conditions. The first set-up that will be presented is used to acquire images when "bright" dust are used (like flour or stainless steel AISI316) and the second set-up is used to acquire images when "dark" dust are used (like carbon or tungsten).

In the first part of the paper results, we will present the images of experiments performed with flour that has been acquired, elaborated and analysed using software developed based on the Lucas-Kanade algorithm used to measure the dust velocity vectors. The velocity modules obtained will be compared with air velocity values (obtained experimentally in previous experimental campaign and numerically with FLUENT). In the second part of the paper results, we will elaborate data coming from experiments done using the 4 different types of dust, (to have 4 different density values) before mentioned changing, for each dust, the height from the bottom of the facility and changing the optical technique used to acquire images. It will be important to analyse three properties in function of the parameter δ and the density of dust: resuspension degree, mean velocity and max velocity. Intensity maps are discussed in order to understand when and how much the re-suspension can be dangerous.

2. Materials and Methods

STARDUST-Upgrade (Figure 1) is a stainless steel cylindrical tank. Thanks to its isolation from the external ambient, it can reach high vacuum values (till 100 Pa).

Figure 1. STARDUST-Upgrade: Windows and inlet valves.

Furthermore, a functional scheme of the facility is presented in Figure 2.

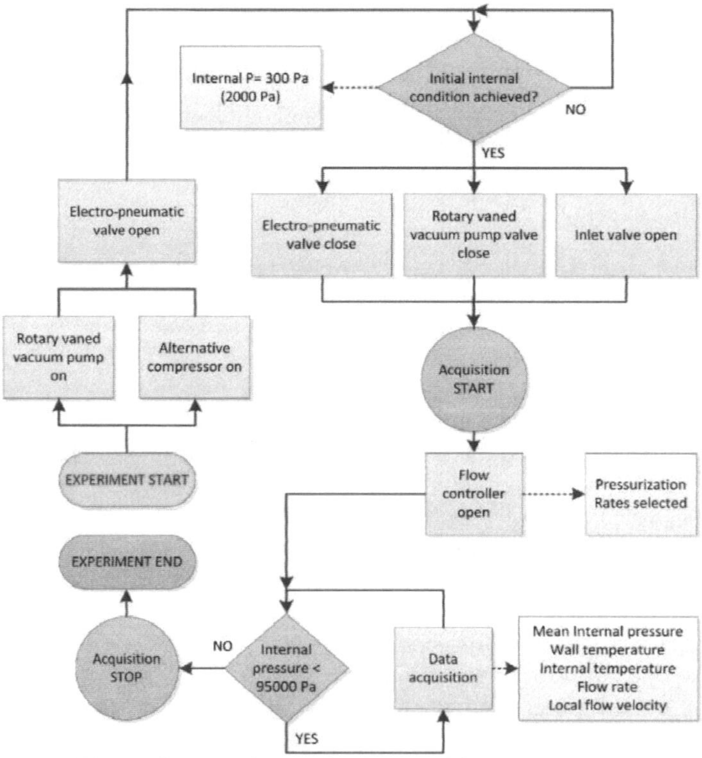

Figure 2. Experimental scheme of STARDUST-U

During each experiment, data acquisition was possible for the following quantities: thermocouples temperatures (T1 [°C], T2 [°C], T3 [°C], T4 [°C]), actual internal pressure (Ps [Pa]), inlet air flow rate [litres/minutes]. Boundary conditions, in terms of initial internal pressure of the vacuum chamber (i.e. "vacuum level") and air flow rate entering the chamber, were achieved using a vacuum pump, and air compressed delivery system respectively, both remote controlled. The internal temperature was measured by four J-thermocouples placed inside the chamber. Internal pressure was measured by two gauges depending on the actual vacuum level. Edwards pressure gauge (ASG-2000-NW16, Edwards, Crawley, UK) covered pressure range from 1000 to 2000 mbar. For higher vacuum levels (pressure below 1000 mbar) a Pirani gauge was used (Alcatel AP 1004 Pirani Gauge, Alcatel Vacuum Technology). Pirani gauge covered a pressure range from 0.0005 to 1000 mbar. Air flow rate was controlled by a mass flow meter (1559A Mass-Flo® Controller, MKS Instruments) covering the range from 20 to 200 slm (standard litres per minute). Acquisition and control system was controlled remotely by a personal computer using a LabVIEW software. Acquisition hardware included an ethernet "CompactDAQ" chassis (NI cDAQ-9188XT) that allowed the introduction of 8 input/output cards.

Before the vacuum generation and loss of vacuum accident reproduction, a dust tray must be prepared and placed inside the camera. The tray has been covered by an opaque cardboard with the aim of decreasing the scattered light noise (Figure 3 a). Tray sizes are shown in Figure 3 b.

(a)

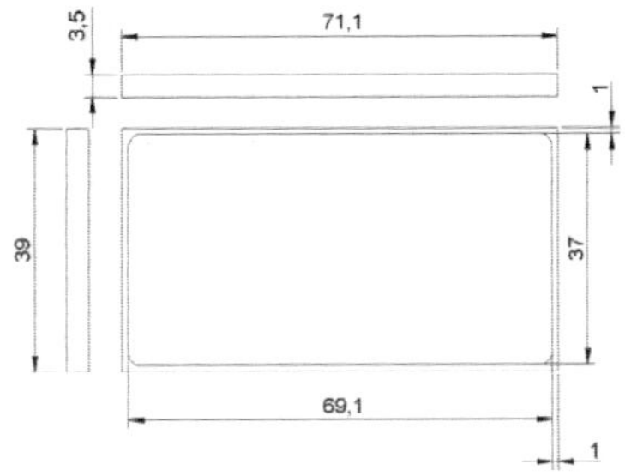

(b)

Figure 3. Dust tray and opaque card board (a); Tray sizes (b).

The dust is deposited on the tray through a sieve, in order to decrease the larger agglomerate particles. In this work, dust samples are located on a tray placed in the center of the camera, along valve E and valve F with the possibility to change its height (Figure 4). An adimensional parameter δ calculated through (1) describes the relative distance between the inlet valve axis and the dust sample position.

$$Valve\ E:\ \ \delta = \frac{h_E - h_{Tray}}{d_{valve}} \quad (1)$$

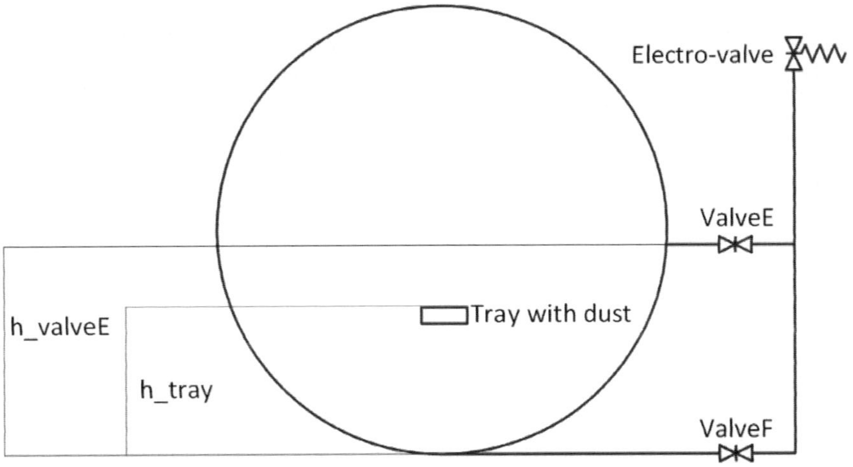

Figure 4. Valves and dust tray position and height parameters used.

STARDUST-Upgrade is provided with glass windows that allow the interface of devices to perform optical measurements. The aim of the work is to measure the velocity vectors of dust during an accident replication and then relate these measurements in function of the parameter δ and the density of dust mobilized. Four types of dust have been used: flour and stainless steel AISI316 (as bright powders), tungsten and carbon (graphite) powder (as dark powders). The particle velocities are measured through a method similar to Particle Tracking Velocimetry [19, 20, 21]. Anyway, since the characteristics of these dust are very different, two different optical setups have been used.
Table 2 describes the main characteristics of the light receiver and the light source of the two experimental setups.

Table 2. Main characteristics of optical experimental setups

Scattering setup		**Absorption setup**	
Light source	Diode Laser	*Light source*	Tungsten lamp
Wavelength	445 nm	*Wavelength*	White
Power	2 W	*Power*	Variable
Light receiver	CCD Fast Camera	*Light receiver*	CCD Fast Camera
Frame rate	5000 Hz	*Frame rate*	5000 Hz
Resolution	1280x128 pixels	*Resolution*	1280x128 pixels

The first experimental set-up (Figure 5) works well with "bright" dust, i.e. particles with a large scattering cross section (like flour of stainless steel in our case). A diode laser generates a laser sheet by means of two lenses, one convex and one cylindrical lens. The relative position of the two lenses allows to have larger or smaller laser sheet. The size of the laser sheet is fundamental to control the intensity of light that interacts with particles and so the intensity of the scattered light. The CCD fast camera is placed orthogonally to the laser sheet. Therefore, the fast camera record images with a dark background and white particles. The dust tracked with this setup is flour and stainless steel dust.

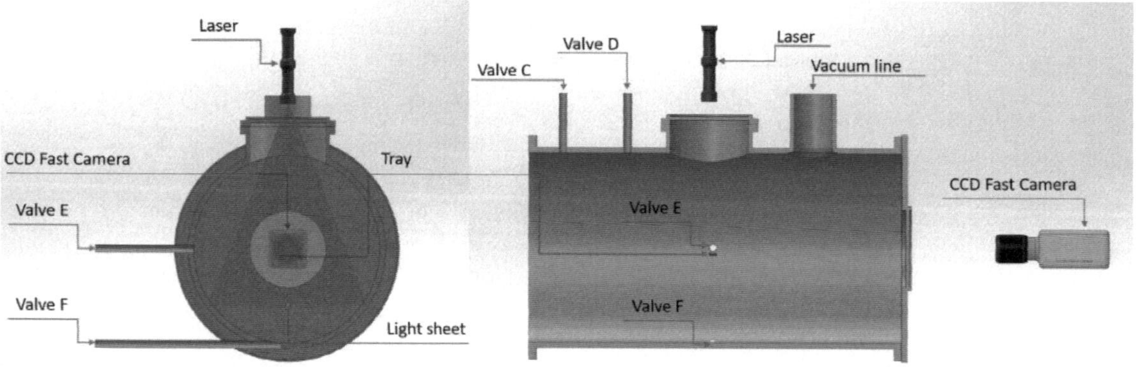

Figure 5. Optical setup to track particles with large scattering cross section.

Since "dark" particles have a too small scattering cross section (like tungsten and carbon in our case), another experimental setup (the second) is needed. In this case, a light source is placed on the same axis of the fast camera. Two convex lenses are used to parallelize the light source inside the camera (Figure 6).

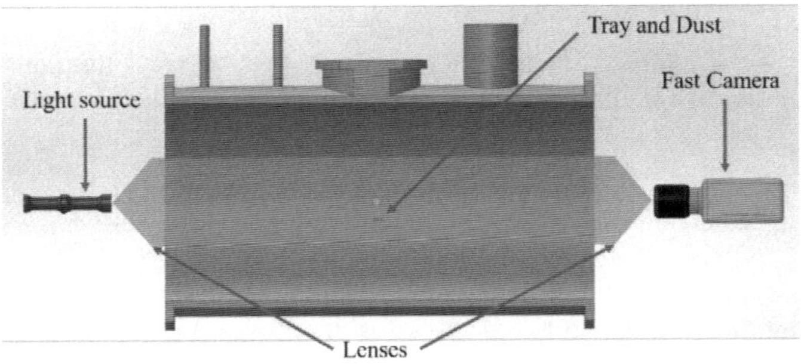

Figure 6. Optical setup to track particles with low scattering cross section and large absorption cross section.

Therefore, this technique record images with white background and black particles. This experimental setup must be used when the scattering cross section of dust particles is too low while absorption cross section is large enough. Anyway, since the software used to track particle works only if the particles are white, pre-processing of these images are needed. Figure 7 shows the main steps for pre-processing the images before velocity computation. Firstly, a background image is built. The background image is computed as the average of 100 frames where the dust is not moved and the inlet valve is not opened. The background image is inverted, therefore the black pixel becomes white and vice versa. Then, each i-image is inverted and the inverted background is subtracted to it. Therefore, the result is an image where only the particles in movement are white and the rest is black.

Figure 7. Pre-processing steps for large absorption cross section optical setup.

IOP Conf. Series: Journal of Physics: Conf. Series **882** (2017) 012011 doi:10.1088/1742-6596/882/1/012011

The software used [14] to track the velocity of the particles has been developed in LabVIEW language, with the aid of the National Instrument Vision tool. It detects the particles in the frame i-1 and looks for the same particles in the frame i. The algorithm used is the Lucas-Kanade algorithm [23, 24, 25].

In order to explain the Lucas Kanade Method we have to start from the basic concepts that a grayscale image is a matrix; The value of each matrix element is delimited between 0 (black) and 255 (White). All algorithms are based on brightness constancy. This hypothesis implies that an object cannot change its intensity along frames, but can vary its position (2).

$$I(x,y,t) = I(x+\delta x, y+\delta y, t+\delta t) \quad (2)$$

Where δx and δy are the pixel displacements and δt is the time difference between two frames. The aim of the optical software is to compute the displacements of each pixel.
Lucas&Kanade algorithm is one of the gradient-based methods. These are founded on developing brightness constancy hypothesis, which can be written as follows (3):

$$\frac{dI(x,y,t)}{dt} = 0 \quad (3)$$

And so, through partial derivations (4):

$$\frac{dI(x,y,t)}{dt} = \frac{\partial I}{\partial x}\frac{\partial x}{\partial t} + \frac{\partial I}{\partial y}\frac{\partial y}{\partial t} + \frac{\partial I}{\partial t} = 0 \rightarrow \frac{\partial I}{\partial x}u + \frac{\partial I}{\partial y}v + \frac{\partial I}{\partial t} = 0 \quad (4)$$

The idea of Lucas&Kanade algorithm consists in assuming a sub-matrix and doing the assumption that each pixel of that sub-matrix have the same velocity of the matrix itself. The algorithm performs (4) in each pixel of each block through (5) [26].

$$\begin{cases} \frac{\partial E}{\partial x}(p_1)\, u + \frac{\partial E}{\partial y}(p_1)\, v = -\frac{\partial E}{\partial t}(p_1) \\ \frac{\partial E}{\partial x}(p_2)\, u + \frac{\partial E}{\partial y}(p_2)\, v = -\frac{\partial E}{\partial t}(p_2) \\ \vdots \\ \frac{\partial E}{\partial x}(p_n)\, u + \frac{\partial E}{\partial y}(p_n)\, v = -\frac{\partial E}{\partial t}(p_n) \end{cases} \quad (5)$$

Furthermore, (5) can be written in matrix form (6):

$$C = \begin{bmatrix} \frac{\partial E}{\partial x}(p_1) & \frac{\partial E}{\partial y}(p_1) \\ \frac{\partial E}{\partial x}(p_2) & \frac{\partial E}{\partial y}(p_2) \\ \vdots & \vdots \\ \frac{\partial E}{\partial x}(p_n) & \frac{\partial E}{\partial y}(p_n) \end{bmatrix}; \; \bar{v} = \binom{u}{v}; d = \begin{bmatrix} -\frac{\partial E}{\partial t}(p_1) \\ -\frac{\partial E}{\partial t}(p_2) \\ \vdots \\ -\frac{\partial E}{\partial t}(p_n) \end{bmatrix} \quad (6)$$

$$C\bar{v} = d \rightarrow C^T C\bar{v} = C^T d \rightarrow \bar{v} = (C^T C)^{-1} C^T d \quad (7)$$

The final form, reached through (7), is (8).

$$\begin{pmatrix} u \\ v \end{pmatrix} = \begin{bmatrix} \sum \frac{\partial E}{\partial x}(p_i)^2 & \sum \frac{\partial E}{\partial x}(p_i)\frac{\partial E}{\partial y}(p_i) \\ \sum \frac{\partial E}{\partial x}(p_i)\frac{\partial E}{\partial y}(p_i) & \sum \frac{\partial E}{\partial y}(p_i)^2 \end{bmatrix}^{-1} * \begin{bmatrix} -\sum \frac{\partial E}{\partial x}(p_i)\frac{\partial E}{\partial t}(p_i) \\ -\sum \frac{\partial E}{\partial y}(p_i)\frac{\partial E}{\partial t}(p_i) \end{bmatrix} \quad (8)$$

In order to compute high velocities, LK algorithm needs a pyramidal implementation, which allows the algorithm to find high-velocity vectors. It is a very popular algorithm in optical method but sometimes heavy filtering function is required because of its high noise sensibility. LK algorithm implementation can be used also for feature point track. Pyramidal implementation consists in creating pyramid representation of each frame and computing point displacement in each representation. Through this expedient, maximum trackable velocity becomes $u_{max,p}=(2^{Lm+1}-1)u_{max}$ [27].

Pyramid images computation has been implemented to allow the computation of large velocity [28]. The output velocity of the algorithm is described in pixels/frame. Spatial conversion is computed measuring the pixel dimension in meters while the temporal conversion is obtained by the frame rate of the fast camera (5000 Hz).
The software has one graphic output that is the frame i with overlaid the particle velocity vectors (Figure 8) and a .txt file with all the velocity vectors information: x and y position, time, magnitude and angle.

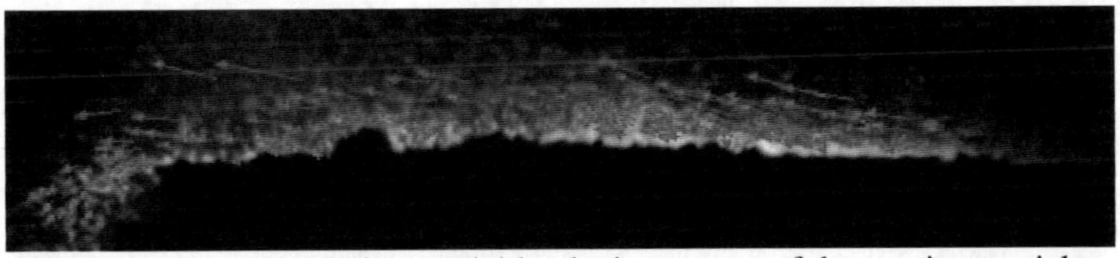

Figure 8. Frame i with the overlaid velocity vectors of the moving particles.

3. Results

In this first part we will analyze only the results coming from an experiment done with flour with the air inlet coming from Valve E and the dust tray placed at the same height of the dust. The software has been used to analyze the dust velocity vectors (modules and direction) and to compare the modules values to those of air flow jet (obtained experimentally and numerically [14-15]). In table 3, there are the experiments parameter:

Table 3. ValveE_0cm – Feature Table.

Valve	E
f_{acq} [Hz]	5000
Exposure time [s]	1/6000
Resolution [Pixel]	1280x128
Spatial res [m/px]	0,0001143
Z (cm to jet axis)	0
Number of Frame	3000
Initial dust weight [g]	2,9607
Final dust weight [g]	0
Dust mobilised [%]	100

IOP Conf. Series: Journal of Physics: Conf. Series **882** (2017) 012011 doi:10.1088/1742-6596/882/1/012011

Total mobilisation time [s]	0,496

Module and direction velocity correlation (Figure 9a) shows how highest velocities are parallel to jet main direction (0°) which is also the main particle direction (Figure 9b). Figure 10 shows the velocity histogram, showing the low number of high velocity particles compared to the low velocity particles.

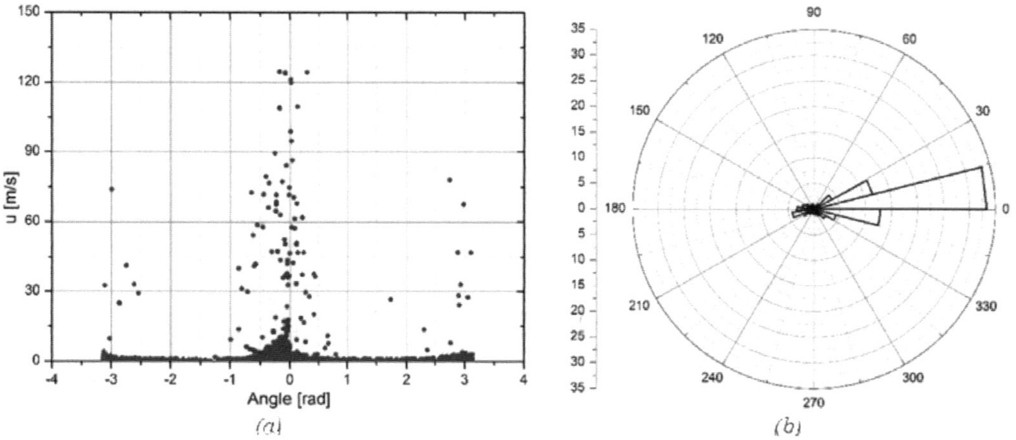

Figure 9. Module vs Direction Dust Particle Velocity (a) and Direction Polar Histogram (b).

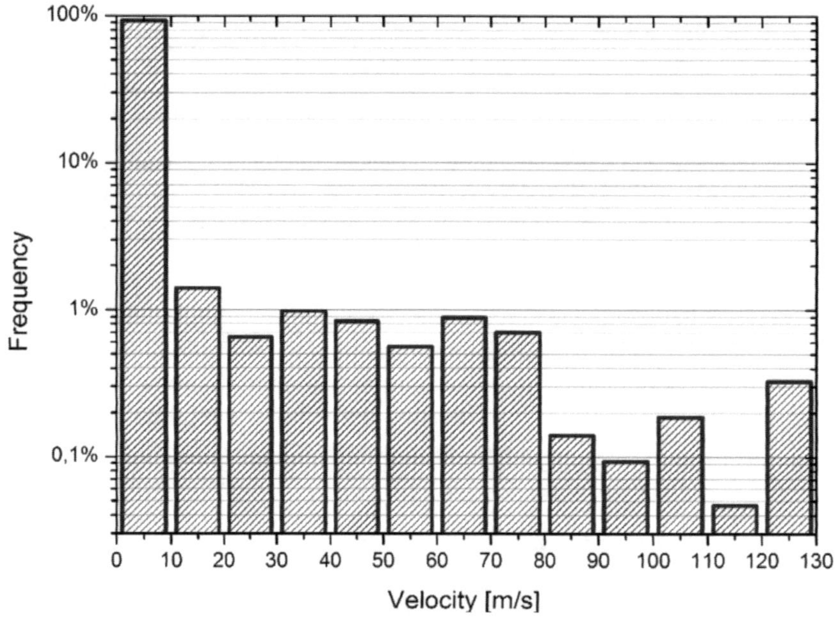

Figure 10. Velocity histogram

Figure 11 shows the correlation between tracked particles and experimental and numerical air jet. It illustrates only the first 0,6 s of the accidents, which is the time required to mobilise all the dust. Only few particles reach the experimental air jet velocity, while the majority does not reach 15 m/s. This occurs because slowest particles are the largest and consequently easier to be tracked. The largest particles are not single particles but agglomerates. The highest velocities are reached at the beginning of the experiment. It may be due to the fact that smaller particles mobilise immediately.

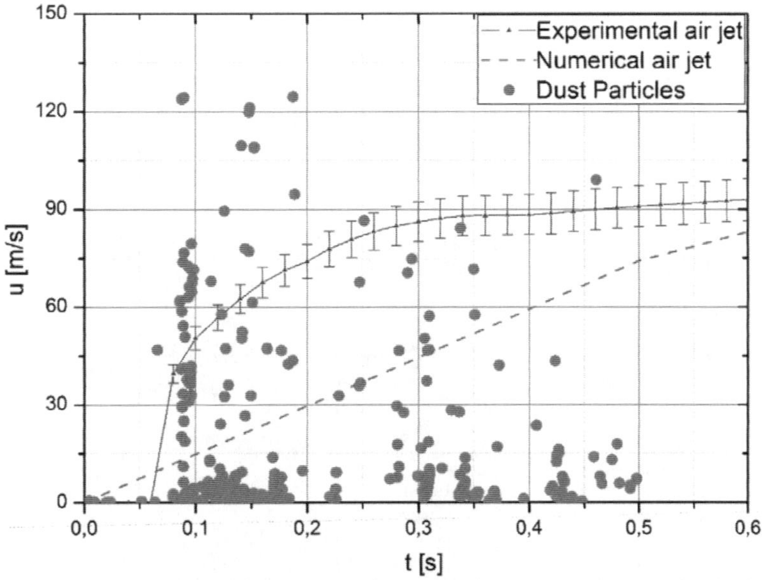

Figure 11. Dust particle velocities compared to experimental and numerical Air Jet velocity.

In Figure 12 there are the velocity vectors at different times.

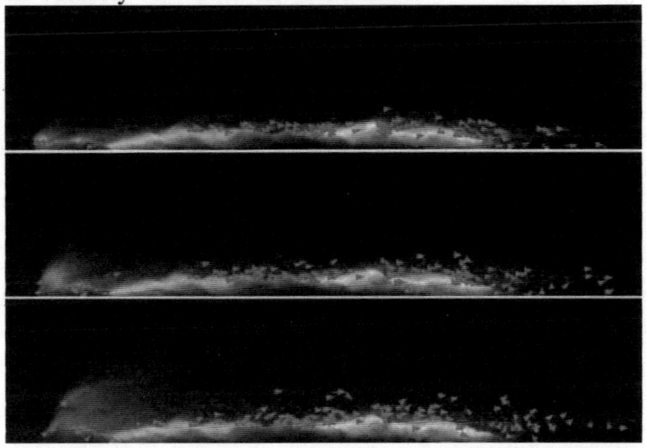

Figure 12. Velocity vectors at t_1=0,0886 s, t_2=0,0896 s and t_3=0,0910s

In the second phase we will elaborate data coming from experiments done using the 4 different types of dusts (to have 4 different density values) before mentioned changing, for each dust, the height from the bottom of the facility and changing the optical technique used to acquire images. It is important to analyse three properties in function of the parameter δ and the density of dusts: resuspension degree, mean velocity and max velocity.

Resuspension degree is calculated as (9):

$$\Delta m_{\%} = \frac{m_i - m_f}{m_i} \tag{9}$$

The mobilization of dust is characterised mainly by two forces: the aerodynamic resistance, drag, which moves particles, and the inter-particles and particle-wall friction forces, which slows down or completely stop the dust mobilization. The drag aerodynamic resistance is a force that is dependent on air-particle relative velocity and fluid density, i.e. the dynamic pressure of the fluid (since the particles are not moving). Frictions forces are a function of dust and wall surface characteristics [29-31]

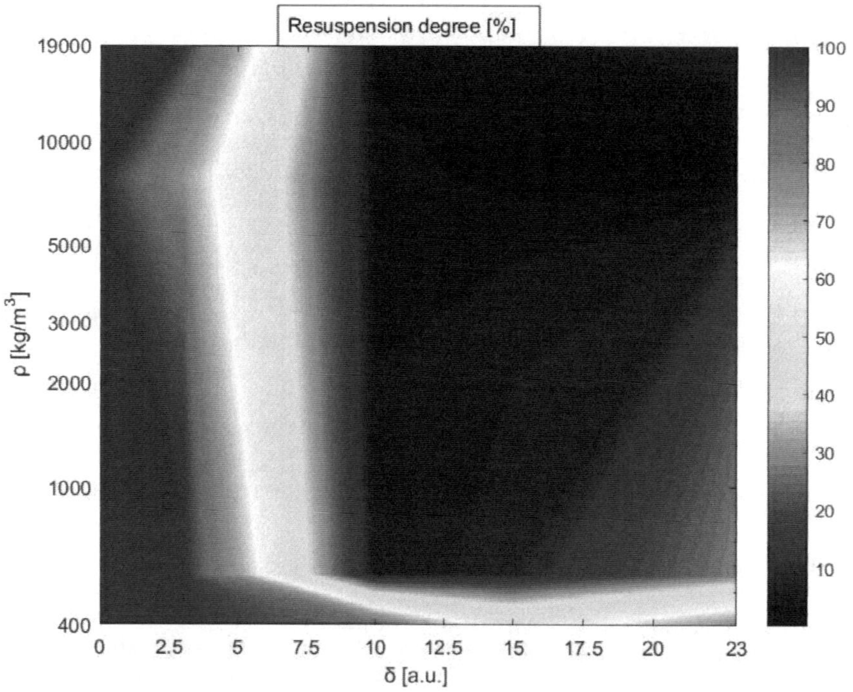

Figure 13. Resuspension degree map.

Figure 13 shows the re-suspension degree in function of density and δ parameter. When the tray with dust is placed at the same height of the valve (δ=0), the dust is completely mobilized, independently of its density. Increasing the distance of the tray from the inlet valve, the re-suspension degree falls down for any type of dust, excluded the lighter one (graphite dust). The minimum of re-suspension is reached for δ=15 while for higher values there is a slight increase. That increase is due to the specific flow inside STARDUST-U vessel. When the air flows inside from the valve E, two counter-vorteices are produced. The center of these vortices are the regions where the dynamic pressure is smaller. Therefore, since the vortex center height is around δ = 10 – 15, the re-suspension degree is minimized. Density of dust plays an important role, since for very small density the re-suspension is always very large, while with its decrease the re-suspension degree falls down.

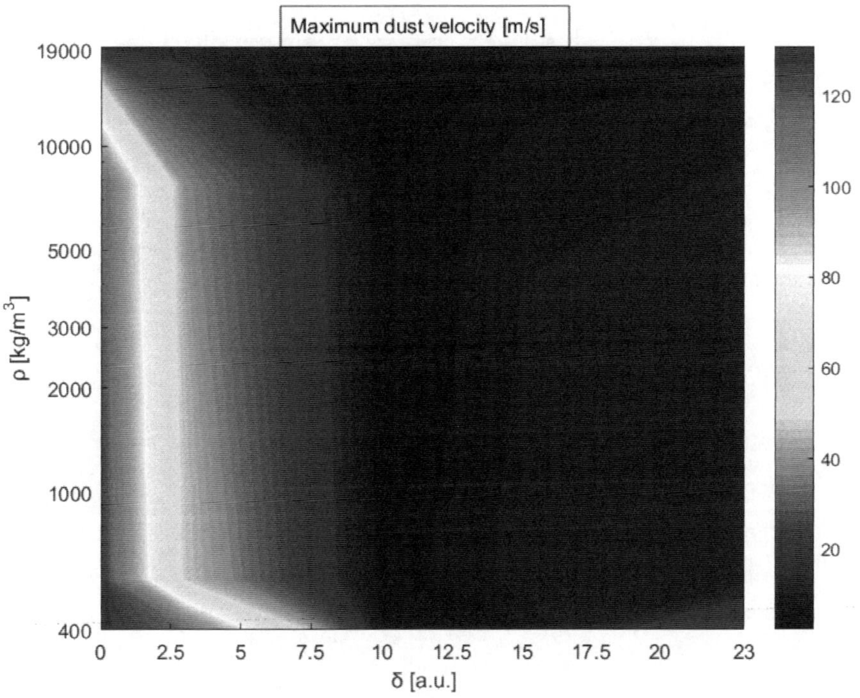

Figure 14. Maximum dust velocity map.

Figure 14 shows the maximum velocity reached by the particles. Still in this case we have a very large velocity only for $\delta = 0$ and for lower density. The velocity map can be explained taking into account a simplified equation that describes the motion of a particle in a flow (9):

$$m_p \frac{du_p}{dt} = D = \frac{1}{2}\rho_f\left(u_f - u_p\right)^2 S_p c_d \qquad (9)$$

From the particle-motion equation is clear that the maximum velocity reachable by the dust particles is the local fluid velocity. Therefore, since the air velocity is very large only along the inlet valve axis, the dust maximum velocity follow the same trend. Moreover, the inertia of the particle is proportional to its mass and so its density. Then, the decrease of velocity in function of density is explained.

Equation (9) describes correctly the motion of a particle when the volume fraction of the disperse phase is very low. Therefore, it can be applied to explain the maximum velocity, since it is reached only by that particles detached from the re-suspension region, where the volume fraction are near to the maximum packing of the dust and so, with large friction due to particle-particle interactions.

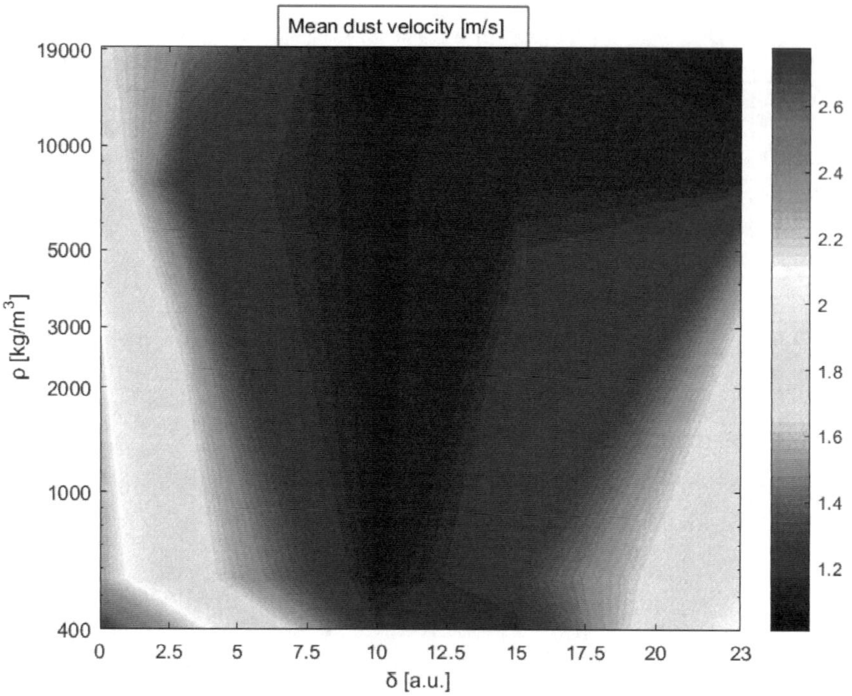

Figure 15. Mean dust velocity map.

Figure 15 shows the mean velocity trend. Still again, the maximum value is reached only for very low density and relative distance equal to zero. The minimum values are placed in the center of the vortex ($\delta \approx 10$) and with the large density of dust (tungsten). Dust density seems to play a larger role in mean velocity, since for each δ, as the density increase the mean velocity decrease.

The resuspension maps demonstrate that strong and dangerous re-suspension occur only if the film of dust is placed very closed to the isolation rupture or where the dynamic pressure of the flow is very large. Furthermore, the study of the continuum phase, in particular of the dynamic pressure field, could be sufficient to understand qualitatively where and how much dust can re-suspend.

The optical measurement techniques could be improved in order to develop a monitoring apparatus to control the concentration and velocity of dust in industrial plants.

4. Conclusion

The authors showed the measurements of dust re-suspension in case of loss of vacuum accident replicated in experimental apparatus called STARDUST-Upgrade. Non-invasive diagnostics involving imaging techniques have been implemented by the authors to perform dust tracking experiments that produced data on dust velocity vectors comparable to the figures for air velocity, that was measured with pressure transducers. Re-suspension degree has been measured through a precision balance. Optical measurements allowed to measure the mean and maximum velocity of mobilized particles. Since dust particles have different optical properties, two experimental setups have been developed. The first one is based on the scattering cross section of the dust while the second one with particles that has large absorption cross section. The experiments have been conducted changing the relative distance between the dust tray and the inlet valve. Then, re-suspension, maximum and mean velocity map have been built in order to study the influence of the parameter δ and the density of dust. Both the two parameters show an influence on the three dispersed phase characteristics. A minimum value of the re-suspension degree and velocity is reached when the dust tray is placed in the center of one of the two counter rotating vortexes, which is characterised with the minimum of the dynamic pressure of the continuum phase. The results demonstrate how the re-suspension problem could be faced, qualitatively, studying the characteristics of the continuum phase. Anyway, if we are interested in accurately results, the best way is the use of multiphase computational fluid dynamic codes. In future, the QEP group will continue to collect experimental results in function of other characteristics, such as flow rate, temperature and initial pressure. Then, numerical simulation with identical boundary condition will be compared with the experimental data in order to validate it. Finally, an optical device able to measure dust characteristics could be developed to perform monitoring of re-suspension inside the industrial plants.

References

[1] R. K. Eckhoff, 2003, *Dust Explosions in the Process Industries.*
[2] Malizia A. et all, 2017, *A novel facility to investigate dust mobilization in confined environments with applications to the security of the pharmaceutical industry* , Materials Science Forum, ISSN: 1662-9760, Vol. 879, pp 1213-1219,
[3] Z. Yuan, N. Khakzad, F. Khan e P. Amyotte, 2015, *Dust explosion: A threat to the process industries,* Process Safety and Enviromental Protection, vol. 98, pp. 57-71
[4] U.S. Chemical Safety and Hazard Investigation Board, 2009, *Investigation Report: Sugar Dust Explosion and Fire*
[5] U.S.Chemical Safety and Hazard Investigation Board, 2005, *Investigation Digest - Dust Explosion at West Pharmaceutical Services*
[6] R. K. Eckhoff, 1996, *Prevention and Mitigation of dust explosions in the process industries: A survey of recent research and development*, J. Loss Prev. Process. Ind., vol. 9, n. 1, pp. 3-20,
[7] R. Eckhoff, 2005, *Current status and expected future trends in dust explosion research*, vol. 18, pp. 225-237.
[8] R. Eckhoff, 2009 *Understanding dust explosions. The role of powder science and technology*, J. Loss Prevent. Proc.
[9] J. Yuan, W. Wei, W. Huang, B. Du, L. Liu e J. Zhu, 2014, *Experimental investigations on the roles of moisture in coal dust explosion*, Journal of the Taiwan Institute of Chemical Engineers, vol. 45, n. 5, pp. 2325-2333,
[10] W. Gao, T. Mogi, J. Sun, J. Yu e R. Dobashi, 2013, *Effects of particle size distributions on flame propagation mechanism during octadecanol dust explosions*, Powder Technology, vol. 249, pp. 168-174.
[11] T. Abbasi e S. Abbasi, 2007, *Dust explosions-Cases, causes, consequences, and control,» Journal of Hazardous Materials*, vol. 140, n. 1-2, pp. 7-44.
[12] M. Abuwser, P. Amyotte, F. Khan e L. Morrison, 2013, *An optimal level of dust explosion risk management: Framework and application*, Journal of Loss Prevention in the Process Industries, vol. 26, n. 6, pp. 1530-1541.
[13] A. Malizia, M. Gelfusa, G. Francia, M. Boccitto, M. D. Vecchio, D. D. Giovanni, M. Richetta, C. Bellecci and P. Gaudio, 2014 , *Design of a new experimental facility to reproduce LOVA and LOCA consequences on dust resuspension*, Fusion Eng. Des.
[14] L. A. Poggi, A. Malizia, J. F. Ciparisse, M. Gelfusa, A. Murari, S. Pierdiluca, E. Lo Re, P. Gaudio. 2016. *First Experimental Campaign to Demonstrate STARDUST-Upgrade Facility Diagnostics Capability to Investigate LOVA Conditions*. J Fusion Energ, 34:1320-1330.
[15] J.F. Ciparisse, A. Malizia et all. 2015, *First 3D numerical simulations validated with experimental measurements during a LOVA reproduction inside the new facility STARDUST-Upgrade*. Fusion Engineering and Design 101, 204–208.
[16] C. Murillo, O. Dufand, N. Bardin-Monnier, O. Lopez, F. Munoz e L. Perrin, 2013. *Dust explosions: CFD modeling as a tool to characterize the relevant parameters of the dust dispersion*, Chemical Engineering Science, vol. 104, pp. 103-116.

[17] A. Malizia, L. A. Poggi, J.-F. Ciparisse, R. Rossi, C. Bellecci and P. Gaudio, 2016. *A Review of Dangerous Dust in Fusion Reactors: from Its Creation to Its Resuspension in Case of LOCA and LOVA*, Energies, vol. 9, no. 8,
[18] J. Ciparisse, A. Malizia, L. Poggi, M. Gelfusa, A. Murari, A. Mancini and P. Gaudio, 2015. *First 3D numerical simulations validated with experimental measurements during a LOVA reproduction inside the new facility STARDUST-Upgrade*, Fusion Eng. Des.,
[19] F. Scarano, 2002. *Iterative image deformation methods in PIV*, Meas. Sci. Technol., vol. 13, pp. R1-R19.
[20] C. Cierpka, N. A., Buchmann, J. Soria and C. J. Kahler, 2014. *Ultra-High-Speed 3D Astigmatic PTV in Supersonic Underexpanded Impinging Jets*, in 17th Int Symp on Applications of Laser Techniques to Fluid Mechanics, Lisbon, Portugal.
[21] L. Souverein, 2006. *PIV based aerodynamic loads determination in supersonic flows*, Delft,
[22] R. Rossi, A. Malizia, L. A. Poggi, E. Peluso e P. Gaudio, , 2016 . *Flow Motion and Dust Tracking Software for PIV and Dust PTV*, Journal of Failure Analysis and Prevention, vol. 2016, pp. 1-12.
[23] B. Schunck, K. P. Horn and B. G. 1980. *Determining Optical Flow*, Artif. Intell., pp. 185-203,
[24] B. D. Lucas and T. Kanade, 1981. *An Iterative Image Registration Technique with an Application to Stereo Vision*, Proceedings of Imaging Understanding Workshop, pp. 121-130.
[25] E. Meinhardt-Llopis, J. Sanchez and D. Kondermann, 2013. *Horn–Schunck Optical Flow with a Multi-Scale Strategy*, Image Processing On Line, pp. 151-172,
[26] B. D. Lucas and T. Kanade, 1981. *An Iterative Image Registration Technique with an Application to Stereo Vision*, Proceedings of Imaging Understanding Workshop, pp. 121-130.
[27] J. Bouguet, *Pyramidal Implementation of the AÆne Lucas Kanade Feature Tracker Description of the Algorithm*, Intel Corporation - Microprocessor Research Labs.
[28] J. Bouguet, *Pyramidal Implementation of the AÆne Lucas Kanade Feature Tracker Description of the Algorithm*, Intel Corporation - Microprocessor Research Labs.
[29] C. E. Brennen 2005, *Fundamentals of Multiphase Flows*, Pasadena, California: Cambridge University Press,
[30] F. Zhanf, M. Reeks e M. P. Kssane, 2012. *Particle Resuspension in Turbulent Boundary Layers and the Influence of Non-Gaussian Removal Forces*, Journal of Aerosol Science, vol. 58.
[31] O. G. Chkhetiani, E. B. Gledzer, M. S. Artamonova e M. A. Iordanskii, 2012. *Dust resuspension under weak wind conditions: direct observations and model*. Atmos. Chem. Phys., vol. 12, pp. 5147-5162.

Acknowledgements
As first author and principal investigator of this work, I want to thank Riccardo Rossi than help me in perform the experimental campaigns and Pasquale Gaudio that help me to mount the optical set-up.

IOP Conf. Series: Journal of Physics: Conf. Series **882** (2017) 012012 doi:10.1088/1742-6596/882/1/012012

On-line 2D monitoring of rolling contact fatigue/wear phenomena in dry tests

I. Bodini[a], C. Petrogalli[a], A. Mazzù[a], M. Faccoli[a], M. Lancini[a], S. Pasinetti[b], G. Sansoni[a], F. Docchio[a]

[a]Department of Mechanical and Industrial Engineering, via Branze, 38, 25123, Brescia, Italy

[b]Department of Information Engineering, via Branze, 38, 25123, Brescia, Italy

ileana.bodini@unibs.it

Abstract. We present the in-field use of a recently developed on-line, real time, optical monitoring system for bi-disk rolling contact tests on railway-wheel- and rail- steel disks, based on laser illumination and a high-speed camera. Qualitative, as well as quantitative information can be obtained by the use of specific indexes derived from the images using blob analysis. Monitoring the tests provides information about the initial transients of turning-crest removal, on the steady state reached, and on the dynamics of the surface in between. Blob anisotropy is a key element to draw information of the mechanisms governing this intermediate state.

1. Introduction

Bi-disk tests on steels are an effective method to characterize the material response to rolling-sliding contact, as they are able to reproduce the damage phenomena of real components at a laboratory scale in controlled working conditions. These testing procedures have been widely used to characterize materials for bearings, gears, railway wheels and rails (see some examples in [1]-[4]).

In bi-disk rolling contact tests different damage mechanisms can interact and affect each other, such as wear, Rolling Contact Fatigue (RCF), strain hardening, ratcheting, etc., as presented in [5], and [6]. Recognizing and monitoring all these phenomena is a key issue for material characterization. However, usually this is performed by means of “gross” parameters, such as weight loss and coefficient of friction, which emphasize the main effects of these phenomena, but do not yield information about the mechanisms leading to them (adhesive or abrasive wear, pitting or spalling, etc.).

A deeper knowledge of the temporal evolution of the surface topology can provide additional key information for identifying damage phenomena, even providing clues of what is happening under the contact surface, which can be directly detected only by means of destructive testing. Surface topology is usually monitored by means of image acquisition systems, able to give a qualitative idea of what is happening at the contact surface [7]-[9].

The major limitations of the present instrumentation for bi-disk tests are (i) the need to periodically dismount the disks to perform ultrasonic cleaning, weight loss ans Barkhausen noise measurements

IOP Conf. Series: Journal of Physics: Conf. Series **882** (2017) 012012 doi:10.1088/1742-6596/882/1/012012

[3], (ii) the need to slow-down the testing apparatus for image analysis, and (iii) the non-quantitative output of the image analysis itself.

In a recent paper [10] we presented a novel approach to bi-disk tests, which allows in-line, time dependent, quantitative 2D and 3D analysis of the disk surfaces during the test. The approach is based on (i) multiple laser source illumination (diffuse and line-shaped), a fast camera, and an analog encoder for on-line disk angular position measurement. The analysis of the images is performed using blob analysis and specific indexes that quantify the effects of wear and of RCF at each stage of the test.

We are currently using the test bench presented in [11] for an extensive series of bi-disk tests on innovative steels to be used for train wheels in the near future. These tests, apart from their obvious importance "per se", are an ideal benchmark to assess the validity of our apparatus in relation to its ability to yield the same information about the sample as derived from the standard approaches, although on a non-invasive, real-time and more quantitative way.

In this paper we show, using four of the steels under tests, how our approach is effective to compare the surface characteristics (i) among the different steels, and (ii) for a single steel, during the test duration, to study the evolution of the disk damage during the test. In particular, we show that the image-analysis indexes are reliable indicators of the state-of-art knowledge on the wear and RCF processes.

2. Materials, instrumentation and procedures

2.1. Materials

Dry rolling contact tests are carried out on disk-shaped samples with diameters of 60 mm and thickness of 15 mm. Four wheel steels have been tested, and in the following, they will be called WS01, WS02, WS03 and WS04, respectively.

Table 1. Hardness [HB] of the four tested wheel steels and Ultimate Tensile Strength (UTS) [MPa].

	Hardness [HB]	UTS [MPa]
WS01	245	623
WS02	286	758
WS03	355	738
WS04	286	758

Due to non-disclosure-agreement, we cannot disclose the chemical composition of the steels, but, in Table 1, we give information about their mechanical properties: WS02 and WS04 have the best mechanical properties among the group and hardness intermediate between WS01 and WS03. This last is characterised by the highest value of hardness. During the test wheel steels samples were coupled with UIC 900A-rail-steel samples, and the friction coefficient was calculated [3]. Figure 1 shows a specimen of WS01 at the end of the test.

Figure 1. Image of a sample under test (WS01 after 130 kcycles)

2.2. The optical layout and the test bench

A detailed description of the optical apparatus for 2D and 3D surface analysis is given in [10].

The apparatus has been integrated in the bi-disk test bench, as shown in Figure 2. The bench has two input windows W1 and W2, which enable the optical access to the disks under test. L1a and L1b are the two defocused laser sources that provide diffuse light illumination of the sample. L2 is the laser stripe used for 3D measurements (not used in the present set of experiments), whereas L3 is the additional laser stripe that is used, in combination to the analogue encoder mounted coaxially with the sample, to measure on-line the angular position of the sample.

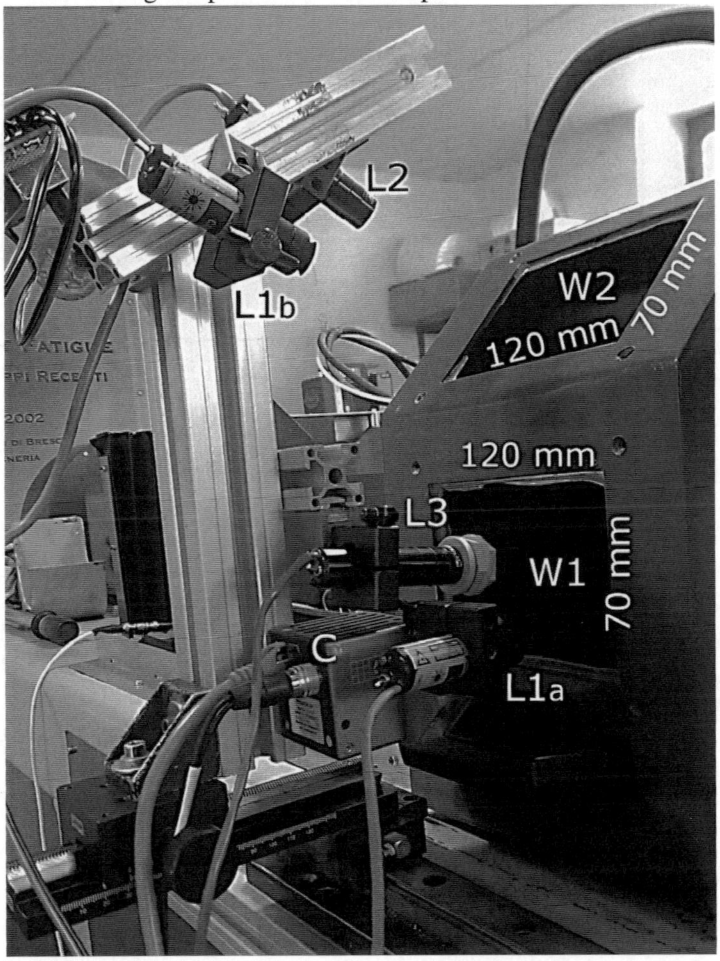

Figure 2. Vision-based measurement system assembled in the RCF test bench.

C is the high-speed camera (Promon 501, AOS Technologies). In these tests it operates at 377 frames per second at a resolution of 1,280x240 px, 8-bit. The images are acquired with an exposure time of 40 μs. The camera lens has the focal length of 50 mm, and this results in a field of view (FOV) of 35 mm × 6.5 mm, with a spatial resolution Rs of 0.027 mm/px.

2.3. Acquisition procedure during the tests

All the samples roll with the same maximum contact pressure of 1.1 GPa, at a speed of 500 rpm and a sliding/rolling ratio of 1%. To comply with the standard testing procedures, the acquisitions are performed before the test (0 kcycles, where a cycle corresponds to a 360° rotation) and after 10, 20, 30, 50, 70, 100 and 130 kcycles. All tests are performed in dry conditions, i.e., air-cooled. Each

IOP Conf. Series: Journal of Physics: Conf. Series **882** (2017) 012012 doi:10.1088/1742-6596/882/1/012012

acquisition consists of a 2 s-long video stream, and it is acquired @ 377 fps, corresponding to 754 frames.

A typical image obtained by the camera on a rotating disk is shown in Figure 3. In Figure 3(a), as a reference, the sample under test is shown as seen from window W2, illuminated by the diffused sources and by the encoder stripe. In Figure 3(b), the image taken by the camera is shown. Two areas are visible: the left area contains the encoder stripe, whose length is proportional to the angular position, whereas the right area contains the speckle field produced by the rough surface illuminated by the two diffused sources. As described in [10], Regions of Interest (ROIs) are defined within the image for analysis: these are shown in Figure 3(b), delimited by the dashed perimeters. ROI3 is used, in each frame of the sequence, to associate the angular position of the disk to the frame, whereas ROI1 is used for the subsequent image analysis.

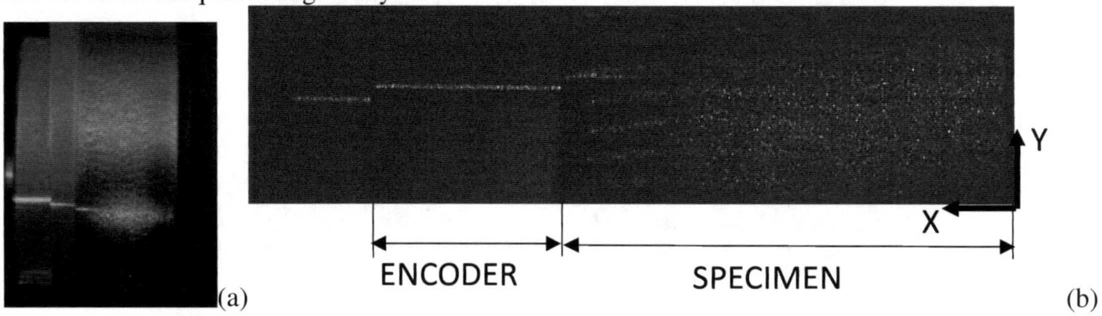

Figure 3. Photograph of the sample under test and acquired image. (a): illumination patterns; (b) corresponding image. The regions of interest ROI1 and ROI3 are drawn in dashed lines.

2.4. Image analysis and quantitative frame evaluation

Each 2-s video produced on the disk rotating at the operating speed of 500 rpm consists of N_T=754 frames equally subdivided over j=1 to 17 revolutions, each in turn formed by k =1 to 46 frames. The situation is depicted in Figure 4. We note that, due to the constraints imposed by (i) the camera settings, and (ii) the field of view deriving from the camera focal length, the frames are partially superimpose. To avoid the redundancy in the analysis, the dimension of ROI1 is set to 4.7 mm (Figure 3(b)).

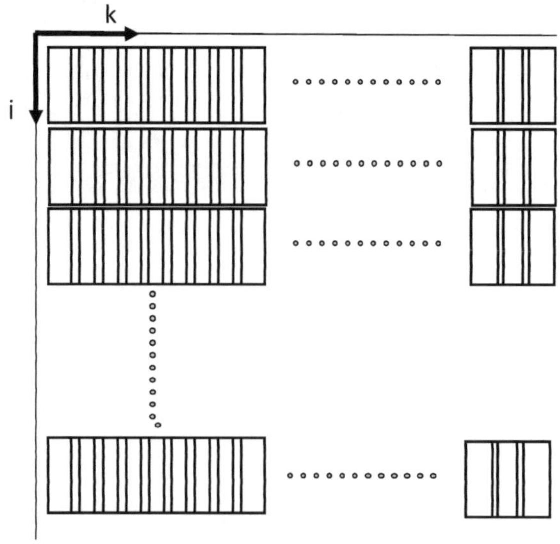

Figure 4. Diagram of the frames of each acquired video: k-th frame in the j-th revolution.

Indexes are obtained in two ways: (i) integral indexes, obtained by averaging on the entire frameset, (ii) angle-dependent indexes, obtained by averaging over frames corresponding to the same angular position of the sample on the shaft.

For every blob in each frame, we derive:

- The blob area, a.
- The blob asymmetry index, a_s, defined as the ratio between the height and the width of the rectangle enclosing the blob (if $a_s > 1$ the blob is oriented along the vertical direction Y, if $a_s < 1$ the blob is oriented along the horizontal direction X, if as = 1 the blob is round-shaped).

For every frame (j,k), we derive the blob number n(j,k), i.e., the number of blobs in the ROI, and hence:

- The average blob area $\bar{a}(j,k)$, in the ROI;
- The damage ratio index, $r(j,k)$, i.e., the ratio of $\bar{a}(j,k)$ divided by the total ROI area;
- The average asymmetry $\bar{a}_s(j,k)$ in the ROI.

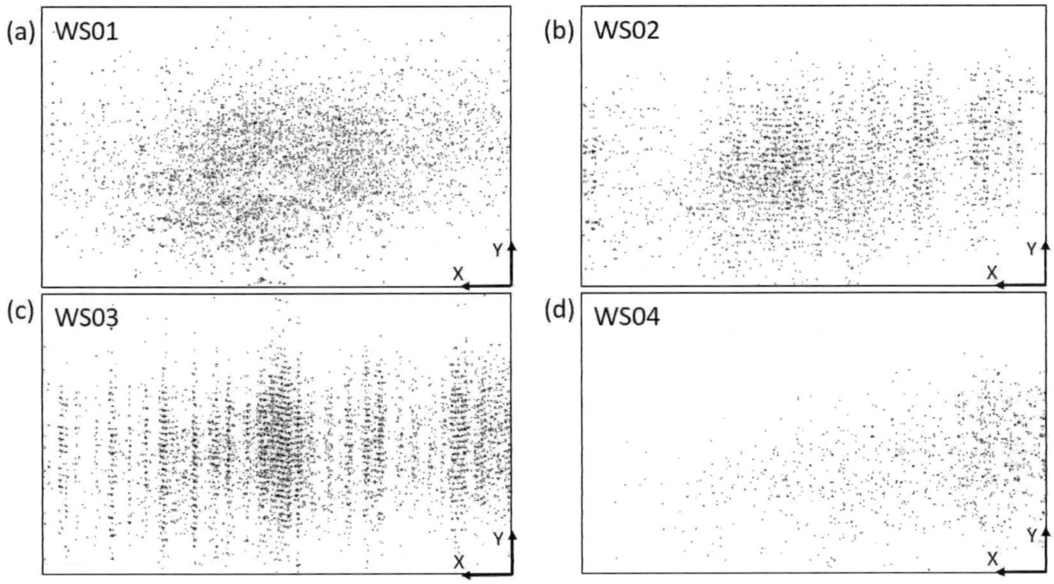

Figure 5. Blob-rendered pictures of the four steels used in the study after 30 kcicles. (a) specimen WS01, (b) WS02, (c) WS03, and (d) WS04.

As an example of the different appearance of the blobs in different situation and/or in different materials, Figure 5 shows the blob-rendered pictures of the four steels used in the study after 30 kcycles (the angular position is 0°, corresponding to j = k = 1). The blobs in the four images differ in size, number, asymmetry (with particular emphasis to Figure 5(b) and Figure 5(c) where they are evidently horizontally oriented) and position in the frame (the blobs in Figure 5(d) are predominantly in the bottom-right part of the frame). This justifies the choice of the above parameters to characterize the phenomenon.

A further level of detail is reached by dividing the blobs into classes, depending on the blob dimensions. Classes are defined as illustrated in Table 2.

For every frame (j,k), we derive the blob number $n_i(j,k)$, i.e., the number of blobs in the ROI belonging to the i-th class, and hence:

- The average blob area $\bar{a}_i(j,k)$, in the ROI
- The damage ratio index $r_i(j,k)$, i.e., the ratio of $\bar{a}_i(j,k)$ divided by the total ROI area;

IOP Conf. Series: Journal of Physics: Conf. Series **882** (2017) 012012 doi:10.1088/1742-6596/882/1/012012

- The average asymmetry $\bar{a}_{s,i}(j,k)$.

Global indexes can be derived by averaging each of the above indexes over all the frames having the same angular position, or over the entire frameset. In particular, in the following, we use:

- R_B, i.e. the average of $r(j,k)$ over the entire frameset;
- N_i, i.e., the average of $n_i(j,k)$ over the entire frameset;
- $A_{S,i}$, i.e., the average of $\bar{a}_{s,i}(j,k)$ over the entire frameset.

The standard uncertainty of each of the above indexes has been calculated as the standard deviation of the populations $r(j,k)$, $n_i(j,k)$, and $\bar{a}_{s,i}(j,k)$, respectively.

Table 2. Classification of the blobs.

Class	Area range [10^{-3} mm^2]
C1	$1.5 < a < 3.7$
C2	$3.7 < a < 10.9$
C3	$10.9 < a < 21.9$
C4	$21.9 < a < 51.0$
C5	$a > 51.0$

3. Results

3.1. Time evolution of the steel disk damage

The efficacy of the on-line derived indexes resulting from the application of the optical setup described above can be demonstrated with the use of examples of time-dependent behaviours of single or multiple steels used in the current tests. An example is given by the time-related behaviour of the damage ratio R_B as a function of the number of cycles in steel W02. The results are summarized in Table 3. From the table we note that R_B has a markedly time-dependent behaviour, starting from 0 kcycles where it is negligible, reaching a peak at 10 kycles, decreasing up to 50 kcycles, and remaining almost constant thereafter until the end of the test (given the standard uncertainties, the values are compatible).

Table 3. Time-evolution of the average damage ratio RB of the surface of the specimen made of WS02.

Number of cycles [kcycles]	R_B [%]
0	0.011 ± 0.019
10	9.23 ± 0.73
20	5.66 ± 0.54
30	4.14 ± 0.61
50	0.85 ± 0.17
70	0.97 ± 0.13
100	0.81 ± 0.11
130	1.46 ± 0.49

These results are consistent with the known mechanism of material removal described in [12]. In fact, it is well-known that, during bi-disk tests, the wear phenomenon exhibits two experimentally recognizable stages: an early stage, with transient wear-rate, and a consequent steady-state stage, with quasi-stable wear-rate [13]. The former stage is mainly caused by the contact between the asperities of the two specimen surfaces. The latter stage (occurring after 50 kcycles) is due to the well-known competition between wear and RCF [4], where wear continuously removes surface material layers and flattens out the product of RCF (that is, the formation of cracks) [10]. This is , to our knowledge, the first time where the evidence of these sequence of phenomena and of their competition is demonstrated quantitatively and located correctly in time with non-invasive way, and online, without the need of examining the sample at different stages of test.

Table 4. Time-evolution of the indexes N_i and $A_{s,i}$, for the specimen made of WS02, grouped into classes of area. The standard uncertainty of the indexes is also presented.

Cycles [kcycles]	Index	C1	C2	C3	C4	C5
0	N_i	26 ± 38	0.10 ± 0.34	0.001± 0.037	0	0
	$A_{s,i}$	1.18 ± 0.44	1.04 ± 0.53	0.78 ± 0.99		
10	N_i	2110 ±80	182 ± 15	41.0 ± 6.4	18.3 ± 4.3	7.9 ± 2.4
	$A_{s,i}$	1.127 ± 0.025	1.073 ± 0.038	1.059 ± 0.071	1.05 ± 0.10	0.99 ± 0.14
20	N_i	1909 ± 80	215 ± 20	42.9 ± 4.3	12.6 ± 4.3	1.4 ± 1.3
	$A_{s,i}$	1.067 ± 0.024	0.724 ± 0.027	0.564 ± 0.086	0.565 ± 0.086	0.63 ± 0.22
30	N_i	1900 ± 87	170 ± 31	17.7 ± 8.1	3.0 ± 2.5	0.19 ± 0.44
	$A_{s,i}$	1.077 ± 0.023	0.728 ± 0.029	0.633 ± 0.095	0.66 ± 0.21	0.66 ± 0.27
50	N_i	970 ± 122	16.1 ± 5.9	0.31 ± 0.57	0.003 ± 0.058	0
	$A_{s,i}$	1.112 ± 0.040	1.045 ± 0.15	1.04 ± 0.44	1.17 ± 0.70	
70	N_i	968 ± 72	23.2 ± 6.6	0.83 ± 0.98	0.04 ± 0.20	0.000 ± 0.013
	$A_{s,i}$	1.080 ± 0.038	0.974 ± 0.097	0.97 ± 0.37	0.94 ± 0.45	0.44
100	N_i	933 ± 69	16.6 ± 5.3	0.39 ± 0.63	0.01 ± 0.10	0
	$A_{s,i}$	1.089 ± 0.040	0.99 ± 0.12	0.94 ± 0.37	0.8 ± 0.35	
130	N_i	1150 ± 155	46 ± 20	3.7 ± 4.0	0.62 ± 1.3	0.07 ± 0.32
	$A_{s,i}$	0.979 ± 0.040	0.825 ± 0.078	0.74 ± 0.24	0.84 ± 0.35	0.90 ± 0.37

The above evidence is further supported by the analysis of the temporal evolution of (i) the class-specific number of blobs N_i, and (ii) their average asymmetry index $A_{S,i}$. This evolution is shown in Table 4. At 0 kcycles the only particles having non-negligible number are the very small (C1) ones. They are predominantly round-shaped, and this is consistent with the machined surface of the new specimens. At 10 kcycles, the blobs of all classes increase in number and are almost all round-shaped: this is consistent with the hypothesis of crest removal during the contact in the early stage.

Between 10 and 30 kcycles, while the smallest (C1) and the largest (C4-C5) decrease in number, the intermediate ones (C2-C3) increase in number. All blobs apart from the smallest are horizontally oriented.

Between 30 and 50 kcycles, all blobs decrease rapidly with respect to their peak value: the smallest blobs (C1) remain round-shaped whereas the others (C2-C5) maintain their orientation. After 50 kcycles, all classes of blobs remain constant in number, with strong prevalence of the smallest ones, and the previously horizontal ones become round shaped again.

This behaviour of the class- and time-dependent blob number and asymmetry is very well consistent with the state-of-art knowledge of the phenomenon. On one side, it confirms the evidence that, after 50 kcycles, all blobs are fairly constant and round-shaped, thus demonstrating the presence of the steady state with quasi-stable wear rate. On the other side, the change of shape between 20 and 30 kcycles towards the horizontal direction is very well consistent with the assumed onset of unidirectional friction of the contact surfaces with an induced plastic flow, which results in the formation of corrugations with a typical wavy aspect [11].

To visually appreciate the data given in tabular form and discussed above, we show in Figure 6 the visual appearance (left) and the acquired frame with a zoomed view of the blobs (right) of WS02 in the most significant sections of the test, i.e., after 10, 20, 30 and 50 kcycles (top to bottom). In the visual sample the onset of the wavy pattern is well evident at 20 and 30 kcycles, with no trace of it at 10 and 50 kcycles. The wavy pattern visible on the sample after 50 kcycles is due to variations of the surface colour related to the oxidation of the material, during the test. Correspondingly, the light pattern acquired from the camera, evidenced by the zoomed blob patterns, clearly show the horizontal elongation of larger blobs, and the round shape of the smaller ones.

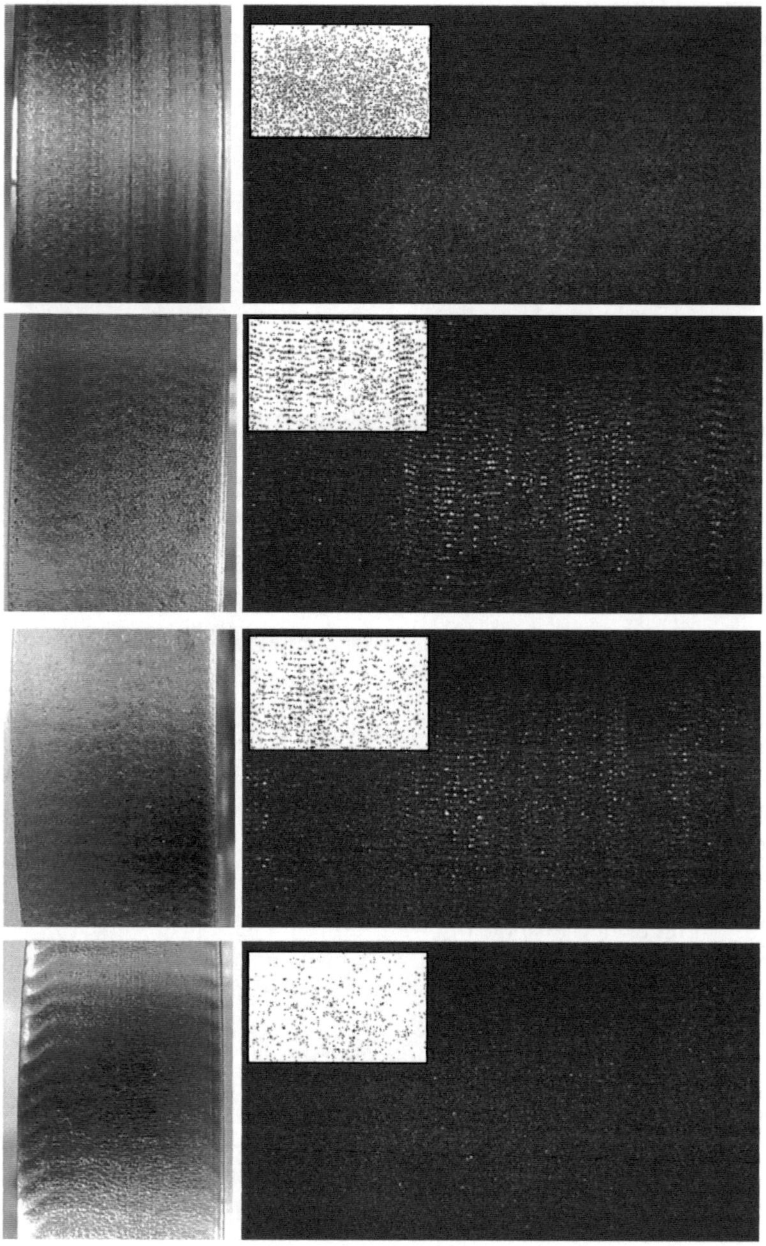

Figure 6. Visual appearance of the sample surface (left), acquired frames with a zoomed view of the blobs of steel W02 (right) from 10 to 50 kcycles (top to bottom).

3.2. Material synopsis and comparison

The on-line, non-contact characterization of disks using our optical setup is able to highlight differences in the time evolution between the steels described in Section 2. As an example, Figure 7 shows the time evolution of the damage ratio R_B, of the four steel disks. All R_B curves present a peak at 10 kcycles, followed by a decrease and a final steady state. However, the peak amplitude, the descent slope and the final state level are evidently material dependent. WS01 exhibits the lowest peak value. The descent slope is very soft until the final value at 130 kcycles, the highest value of all samples.

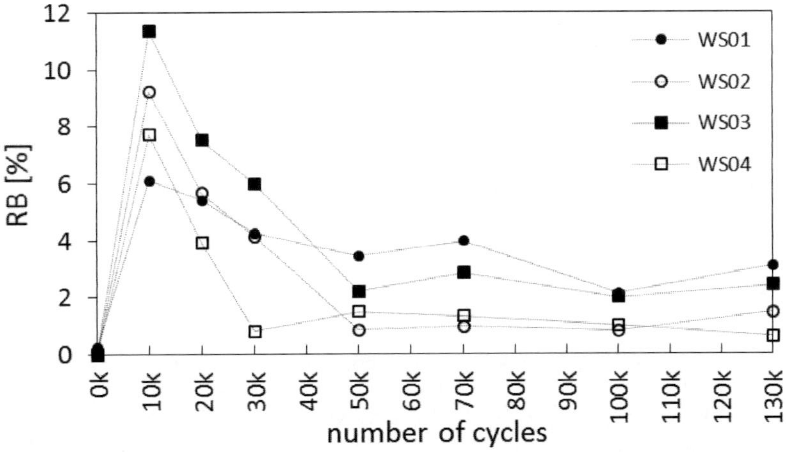

Figure 7. Time evolution of the damage ratio R_B, calculated for the four tested steels. Dotted lines are inserted to increase the clarity of the graph, they do not represent data.

WS04 exhibits a peak value higher than that of WS01, and the highest slope after the peak, with the steady state starting already after 30 kcycles at very low R_B values. WS02 and WS03 show the highest peak values, together with intermediate descent slopes.

The above evidence is well justified by considering steel hardness (Table 1): disks with higher hardness exhibit higher values. Depending on hardness, different mechanisms are responsible for the initial turning-crest removal: in hard materials, crests are removed mainly by brittle micro-fracture, leaving more micro-pits on the contact surface (higher R_B). In softer materials, the crests are mostly flattened by plastic flow, leaving a smoother surface (lower R_B) [4], [6], and [8].

The behaviour of R_B for WS03 and WS04 is similar to the one observed in steel WS02. In fact, these three steels present well-separated early stage and steady stage. However, each steel is characterised by a different duration of the early stage, which is typical of the material.

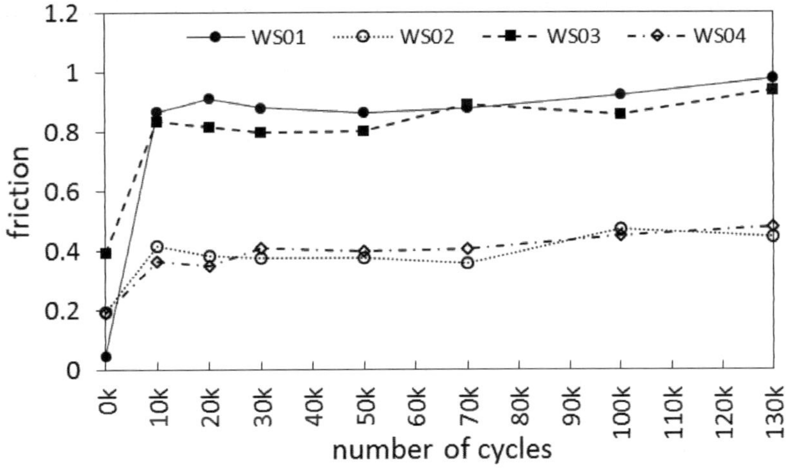

Figure 8. Friction calculated for the four studied steels.

Equally enlightening is the examination of the end-test R_B values of the four disks: WS01 and WS03 exhibit similar R_B values, higher than the other two. This is well in agreement with the friction coefficient measured during the test and shown in Figure 8. Friction calculated for the four studied steels.. WS01 and WS03 have similar friction coefficients and the same holds for WS02 and WS04. The steels with higher R_B in the steady state are the ones with higher friction coefficient.

3.3. Detection of possible discontinuous damage over a specimen

A unique advantage of the online optical system stems from its ability to monitor the temporal evolution of a disk under test on an angle-dependent basis. This makes it possible to quantify and follow in time the so called "polygonal" effect [14]. Figure 9 gives a clear example: the specimen in Figure 9(a) is made of UIC 900 A, which is a rail steel. It underwent an 800 kcycles-long dry test against a wheel-specimen made of Superlos®, and became polygonal during the test, i.e. portions of surface that appear very rough alternated with portions that appear very smooth characterize it.

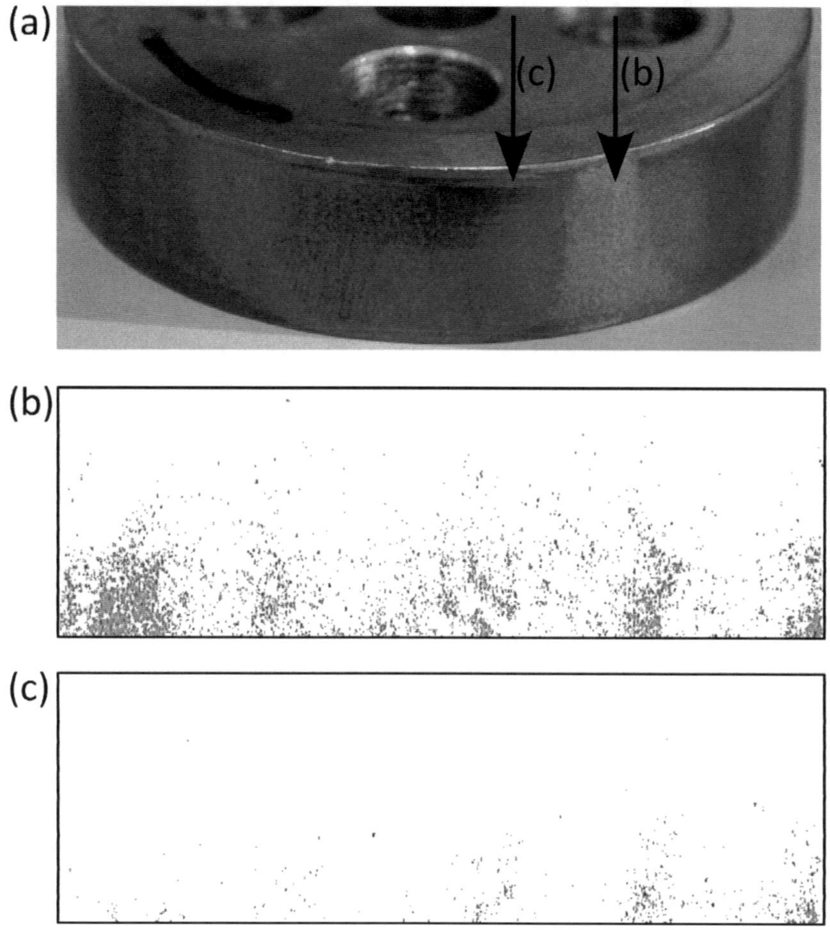

Figure 9. Detection of damage discontinuity. (a) Polygonal specimen; (b) particles on a portion of the specimen; (c) particles on another portion of the same specimen.

Considering measurement points evenly spaced, 15° apart one from another, the vision-based measurement system recognizes these two situations in Figure 9(b) and in Figure 9(c), respectively. The former shows a portion with many particles, to which correspond an R_B value of

(10.28 ± 0.18) %, the latter shows the consecutive portion, which has a small number of particles and R_B equal to (0.880 ± 0.072) %.

The ability of the system to recognize, on-line, a non-uniformity in the damage of different portions of the same surface can be very useful because it is an indicator of incorrect specimen mounting, which should lead to stop the test.

4. Conclusions

In this paper, we have presented, with a number of examples taken by extensive tests in progress in our laboratories, the performance of the on-line, real-time optical monitoring system described in [10]. We have also presented the ability of the indexes derived to provide, in a quantitative way, indications on the temporal evolution of the specimen under test without altering the test flow (e.g. by slowing it or stopping it), and especially without removing the disk for weighing or inspection.

In particular, the optical system can accurately follow the evolution of the surface by identifying the early transient stage and the subsequent steady-state stage. The synthetic indexes introduced are effective in quantitatively describing the surface topology, and can be correlated to quantities used in the common practice to define the steel behaviour in rolling contact tests (such as weight loss and friction).

One of the clear advantages of using the optical system in bi-disk tests is the ability to predict, based on the value of the various indexes, when the specimen is in a steady state, therefore to define the total duration of the test, suggesting to stop it as further monitoring is not important.

The optical system has now firmly become a monitoring companion in bi-disk tests, with a high added value in combination with the measurement of other parameters such as torque, friction, vibrations, and weight loss.

References

[1] Wang W J Lewis R Yang B Guo L C Liu Q Y and Zhu M H 2016 Wear and damage transitions of wheel and rail materials under various contact conditions *Wear* vol 362-363 p 146

[2] Wei J Zhang A Q and Gao P 2016 A study of spur gear pitting under EHL conditions: Theoretical analysis and experiments *Tribology International* vol 94 p 146

[3] MazzùA Solazzi L Lancini M Petrogalli C Ghidini A Faccoli M 2015 An experimental procedure for surface damage assessment in railway wheel and rails steels *Wear* vol 342-343 p 22

[4] Donzella G Faccoli M Ghidini A Mazzù A Roberti R 2005 The competitive role of wear and RCF in a rail steel *Engineering Fracture Mechanics* vol 72 p 287

[5] Ekberg A and Kabo E 2005 Fatigue of railway wheels and rails under rolling contact and thermal loading – an overview *Wear* vol 258(7-8) p 1288

[6] Mazzù A Petrogalli C and Faccoli M 2015 An integrated model for competitive damage mechanisms assessment in railway wheel steels *Wear* vol 322-323 p 181

[7] Zhang J Korsten M and Regtien P 2003 A vision system for online wear detection *Proceedings, XVII Imeko World Congress* June 22 – 27 Dubrovnik Croatia p 1960

[8] Soleimani S Sukumaran J Douterloigne K De Baets P and Philips W 2016 Online wear detection using high-speed imaging. *Microscopy and Microanalysis* vol 22 p 820

[9] Li L and An Q 2016 An in-depth study of tool wear monitoring technique based on image segmentation and texture analysis *Measurement* vol 79 p 44

[10] Bodini I Sansoni G Lancini M Pasinetti S and Docchio F 2016 A novel optical apparatus for the study of rolling contact wear/fatigue based on a high-speed camera and multiple-source laser illumination, *Review of Scientific Instruments* vol 87 n. 083701

[11] Donzella G Faccoli M Mazzù A Petrogalli C and Roberti R 2011 Progressive damage assessment in the near-surface layer of railway wheel-rail couple under cyclic contact *Wear* vol 271 p. 408

[12] *Modern Tribology Handbook* vol 1 CRC cap 7

[13] Archard J F 1953 Contact and rubbing of flat surfaces *Journal of Applied Physics* vol 24 p 981
[14] Meywerk M 1999 Polygonalization of railway wheels *Archive of Applied Mechanics* vol 69 p 105

Acknowledgments
We wish to thank Mr. Gabriele Coffetti and Mr. Silvio Bonometti for their precious support in the experimental activities.

RANDOM SEQUENCE FOR OPTIMAL LOW-POWER LASER GENERATED ULTRASOUND

D. Vangi[a], A. Virga[a], M. S. Gulino[a,b]

a. Università degli Studi di Firenze, Department of Industrial Engineering

b. michelangelo.gulino@unifi.it

Abstract. Low-power laser generated ultrasounds are lately gaining importance in the research world, thanks to the possibility of investigating a mechanical component structural integrity through a non-contact and Non-Destructive Testing (NDT) procedure. The ultrasounds are, however, very low in amplitude, making it necessary to use pre-processing and post-processing operations on the signals to detect them. The cross-correlation technique is used in this work, meaning that a random signal must be used as laser input. For this purpose, a highly random and simple-to-create code called T sequence, capable of enhancing the ultrasound detectability, is introduced (not previously available at the state of the art). Several important parameters which characterize the T sequence can influence the process: the number of pulses N_{pulses}, the pulse duration δ and the distance between pulses d_{pulses}. A Finite Element FE model of a 3 mm steel disk has been initially developed to analytically study the longitudinal ultrasound generation mechanism and the obtainable outputs. Later, experimental tests have shown that the T sequence is highly flexible for ultrasound detection purposes, making it optimal to use high N_{pulses} and δ but low d_{pulses}. In the end, apart from describing all phenomena that arise in the low-power laser generation process, the results of this study are also important for setting up an effective NDT procedure using this technology.

1. INTRODUCTION

Ultrasounds (USs) are commonly used in the Non-Destructive-Testing (NDT) field to investigate upon the structural integrity of materials [1,2]. The more utilized instruments for this kind of inspection are narrow band, piezoelectric probes which need the contact with the inspected object to be guaranteed. Non-contact applications use high-power pulsed lasers as a source of elastic waves [3-6], mainly in order to allow an on-line inspection [7]. These waves are generated by the material expansion and contraction due to the laser quick heating and the subsequent sudden cooling. The ultrasonic generated frequencies reach hundreds of MHz, since the laser pulse is typically few nanoseconds long. In order to carry on a definitely NDT procedure while obtaining the highest ultrasonic amplitude possible, the power density is usually kept close to the material's ablative threshold (around $20\,MW/cm^2$ for steel [3]). The ultrasonic amplitudes thus generated, often higher in comparison to the piezoelectric probes ones, make the US detection easy without particular signal manipulations. In some other cases, post-processing techniques can be used, the most recent ones featuring cross-correlation [8] and Laser Nonlinear Wave Modulation Spectroscopy (LNWMS) [9].

IOP Conf. Series: Journal of Physics: Conf. Series **882** (2017) 012013 doi:10.1088/1742-6596/882/1/012013

Pulsed lasers, such as Nd:YAG, are extremely expensive and particularly bulky; in addition, pulsed lasers spread the energy in a wide range of frequencies, and thus a large energy amount is outside of the receiving probe band too. Recent advances in low-power Continuous Wave (CW) laser diodes technology, which is very important in the medical research field, opened the path to NDT procedures which can guarantee the integrity of the materials, even low melting point ones as composites, making use of a simple, small and low-cost testing layout.

Low-power CW laser diodes have some disadvantages, the main one being a low Signal-to-Noise Ratio (SNR) in the generated US due to the low power used. In order to enhance the SNR, particular modulation sequences are usually employed, which have 2 principal features:

- the sequence consists of a large amount of pulses, since one single pulse cannot generate an ultrasonic oscillation detectable by the instruments (less than 1 pm [10]);
- the sequence is extremely random to accurately characterize the US in Time-Of-Flight (TOF) and Peak-to-Peak (PtP) terms using cross-correlation techniques [11,12].

Low-power laser US are generated feeding (driving) the diode with TTL signals, especially Pseudo-Noise (PN) or pseudo-random sequences [10-15], consisting of several pulses with different lengths. PN codes have a perfectly centred auto-correlation, because of their high randomness characteristic; however, even if this can be an advantage considering only the cross-correlation post-processing purpose, pulses with different durations generate an US with different frequency content [16].

The purpose of this study is to analyse low-power laser generated USs from the thermo-elastic and experimental point of view, considering longitudinal waves propagation inside a 3 mm steel disk. Finite Element (FE) models allow to indagate the features of low-power laser US as can be done in those cases in which high power lasers are used [17-19]. FE analysis does not cover the experimental practice, where a high number of pulses and high randomness in the TTLs driving the laser are necessary [15, 20]: an optimum in SNR can be reached adjusting parameters directly connected to the ultrasonic PtP value and the input sequence's randomness.

PN codes are not flexible because, once their length has been set, there is no way to change the sequence. For this reason, a new highly random sequence has been created, called T-sequence (T stands for "Train of pulses"). It is not present at the state of the art and it gives the researchers the possibility to change as many TTL's parameters as possible.

It will be shown that a proper choice of the influencing parameters can lead to the detection of the US even without using PN codes, which many past studies considered as a basis. The change in the T-sequence's fundamental parameters will be analysed; the best combination of these parameters will be the one which allows to increase SNR the most. Even if the results of a testing campaign (and of FE simulations) on a 3 mm thick steel disk are reported, results relatively close to these can be achieved for a 20 mm thick disk; because of this and other arising phenomena that will be described, it can be assessed that resonances inside the specimen are not always important but only in some particular circumstances.

2. MATERIAL AND EXPERIMENTAL METHOD

2.1. Material

The experimental layout, visible in figure 1, consists of:

- a modulable CW laser diode (TOPTICA iBeam Smart 640 S, 3B class) with a power of 150 mW and wavelength equal to 639 nm;
- a 75 mm focal length spherical lens with reflectivity coefficient R≅0.5%;
- 2 steel disks– with 3 mm and 20 mm thickness- are considered throughout the study (the 20 mm one is shown in the picture);
- a broadband (from 100 kHz to 1 MHz) Brüel & Kjær acoustic probe attached to the steel disk's surface which is opposite to the one irradiated by the laser.

The obtainable focus for the laser was measured by means of a CMOS sensor and resulted in a 30 μm beam waist. The power density is equal to 0.02 MW/cm^2 in this configuration, being under the material's ablative threshold [3].

IOP Conf. Series: Journal of Physics: Conf. Series **882** (2017) 012013 doi:10.1088/1742-6596/882/1/012013

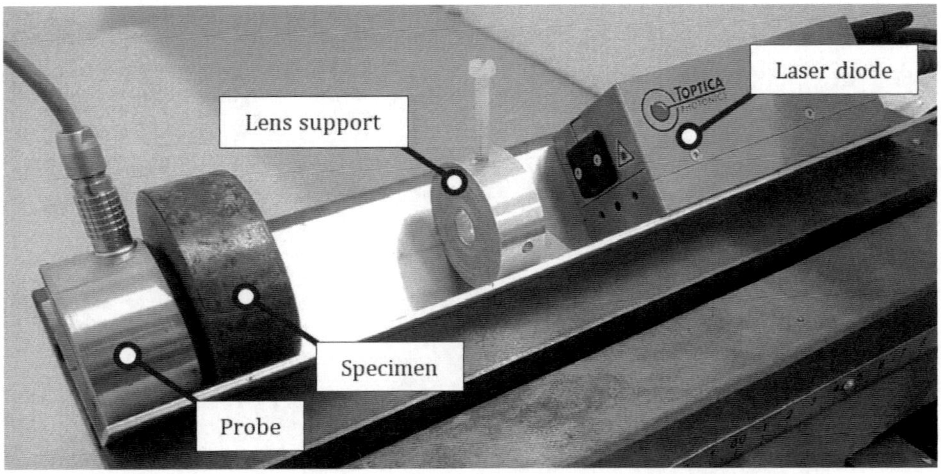

Figure 1. Experimental layout used in the testing procedure.

2.2. T sequence

The US detection in this kind of application mostly depends on signal post-processing techniques: in this context, the cross-correlation algorithm was chosen because of its versatility and use in many other studies on the topic [10-13]. Its application can be effective only if highly random signals are used as laser inputs [15]: the most widely used sequences for this application are PN codes [10-14]. PN codes are extremely random but the thermo-elastic effect is not optimized, because of the difference in each laser pulse length. This is evident considering that they were created for spreading the spectrum in a system [16].

In the ultrasonic field, piezoelectric contact or air-coupled narrow band probes are commonly employed, meaning that a spread spectrum is inefficient for this NDT purpose. PN codes (e.g. Golay codes and M sequences [10,14]) have a precise order in ON/OFF lengths too, so that the only sequences' parameters that can be modified are the length in terms of point and the length in terms of time [15].

To conjugate the necessity of having an output narrow band and a flexible input sequence, a particular sequence was created which is not present in other studies at the state of the art: the T sequence (where T stands for "Train of pulses"). Basically, the T sequence is a signal which possesses equal ON lengths for all pulses, while OFF lengths are created randomly. Every time a T sequence is created through this algorithm, the result can be different. The T sequence in fact represents a category of signals, rather than a single sequence.

The first 50 points of a 6500 points long T sequence are shown in figure 2a. The randomness of the T sequence is visualized through the histogram in figure 2b, where the OFF lengths statistical frequency $p(OFF)$ has a trend comparable to the distribution for the white noise (flat histogram). While PN codes possess an autocorrelation perfectly "centred" (different from 0 only at point 0), the T sequence's auto-correlation (figure 2c) has a minimum adjacent to the maximum. The T sequence, so, is random but also less performing in respect to the PN codes from a randomness point of view. In the next paragraphs, only the cases where this particular T sequence is applied are considered, if not differently specified.

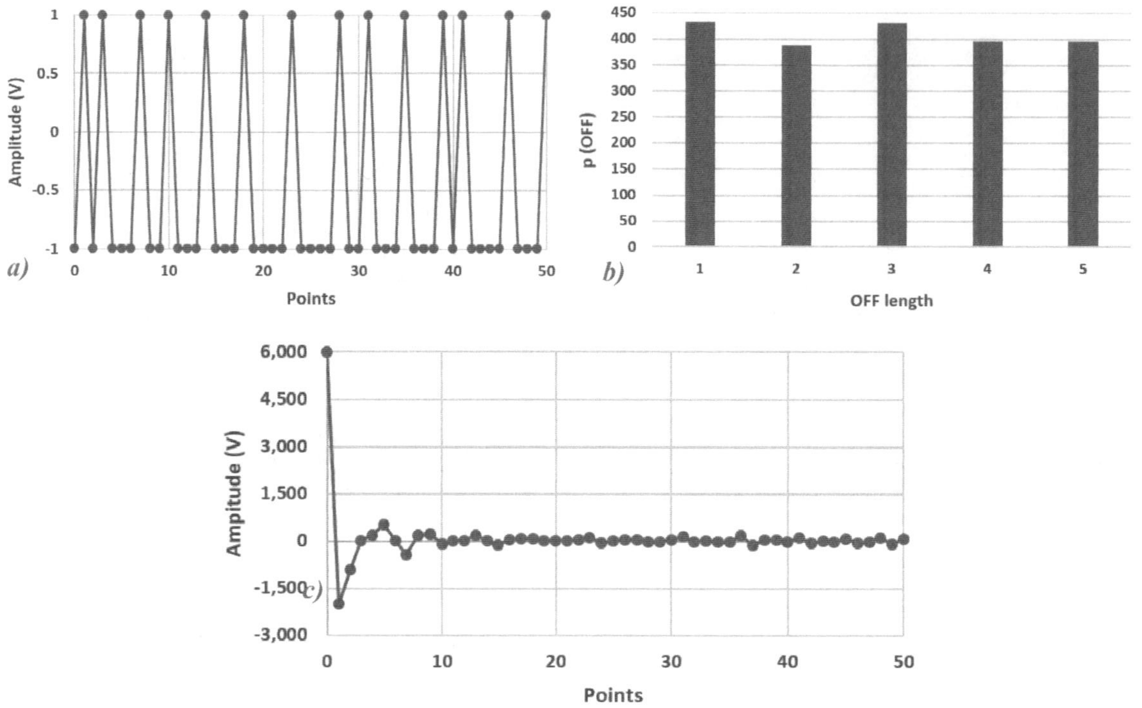

Figure 2. Portion of the T-sequence (a); its randomness is highlighted through the OFF lengths statistical frequency $p(OFF)$ (b) and part of the related auto-correlation (c).

2.3. Method

As a reference, figure 3 shows the steps taken to carry out the testing campaign. First of all, the T-sequence's parameters must be chosen (a); the sequence is generated through a programmable pulse generator (b) and imposed to the laser. The input sequence and the output response to the laser excitation, in which the US is hidden, are displayed on an oscilloscope. The instrument then applies a defined number of ensemble averages on these two interesting signals, in order to partially mitigate the noise influence (c). The probe output is amplified of 60 dB. After these two signals acquisition, the cross-correlation operation (d) is applied between them [8,11,12], and the ultrasonic TOF and PtP can be evaluated from the cross-correlation output (e).

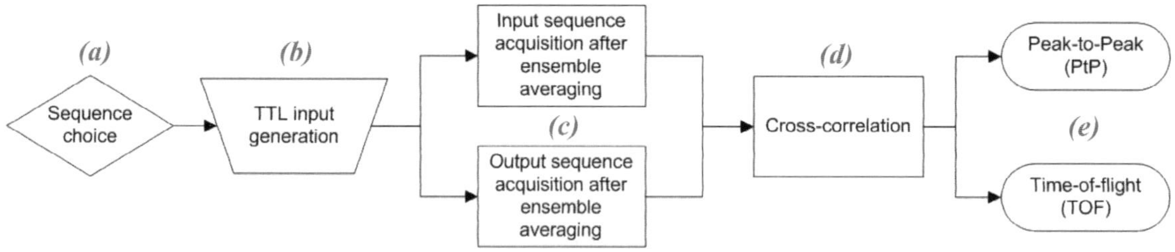

Figure 3. Flow chart regarding the procedure.

Even if SNR can be calculated through a large variety of methods [4,5,8,10,11], it is always represented by the formula:

$$SNR = \frac{Y_{signal}}{Y_{noise}},$$

where Y_{signal} is a generic quantity which considers only the portion of acquired signal interested by the US after the cross-correlation application, while the same applies for the noise with Y_{noise}. The noise is coherent with the generated US, which makes the cross-correlation technique more efficient than other time-frequency methods like Wavelet Transforms [20]. A simple formulation has been considered:

$$Y_{signal} = max(signal) - min(signal) = PtP_{US},$$

$$Y_{noise} = max(noise) - min(noise) = PtP_{noise}.$$

PtP_{US} is calculated on the cross-correlation portion interested by the US and PtP_{noise} on the cross-correlation remaining part.

3. FINITE ELEMENT CHARACTERIZATION

In order to properly study the ultrasonic waves propagation inside a steel disk, a FE analysis regarding the thermo-elastic effect has been carried out through COMSOL© Multiphysics. The laser effect on the material is simulated by a thermal pulse on the disk surface. As the thermal pulse duration is varied, the resulting oscillation's key-features are highlighted.

The pulse duration in the FE analysis will be called δ from now on. Referring to similar works [17-19] which consider high power laser sources, a 2D axisymmetric model has been created (represented in figure 4), allowing to analyse a 3 mm thick revolution solid model without meshing its complete shape. The region close to the axis is the most defined in the r and z directions, with 5 μm wide elements: a particularly high accuracy is needed here, the ipocentre, because of the presence of the laser irradiation with a 15 μm spot radius (the experimentally measured beam-waist). The peripheral one, on the other hand, is coarser and made up of 150 μm wide elements: at the edge, a fixed-end constraint is imposed, as in the experimental layout (disk fixed in respect to a L-shape bar). In the z direction, dimensions vary linearly with the increasing distance. These sizes allowed to minimize the calculation time.

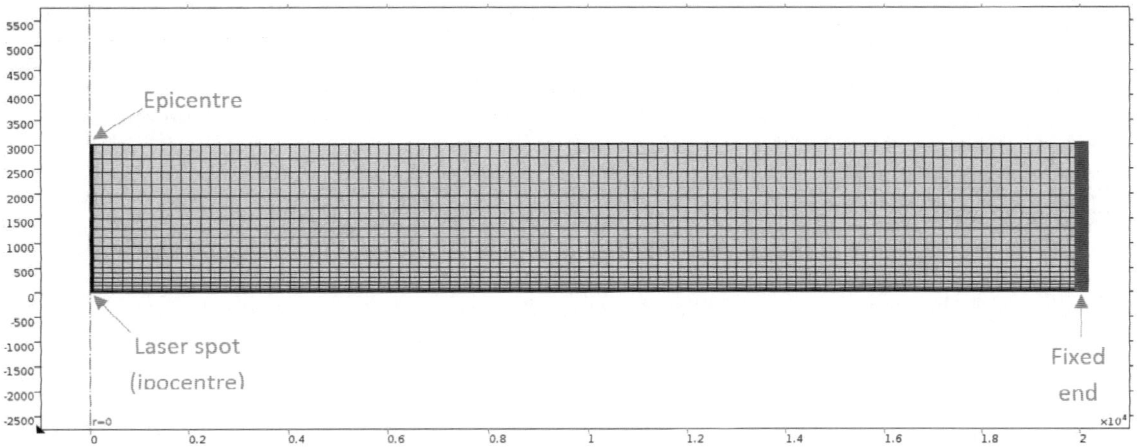

Figure 4. Mesh of the revolution solid with r=20000 μm and z=3000 μm. Along the axis, the laser spot irradiation area (ipocentre) and the epicentre can be seen. A fixed-end constraint is imposed at the edge of the steel disk.

IOP Conf. Series: Journal of Physics: Conf. Series **882** (2017) 012013 doi:10.1088/1742-6596/882/1/012013

A thermo-elastic coupling for this kind of application is necessary. Above, all letters in bold indicate vectors. The software initially resolves the transient heat-exchange problem through the Fourier equation:

$$\rho c_p \frac{\partial T}{\partial t} + \rho c_p \mathbf{u} \nabla \mathrm{T} = \nabla(\mathrm{k}\nabla\mathrm{T}) + \mathrm{Q}$$

where T is the temperature, t the time, Q is the heat flux (Gaussian profile on the spot), ρ, c_p and k are the material's density, heat capacity at constant pressure and thermal conductivity, respectively. After the determination of temperature distribution in the solid, a generic structural strain ε is calculated as:

$$\varepsilon = \alpha\left(T - T_{ref}\right)$$

where α is the material's linear expansion coefficient and T_{ref} the reference temperature (293 K). The structural mechanics equations used to find the displacement vector $\boldsymbol{u}$ (which represents the elastic waves in the sample) are:

$$\rho \frac{\partial^2 \boldsymbol{u}}{\partial t^2} - \nabla \boldsymbol{\sigma} = \boldsymbol{F_v}, \qquad \boldsymbol{\sigma} = \boldsymbol{S},$$

$$\boldsymbol{S} - \boldsymbol{S_0} = \boldsymbol{C}{:}\,(\boldsymbol{\varepsilon} - \boldsymbol{\varepsilon_0}),$$

$$\boldsymbol{\varepsilon} = \frac{1}{2}[(\nabla \mathbf{u})^{\mathrm{T}} + \nabla \mathbf{u}],$$

where **C** is the elasticity tensor, '**:**' represents the double dot product, $\boldsymbol{\varepsilon}$ is the strain tensor, $\boldsymbol{\varepsilon_0}$ its initial value (**0**), **σ** the stress tensor and $\boldsymbol{F_v}$ the volume forces vector. In order to obtain a wave propagation which is congruent with the real one, it is necessary [17] to use:

$$t_{max}^{s} = \frac{1}{180 f_d} \qquad , \qquad l_{max}^{e} = \frac{\lambda_{min}}{12}$$

where t_{max}^{s} is the maximum simulation time-step, f_d represents the design frequency (the maximum frequency of interest), l_{max}^{e} is the maximum length for the finite elements and λ_{min} the minimum ultrasonic wavelength; the latter can be calculated using the minimum speed for the various wave modes v_{min}:

$$\lambda_{min} = \frac{v_{min}}{f_d}.$$

Choosing a f_d value equal to 1 MHz (which is the maximum perceived frequency for the broadband probe), the resulting maximum allowed time-step results in $t_{max}^{s} = 5.5\ ns$ and then t^s has been set to 5 ns. Since shear waves are the slowest (3100 m/s for steel), $l_{max}^{e} \cong 234\ \mu m$ and thus the longest element is 200 μm. Only longitudinal waves (L waves) have been studied in this work despite the necessity to represent shear waves.

Ultrasonic probes are sensitive to pressure variations for waves which propagate inside the material; at the exterior, a global displacement is generated with frequency modulations over it. The displacement curves obtained through FE simulations have been filtered considering a low cut-off frequency equal to 100 kHz, so only modulations with frequencies which are inside the probe band are displayed. As can be seen in figure 5 for the epicentre (the disk's opposite point to laser spot), a single laser pulse with a δ equal to 400-800 ns can lead to a first oscillation with high amplitude and a second one which is yet evident. On the other hand, 1200 or 1600 ns long pulses can generate a first oscillation, but there is not a second one. An increase in the oscillation can be highlighted while δ increases, but the maximum oscillation is of the order of $0.1\ pm$. The arrow indicates the ultrasonic time of flight:

$$TOF = \frac{b}{v_l} \cong \frac{0.003\ m}{6000\frac{m}{s}} = 500\ ns$$

where b is the thickness of the disk (the ultrasonic path) and v_l the longitudinal wave speed (average value for common steel). In the graph, the point where the displacement slope begins demonstrates a good agreement with the calculated TOF value.

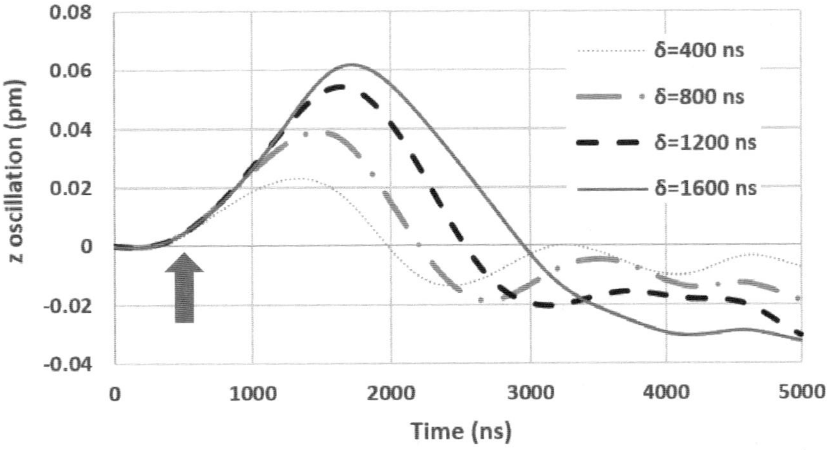

Figure 5. Axial displacement (L wave) at the epicentre as a function of δ (400, 800, 1200, 1600 ns in duration).

A modal analysis has been carried out to find the resonances of the constrained system, resulting in the epicentre maximum displacement for a load frequency of around 300 kHz.

The displacements' frequencies of interest (probe band 0.1-1 MHz) are shown in figure 6. Fast Fourier Transform FFT coefficients are presented as a function of δ, without the zero-frequency term and normalized with respect to the maximum value for the various durations. As can be seen, the normalized coefficients shift towards the low frequencies with increasing δ: the longer the heating period, the lower the value of high frequencies coefficients and vice versa. The change in FFT coefficients for different frequencies as the pulse duration varies means that resonance effects only partially arise due to the thermo-elastic excitation.

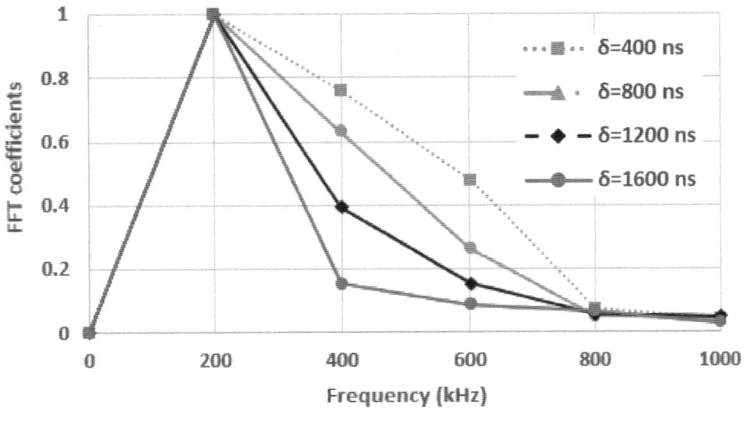

Figure 6. FFT normalized coefficients as a function of frequency and pulse length δ.

In figure 7, the period of the generated US is shown as a function of δ; tracing a $P = \delta$ straight line, it can be shown that the generated ultrasonic wave has a period which is more than double the heating duration δ. In addition to this, P greatly increases as the δ value rises. So, for example, to create an ultrasonic oscillation period under 2000 ns (US complete sinusoid), a heating period of 1000 ns is not sufficient (laser ON length only).

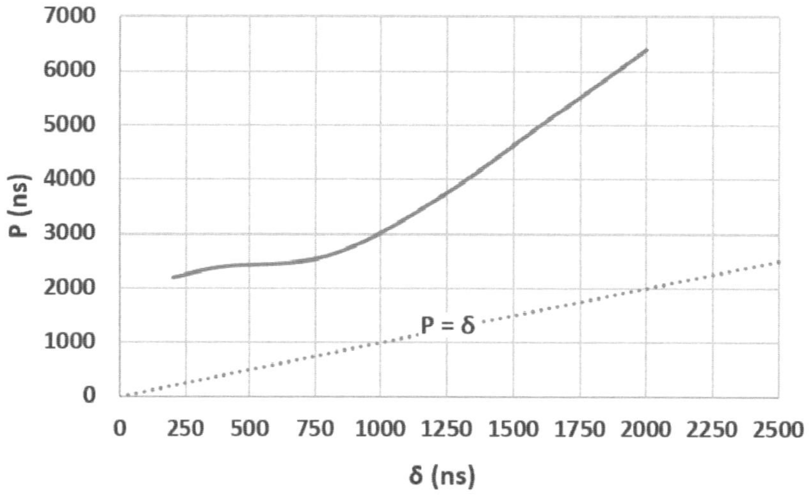

Figure 7. The ultrasonic period P as a function of the pulse length δ.

Multiple pulses are commonly used in experimental testing to increase the amplitude of the generated displacement and to apply post-processing SNR enhancing techniques (i.e. the cross-correlation) [10-12]. As an example, the influence of 3 pulses on the epicentre displacements is shown in figure 8. The pulse durations δ are all the same and equal to 800 ns, but the relative distances are 3200 ns. The TOFs are indicated, as in the single pulse case, with an arrow: the first and third calculated TOFs coincide with the reported ones; the second, though, differs from it: referring to figure 6, an existing oscillation is present in correspondence of this TOF for a single pulse, thus the second generated US interfere with it. If the relative distance between pulses is adequately chosen (i.e. reheating at the minimum of an oscillation), no amplification nor decrease in the displacement is produced, but the US is generated approximately in the same way for every laser pulse.

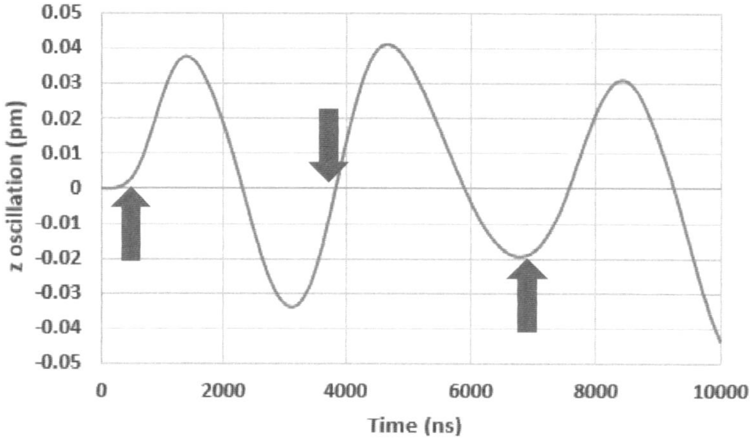

Figure 8. Displacement at the epicentre for a 3 pulses excitation (δ=800 ns).

For the considered 3 pulses excitation, the ultrasonic frequencies generated are reported in figure 9 (not normalized). It can be seen, comparing this curve to the $\delta = 800\ ns$ one in figure 5, a narrowing of the band around 300 kHz, which is also the steel disk resonance frequency.

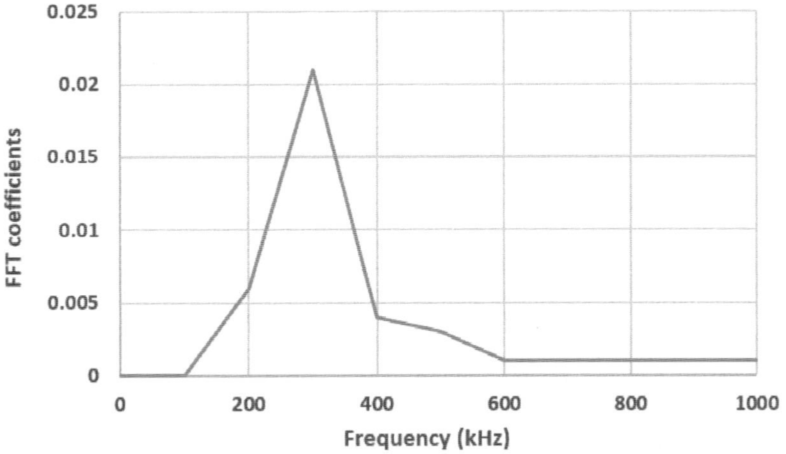

Figure 9. FFT normalized coefficients for a 3 pulses excitation (δ=800 ns) as a function of frequency.

The FE analysis has demonstrated that, effectively, generating a high number of pulses does not increase the oscillations amplitude: preliminarily, it can be assessed that the experimental need to use a high number of pulses is due to post-processing purposes and not for the amplification of oscillations. The use of 3 pulses allows to obtain oscillations around 0.05 pm. If the distance between pulses is adequately imposed, a non-interference condition can be reached between them, thus generating USs with the same characteristics (in particular the main frequency). For the chosen example, though, the main frequency obtained correspond to the resonance one, implying that many conditions regarding different values of δ and distance between pulses should be considered to determine if resonance conditions arise in every case. Simulating them all is not a practical choice. In addition, it was not possible to simulate entire sequences with hundreds or thousands of pulses with HIGHs and LOWs random distribution because of the low time-step, meaning that the influence of interactions between many ultrasounds could not be fully investigated. For all these reasons, the importance of parameters like the pulses lengths, the relative distance between pulses and their number is then determined through experimentation.

4. RESULTS

Experimental tests were carried out to establish how the SNR is influenced by the change in some T sequence parameters. SNR represents how much higher the ultrasonic amplitude is in respect to the noise, thus highlighting how much the US is evident after a post-processing (cross-correlation) procedure.

As displayed in figure 8, the FEM analysis allowed to preliminarily study USs generated through 3, 800 ns long and 3200 ns spaced in time laser pulses. To characterize low-power laser generated US, thus, 3 main parameters must be studied:

1. δ: pulses duration;
2. N_{pulses}: number of pulses contained in the T sequence;
3. d_{pulses}: relative mean distance between pulses.

Several T sequences with different randomness have been generated and studied: the number of LOW levels between pulses ranged between 1 and 5 points for each sequence, but the random spacing was always different for each run. The use of different sequences led though to the same results: hereinafter, thus, only the results for one T sequence will be reported without losing generality.

4.1. Pulse length δ

The pulse length δ determines how much the specimen is heated: the more the heating period, the more the displacement amplitude (but not necessarily the oscillation). As can be seen in figure 10, the SNR increases as δ rises. A T sequence consisting of 2000 pulses has been chosen as a reference. For the highest values of δ, an asymptotic behaviour can be highlighted, probably because the generated US main frequency moves away from the probe band (under 100 kHz). This phenomenon mainly depends on the material characteristics.

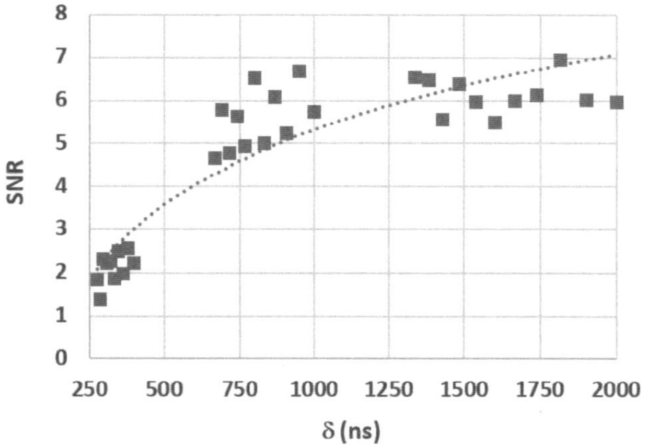

Figure 10. SNR for the T sequence as a function of δ.

Since the change in the ultrasonic frequencies can greatly affect SNR, it is necessary to analyse the frequency composition of the ultrasonic signal resulting from a δ long laser excitation. Figure 11 shows the Fourier coefficients' values corresponding to frequencies between 0 Hz and 1 MHz as a function of δ. Low values of δ imply the US generation with a main frequency band (dark in the figure) with some modulations (more clear and next to it), capable of covering almost the entire probe frequency band; as δ raises, the ultrasonic frequencies reduce, with a band which is very narrow for $\delta = 2000\ ns$: this is perfectly in line with FE simulations for a single pulse.

The ultrasonic main frequencies for $\delta = 2000\ ns$ are extremely close to the lower limit of the band probe, justifying the SNR asymptotic behaviour shown in figure 10. Points in the graphs with amplitude values above 2 can be approximately fitted with a linear trend: it can be an important feature for studies like NDT or material characterization where the suitable ultrasonic frequency choice has a great impact. The frequency composition obtained from the FE model when multiple pulses are used (figure 9) is very similar to the one obtained experimentally. The decrease in the main frequency as δ rises suggests, though, that the correspondence with the resonance frequency is not applicable for all δ values. The resonance, thus, influences only partially the generated US frequency.

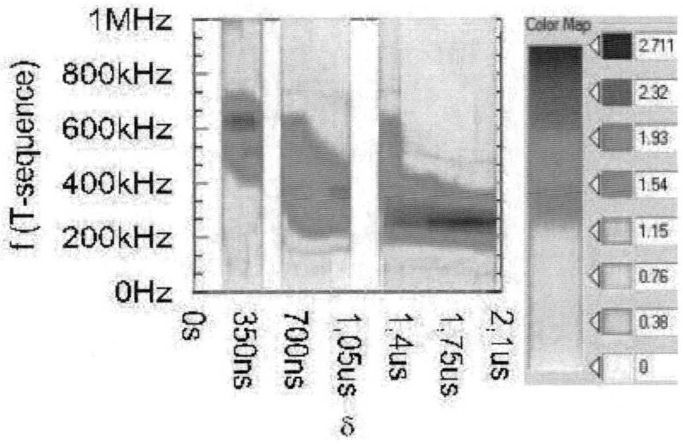

Figure 11. Frequencies and related Fourier coefficients at various δ.

In the end, the use of a high δ is recommended to maximise the SNR, but always keeping in mind that a very high value of this parameter can generate ultrasonic frequencies out of the probe band. If ultrasonic probes with an inferior lower band limit are available (20 kHz), δ can be further increased.

4.2. Number of pulses N_{pulses}

As previously mentioned, the number of pulses is the main parameter which can lead to ultrasonic individuation in real testing campaigns, since a single pulse creates oscillations which are not detectable (tenth of pm). The influence of the pulses number N_{pulses} on the SNR is shown in figure 12, with an imposed δ value equal to 800 ns. The T sequence application allows to obtain a SNR which tends to increase as N_{pulses} raises: this increase is not quick, coherently with FE simulations (figure 8). In fact, the number of pulses is not that important since USs with the same amplitude and frequency are generated at each laser pulse; the increase is mainly due to a major effectiveness of the cross-correlation operation when a large amount of pulses is applied. So, a high number of pulses is desirable to obtain a high SNR. It is however important to underline that the inspected time window widens if many pulses must be used, so also this facet must be considered.

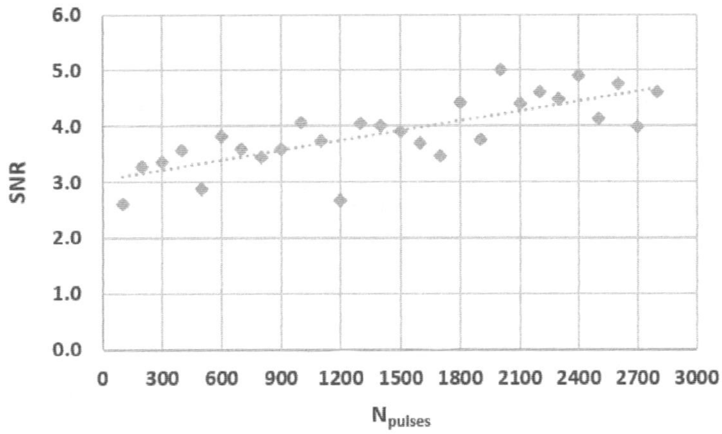

Figure 12. SNR as a function of the number of pulses N_{pulses} with δ equal to 800 ns (inspected time window 8 ms long).

IOP Conf. Series: Journal of Physics: Conf. Series **882** (2017) 012013 doi:10.1088/1742-6596/882/1/012013

4.3. Distance between pulses d_{pulses}

Another topic which could not be investigated through a FE approach is the influence of the relative distances d_{pulses} for a high number of pulses. Because of the sequences randomness which is necessary to obtain only one and easily characterizable peak in the cross-correlation [15], d_{pulses} represents a mean distance between pulses.

Considering $N_{pulses} = 2000$ and $\delta = 800\ ns$, the trend of SNR is shown in figure 13. An absolute minimum is present around $d_{pulses} = 10\ \mu s$: this is mainly due to the interference between ultrasonic waves generated and reflected at the epicentre (same phase and frequency) caused by the low specimen thickness; a successive minimum (but with a higher SNR) indicates another ultrasonic interference, but between the generated US and the one which has been reflected twice. SNR is high when $\boldsymbol{d_{pulses}}$ is low, meaning that generated and reflected waves do not have the same phase (no ultrasonic interaction). The graph suggests that using pulses relatively close to each other can give better results from a SNR maximization perspective.

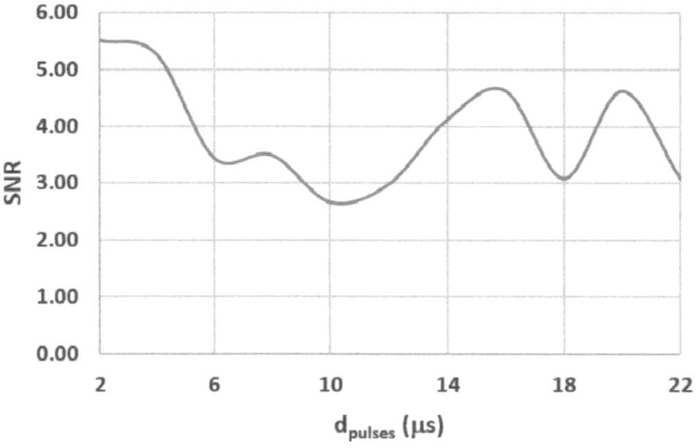

Figure 13. SNR as a function of the mean distance between pulses d_{pulses}, considering $\delta = 800$ ns.

5. CONCLUSIONS

A FEM analysis has been carried out to highlight the main characteristics of low-power laser generated US. Some important parameters' influence was studied preliminarily, as for the pulse length and the application of multiple pulses. Data regarding the generated US show: 1) the low displacements entity, even for high heating period, 2) the decrease in high-frequencies components inside the displacement itself. The FE analysis also showed that a large amount of pulses (with same length) used in testing campaigns increases the number of ultrasonic oscillations generated but not their amplitude, leading to a more easily detectable peak when post-processing techniques (i.e. cross-correlation) are applied.

Starting from the FE results, experimental procedures were applied to highlight the importance of the parameters considered in simulations. Highly random codes are meant to be used as input of the laser modulated diode, to improve the efficiency of the cross-correlation post-processing technique. So, a family of sequences called T sequence (which is not present at the state of the art) has been created.

The results of the application of one particular T sequence are reported for every important parameter considered:

1) δ - The pulse length varies the generated US in two different ways, i.e. changing the US amplitude and frequency content. An increase in δ increases the thermal oscillation and so it allows to obtain a high SNR in general too. However, it is true if the US frequency band is entirely inside the probe band. So, while the carried-out analysis has made use of a broad-band probe which is near the ultrasonic lowest frequency (20 kHz), the use of probes with an inferior lower band limit can lead to an increase in SNR if higher δ are used.
2) N_{pulses}- A very high number of pulses implies a wider inspected time window. Apart from this, high values of this variable are necessary in order to detect the ultrasound and are recommended to obtain good SNRs.
3) d_{pulses}- The T sequence allows to obtain a maximum in SNR when d_{pulses} is very low, while an increase creates a minimum in SNR because of the interference between counter-phase generated and reflected ultrasonic waves; the point where the minimum is found depends on the geometrical and thermo-elastic features of the inspected object.

In conclusion, the FE analysis allowed to adequately comprehend the US thermo-elastic generation phenomenon inside the material. The successive experimentation expanded the study and the phenomenon comprehension evaluating the influence of other parameters through which the CW laser is modulated. The features of the generated US are defined once the parameters defining the input T sequence are chosen and imposed. In the end, the T sequence which allows to obtain the highest SNR is the one with the highest values of N_{pulses} and δ (with ultrasonic frequency inside the probe band) and with the lowest d_{pulses}.

Since input PN codes' influence has not been considered, future studies will take on a comparison between their use and the T sequence one. In addition, only longitudinal waves have been investigated so other works will naturally deal with low-power laser generated surface US and their characteristics.

References

[1] Withers P J et al. 2008 Recent advances in residual stress measurement *International Journal of Pressure Vessels and Piping* **85-3** 118-127

[2] Shaswary E et al. 2015 A new algorithm for time-delay estimation in ultrasonic echo signals *Transactions on Ultrasonics, Ferroelectrics, and Frequency Control* **62-1** 236-241

[3] Scruby C B, Drain L E 1980 *Laser Ultrasonics - Techniques and applications* Ed. Adam Hilger

[4] Singhal M et al. 2006 Characterization of laser generated bulk waves using wavelet transforms and pattern recognition *A-PCNDT*

[5] Cavuto A et al. 2015 Experimental investigation by laser ultrasonics for high speed train diagnostics *Ultrasonics* **55** 48-57

[6] Lee J R et al. 2013 Comparative analysis of laser ultrasonic propagation imaging system with capacitance and piezoelectric air-coupled transducers *Journal of Intelligent Systems and Structures* **3** 551-562

[7] Park, B. et al. 2013 Laser ultrasonic imaging and damage detection for a rotating structure *Structural Health Monitoring* **12-5/6** 494-506

[8] Vangi D et al. 2013 Laser-ultrasound application for train axles monitoring (Applicazione Laser-Ultrasuoni per il Monitoraggio degli Assili Ferroviari) *Proc. of 42nd AIAS Nat. Conf.* 215

[9] Sohn H. et al. 2014 Fatigue crack localization using Laser Nonlinear Wave Modulation Spectroscopy (LNWMS). *Journal of the Korean Society for Nondestructive Testing* **34-6** 419-427

[10] Veres I A et al. 2013 Golay code modulation in low-power laser-ultrasound *Ultrasonics* **53-1** 122-129

[11] Pierce S G, Culshaw B 1998 Laser generation of ultrasonic Lamb waves using low power optical sources *Science, Measurements and Technology* **145-5** 244-249

[12] Anastasi R F, Madaras E I 1999 Pulse Compression Techniques for Laser Generated Ultrasound *Ultrasonics Sym., Proc. of the IEEE* **1** 813-817

IOP Conf. Series: Journal of Physics: Conf. Series **882** (2017) 012013 doi:10.1088/1742-6596/882/1/012013

[13] Kohanzadeh Y et al. 1974 Thermoelastic waves generated by laser beams of low power *The Journal of Acoustical Society of America* **57-1** 67-71
[14] Cleary A et al 2011 Low power laser generated ultrasound: signal processing for time domain data acquisition *Journal of Physics: CS* **278-1**
[15] Vangi D et al. 2017 Study on the most influential parameters in low-power laser generated ultrasound *FME Transactions* **45** 323-330
[16] Dixon R C 1984 *Spread Spectrum Systems* Ed. J. Wiley
[17] Wang J et al. 2007 Numerical simulation of laser-generated surface acoustic waves in the transparent coating on a substrate by the finite element method *Optics & Laser Technology* **39-1** 21-28
[18] Mineo C et al. 2014 Numerical study for a new methodology of flaws detection in train axles *Ultrasonics* **54-3** 841-849
[19] Liu P et al. 2016 Numerical simulation of damage detection using laser-generated ultrasound *Ultrasonics* **69** 248–258
[20] Vangi D et al. 2016 Noise reduction in ultrasonic signals through cross-correlation algorithms for mechanical components' on-line monitoring (Riduzione del rumore in segnali ultrasonori tramite algoritmi di cross-correlazione per il monitoraggio di componenti in servizio) *Proc. of 45th AIAS Nat. Conf.* 673

Non-invasive dynamic measurement of helicopter blades

J Serafini[1], G Bernardini[1], L Mattioni[1], V Vezzari[2] and C Ficuciello[3]

[1] Department of Engineering, University Roma Tre, Via della Vasca Navale 79, Roma, 00144, Italy

[1] Globalsensing s.r.l., Via Portuense 956, Roma, 00148, Italy

[3] I.R.I. "Italian Rotors Industries s.r.l.", Via delle Valli 44/C, Aprilia, 04011, Italy

E-mail: serafini@uniroma3.it

Abstract. This paper presents the development and the application on helicopter blades of a measurement system based on FBG strain gauges. Here, the main goal is the structural characterization of the main rotor blades, with the aim of showing the potentialities of such a system in blades quality check applications, as well as in the development of structural health monitoring and rotor state feedback devices. The device has been used in both non-rotating and rotating tests, and does not require the presence of slip rings or optical joint since it is completely allocated in the rotating system. It has been successfully applied to characterize the frequency response of blades lead-lag, flap and torsion deformations, up to 250 Hz.

1. Introduction

Aircraft structures integrity is subject to several threats due to normal and abnormal operating conditions. Moreover, the production phase is not always sufficiently monitored, especially when considering processes with low quality control standards. Structural quality and integrity checks and structural health monitoring (SHM) are then areas of paramount importance in aircraft industry to reduce failures and performance degradation of aircraft structures. Nowadays, structural monitoring is mainly based on the traditional damage tolerance philosophy [1], and it is performed through the combination of visual checks and off-line non-destructive testings (NDTs). They are based on different techniques, including ultrasonic, liquid penetrant, magnetic particle, electromagnetic, laser, radiography [2]. Although this methodologies are reliable and consolidated, being offline, they rely on periodical control protocols and then present some obvious disadvantages: (i) time granularity: between one check and another, there is a non monitored period; (ii) high maintenance cost, due to the necessity of having limited amount of time between checks; (iii) decreasing of time-on-wing. Moreover, they require dedicated and usually costly equipment and dedicated spaces. This is especially true in helicopter applications, where helicopter blades also require periodic track and balance to keep low vibration levels, generated by differences of blade characteristics due to the fabrication process and use.
In-flight systems for structural monitoring are receiving more and more attention as they may increase flight safety reducing, at the same time, maintenance cost [3, 4]. This could be achieved thanks to the continuous observation of many critical aircraft components. The sensors to be used must have specific characteristic in terms of low intrusiveness and weight and the potential candidates include, among the others, micro electromechanical systems (MEMS)

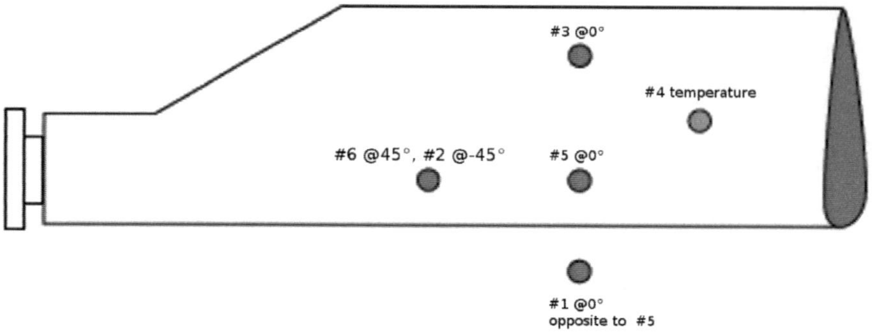

Figure 1. Sensors setup.

and FBG strain gauges. Moreover, their applications to helicopter main rotor blades, present additional constraints due to the relative motion between rotating and non rotating frames. The transmission of signal and power between the two frames via slip-ring or optical joints is particularly critical, making a system entirely housed in the rotating frame preferable. A drawback of such a system lies in the fact that it requires an autonomous power source directly located on the rotating system. To alleviate this problem a low-power system processing and exchanging few data is desirable. In terms of measurement capability, the in-flight health monitoring system must be able to deal with moderate lead-lag, flap and torsion deflections, typical of helicopter blades. Moreover, the monitoring system has to be able to measure and resist signal up to 50 Hz, caused by the periodic loads acting on the rotor system and by low damping values of its aeroelastic modes. Finally, for inertial (centrifugal forces) and aerodynamic (to produce small interferences) issues, optimal positioning of the sensors would be the inner part of the blade, near the root section. This makes the strain-based measurements preferable with respect to acceleration-based measurements, as in this region the acceleration due to the deformation is small, whereas the strain is maximum there.

In this paper, a quality check system based on the use of a limited number of Fiber Bragg grating (FBG) strain gauges entirely housed in the rotating frame is proposed and tested. The system is used for the structural characterization of ultra-lightweight helicopter blades, through experimental tests in non-rotating and rotating conditions. The work done in this paper has to be seen as a fist step toward the development of a fully sensorized blade, allowing online Structural Health Monitoring, Rotor-State Feedback and Track and Balance.

2. Measurement setup

The measurement setup consists of: (i) a battery-powered interrogator, located at the top of rotor mast; (ii) six FBG measurement points along a single fiber, arranged following the scheme depicted in figure 1

The sensors are fixed on the blade surface through cyanoacrylic glue, and subsequently protected along with the fiber between them with aluminum tape (figure 2). Only five sensors out of six (1,2,3,5,6) are actually used for strain measurement, whereas sensor 4 works as temperature probe to compensate thermal expansion effects on strain. Among the five sensors dedicated to strain measurements, sensors 1, 3 and 5 are parallel to the blade span, whereas sensors 2 and 6 are angled $\pm 45^\circ$ with respect to it. Sensors are used in pair (*i.e.,* evaluating the difference between the signal from two of them): (i) sensors 1 and 5 are used for flap banding measurement; (ii) sensors 1 and 3 are used for lead-lag banding measurement; (iii) sensors 2 and 6 are used for

Figure 2. Final setup for flight testing.

Figure 3. Flange for non-rotating tests.

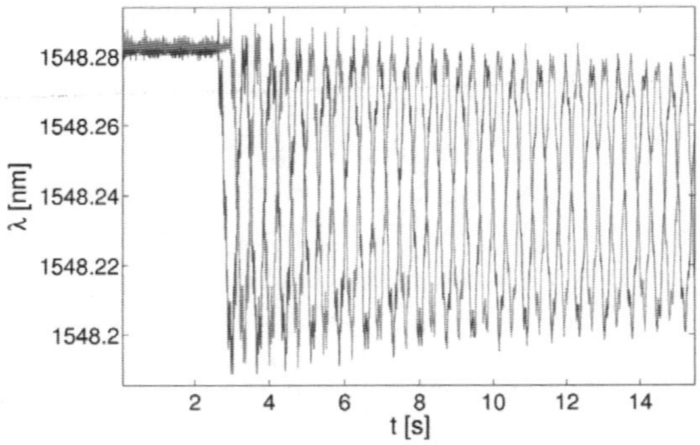

Figure 4. Output of two tests in response to step relaxation excitation.

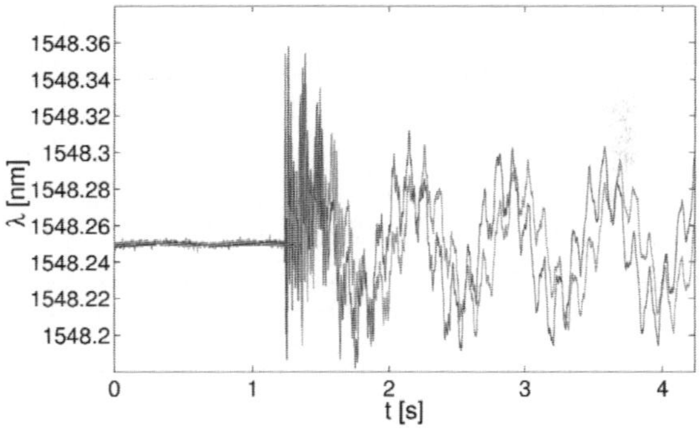

Figure 5. Output of two tests in response to hammering excitation.

torsion measurement. In order to improve the selectivity of individual pairs (*i.e.*, the capability of each pair to identify the natural frequencies pertaining the deformation they are dedicated to), a weighted difference between sensor signals may be used, on the basis of the knowledge of the sensors position with respect to the blade elastic axis [5].

3. Non-rotating tests

To perform non-rotating tests a suited flange to constraint the blade root has been designed and manufactured (figure 3). This flange has been especially designed to allow the blade rotation around its feathering axis, in order to have the possibility of analyzing blades at different pitch angles.

First of all, the proposed technique has been used to verify the accuracy of the structural properties of rotor blades identified through the experimental characterization procedure described in [6, 7]. This has been performed comparing the blade non-rotating natural frequencies, numerically predicted by using the identified experimental structural properties [8, 9], with those experimentally determined from the blade transient response to an initial excitation. Specifically, three different excitation techniques have been employed, namely: step relaxation, hammering and shaking.

Note that, since the goal of these tests was the analysis of the free blade response, an accurate measure of the excitation signal was not necessary, thus avoiding the need of expensive tools

IOP Conf. Series: Journal of Physics: Conf. Series **882** (2017) 012014 doi:10.1088/1742-6596/882/1/012014

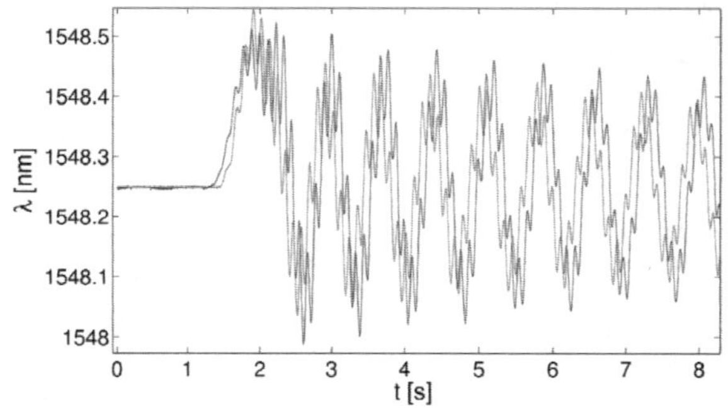

Figure 6. Output of two tests in response to shaking excitation.

specifically dedicated to this purpose (like instrumented hammer); it was sufficient to analyze the signals once the transient response was over. It is worth noting that the above experimental approach may be easily used for rotor blades quality check.

figures 4 to 6 show the time histories of the FBG output due to two consecutive tests per each excitation procedures. It is possible to notice that the time histories are very similar to each other when the same input signal is used, whereas they exhibit different harmonic contents when different excitation methodologies are applied, essentially due to the different natural modes excited by each of them. The step relaxation, obtained releasing a weight tied to the blade, mostly excites low frequency modes. The hammering excitation yields a richest response in terms of frequency content. The shaking excitation gives a spectrum strongly dependent on the spectrum of the input, as expected. These consideration are confirmed by figure 7, which depicts the spectra of the difference between the sensors signals from the three different excitation techniques. Note that the flap spectrum is here also evaluated by using the sum of signals from sensors 5 and 3 (top-left figure), with similar results with respect to the difference between sensors 5 and 1. This is expected, since most of the deformation energy is associated to flap modes, due to their relatively small stiffness. However, although the high frequency modes are captured only through the hammering, the three excitation techniques give equivalent results on the low-to-mid-frequency range.

In figures 8 and 9 the effect of moving the application point of the excitations on the PSD of the output signals is presented, for both step relaxation and hammering. Generally, moving the excitation point from the inner to the outer portion of the blade gives an increase of low-frequency content of the spectrum. Moreover, the higher frequency content is only weakly influenced by the excitation location, making the identification of high frequency modes a very hard task. Contrary to what expected, this is the case even when the excitation input is applied at the inner part of the blade: this behavior is probably due to the low level of deformation energy given to the blade in this test condition. Nevertheless, performing several tests with different excitation points is useful to reduce the possibility of having hidden modes, due to the eventual positioning of the excitation device in one of their nodes.

Next, figure 10 shows the natural frequencies of the blades, obtained through a statistical analysis on several set of data obtained from several tests. Note that, for graphical reasons, frequencies are associated to a set of curves that are specific of a single test; in this sense, this picture has to be considered only as an illustrative one. It is worth noting that, independently from the aforementioned problems in dealing with high frequency modes, it has been possible to identify modes up to 250 Hz. The outcome of the statistical analysis of the test data is reported in Table 1. Results from step relaxation and hamming excitation seem to be quite similar, except for the third torsion frequencies which differ of about 4%, whereas the shaking

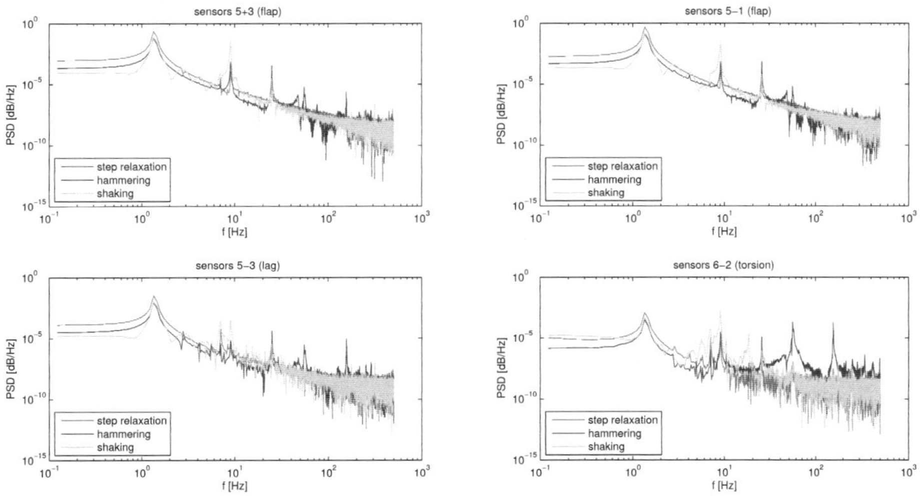

Figure 7. Comparison of the spectra for different excitation techniques.

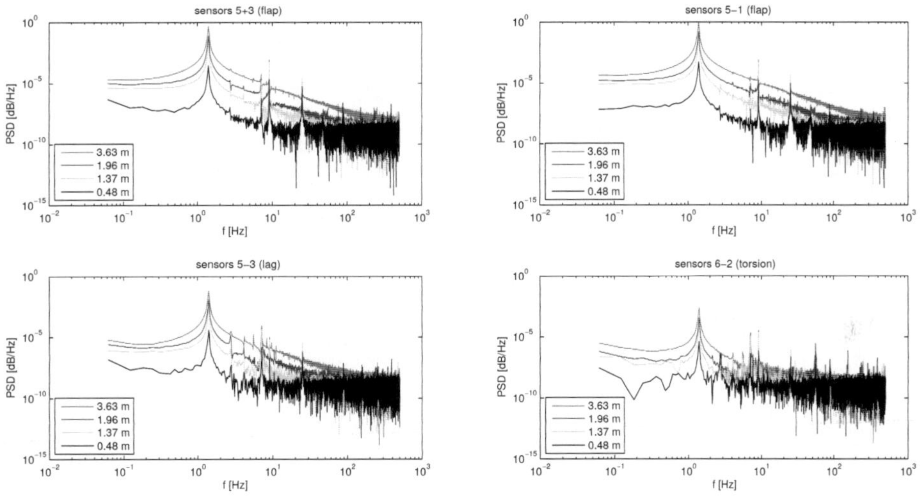

Figure 8. Effect of step relaxation excitation location on identified spectra.

tests completely miss the high-frequency modes. Moreover, since the third flap frequency has been identified by the shaking only in one test, its standard deviation is not shown. The results repeatability is quite good, with the only exception pertaining the torsional modes, for which a slightly deterioration has been detected. Finally, it is worth noting note that, due to the significant difference between flap and lead-lag bending stiffnesses, it was impossible to identify higher lead-lag modes, essentially for the practical difficulty to transmit to them a significant amount of deformation energy.

The comparison with the numerical eigenfrequencies evaluated by using the structural properties from the characterization process described in [7] is shown in table 2, in terms of their relative error. The torsion modes are missing, in that it was impossible to determine the sectional moment of inertia during the characterization process. The other modes are in quite good agreement with those experimentally identified.

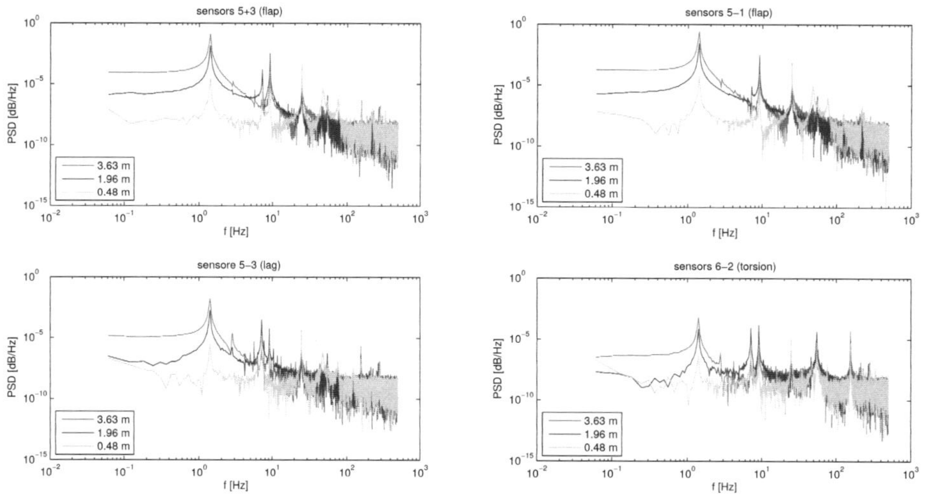

Figure 9. Effect of hammering excitation location on identified spectra.

Table 1. Mean value and standard deviation of identified frequencies.

Mode		Step Relaxation	Frequency [Hz] Hammering	Shaking
1st flap	Mean	1.343	1.343	1.343
	Std Dev	0	0	0
2nd flap	Mean	7.751	7.777	7.775
	Std Dev	0	0.012	0.025
3rd flap	Mean	20.905	21.204	21.240
	Std Dev	0.037	0.148	—
1st lag	Mean	5.219	5.219	5.219
	Std Dev	0	0	0
1st torsion	Mean	48.691	48.229	48.693
	Std Dev	3.181	0.798	—
2nd torsion	Mean	156.2	155.5	—
	Std Dev	0.462	0.967	—
3rd torsion	Mean	238.6	249.3	—
	Std Dev	7.27	7.38	—

Finally, using the moving block technique [10, 7], it has been also possible to identify the structural damping of the low frequency modes (see table 3), which is an important parameter for structural stability. Moreover, its change may be an indicator of some kind of damage [11], making it a possible candidate in detecting blade damages (*i.e.*, a possible index in the definition of health monitoring criteria).

4. Rotating tests

After the non-rotating tests, the measuring system has been set up on the main rotor of a T-22 helicopter, in order to assess its in-flight reliability and performance. The flight test has been a

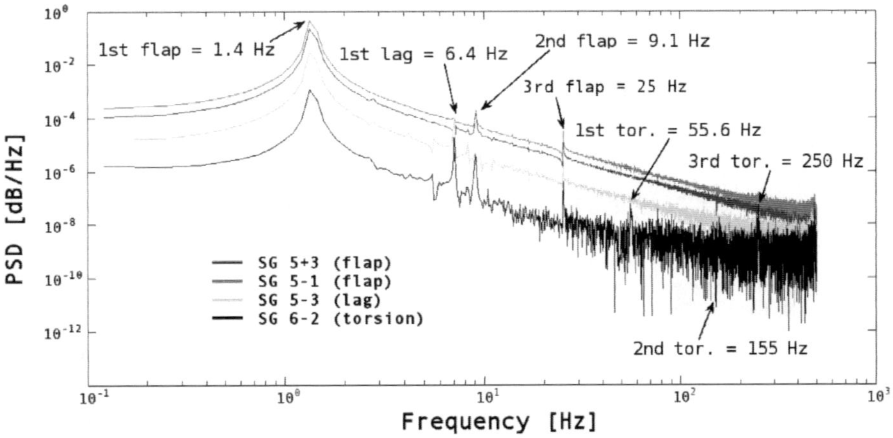

Figure 10. Identified non-rotating eigenfrequencies.

Table 2. Relative error on measured and predicted blade eigenfrequencies.

Mode	Relative error
1st Flap	$< 0.5\%$
1st Lag	$< 0.5\%$
2nd Flap	6.6%
3rd Flap	6.8%

Table 3. Mean value and standard deviation of identified damping.

Mode		Step Relaxation	ζ Hammering	Shaking
1st flap	Mean	0.022	0.016	0.023
	Std Dev	0.008	0.004	0.004
2nd flap	Mean	0.047	0.051	0.068
	Std Dev	0.003	0.007	0.002
3rd flap	Mean	0.226	0.224	—
	Std Dev	0.011	0.013	—
1st lag	Mean	0.057	0.058	0.061
	Std Dev	0.005	0.003	0.001

low level hovering of about eight minutes from ignition to shutdown. Figure 11 shows the time history of the output of sensors 1,2,3,5 and 6 during the whole test. In order to excite blade modes, the pilot performed simple maneuvers, slightly modifying the trim controls. Figure 12 depicts a detail of the sensors signal time history after the application of a collective double-step input. Note that, from Figure 11 it is possible to clearly recognize all the flight phases, as well

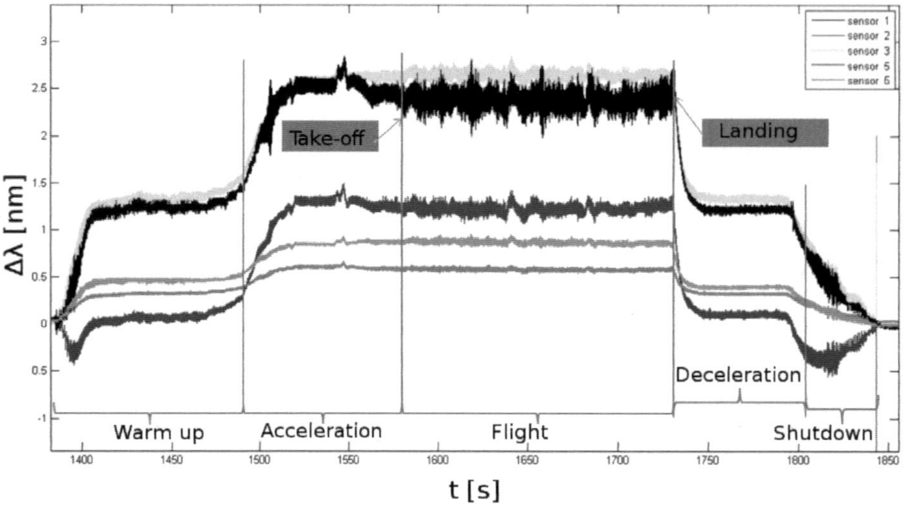

Figure 11. Sensors output during flight.

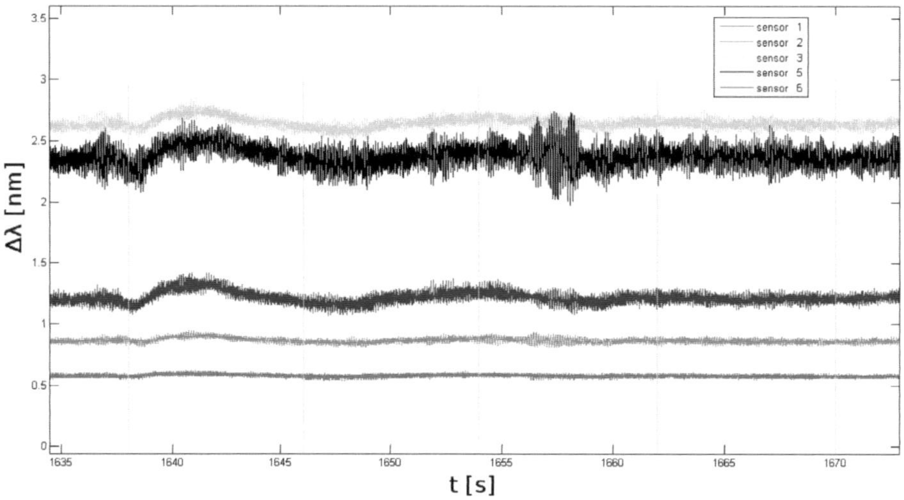

Figure 12. Sensors output during flight, detail.

as the response to the variation of the control inputs.

To highlight the harmonic content of the signal, it has been analyzed through a Fast Fourier Transform. First, three 8-seconds chunks of the signal, 1638 s - 1646 s, 1646 s - 1654 s and 1662 s - 1670 s, have been processed. The first chunk (green line in figure 13) concerns the double-step collective variation phase, the second chunk (blue line in figure 13) is immediately subsequent to this phase and the third one (red line in figure 13) is representative of the stationary hovering phase. As expected, the spectra are significantly richer than those in the static test. Several peaks, corresponding to rotating modes frequencies and to the periodic response are present. Note that, the latter are caused by the non perfect hovering flight condition, and by the aerodynamic interference between the main rotor blades and the fuselage. The spectra are almost coincident for $\omega \geq 10$ while significant discrepancies arise in the low-frequency range,

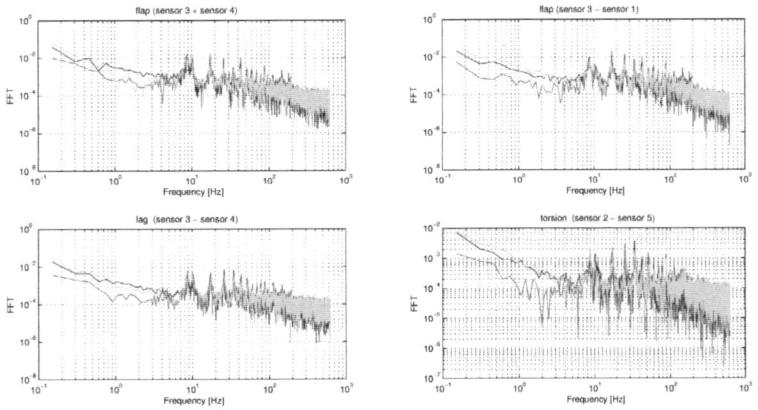

Figure 13. Fast Fourier Transform of signals during flight. —: signal during collective excitation; —: immediately after collective excitation; —: steady hovering.

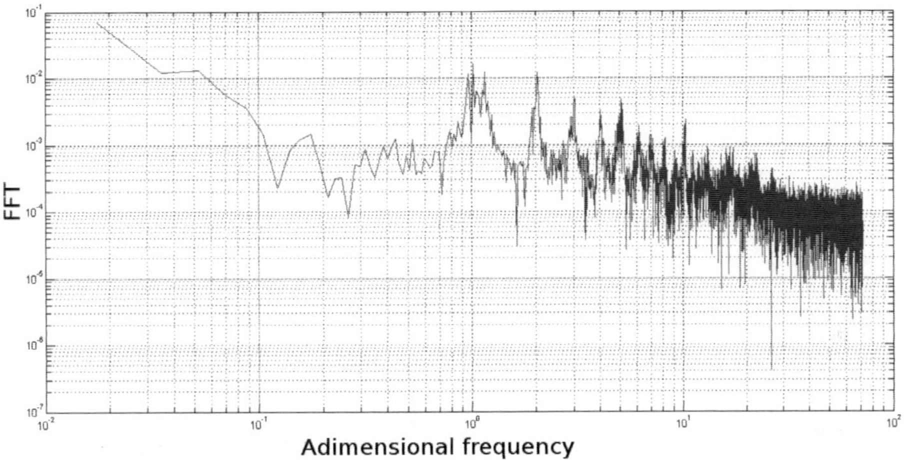

Figure 14. Example of spectra in terms of adimensional frequency.

which is dominated by the rigid body motion induced by collective pitch actuation. A better insight of the spectra may be obtained from the analysis of figures 14 and 15. The former shows the output of a sensor in terms of the adimensional frequency $\nu = \omega/\Omega$, where Ω is the rotor angular speed. Several peaks are evident at integer values of ν (*i.e.*, at frequencies multiple of Ω), due to the steady periodic elastic response. The latter shows a detail of the low frequency spectrum of the signal from the flapping sensors. In addition to the peaks at frequencies Ω and 2Ω, other two peaks are present: one associated to the first elastic flap, and the other one to the rigid flap motion. The latter is only visible during the collective variation phase, and it is probably hidden by Ω peak during the other two flight conditions. Both flapping frequencies are coherent with the rotating ones numerically evaluated (error below 1%). The rigid flap peak is very difficult to identify in steady hovering conditions, in that it is barely visible in some cases and completely hidden in the other ones. This is particularly evident in figure 16, which shows the output of the flapping sensors regarding four phases of steady hovering flight.

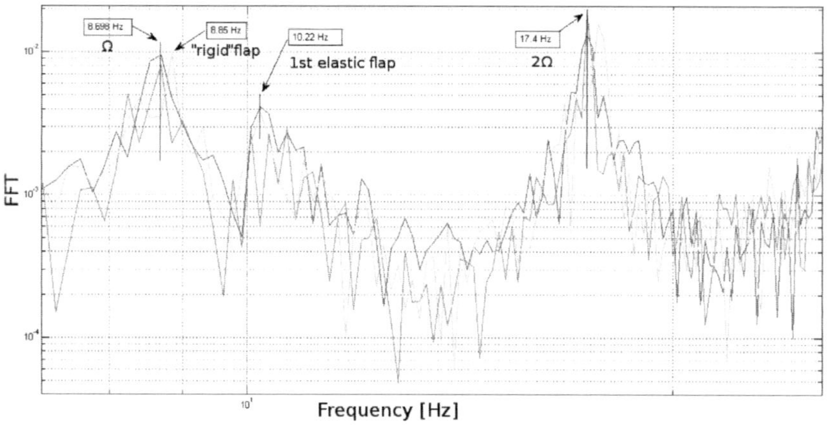

Figure 15. Fast Fourier Transform of flap signals during flight, detail. —: signal during collective excitation; —: immediately after collective excitation; —: steady hovering.

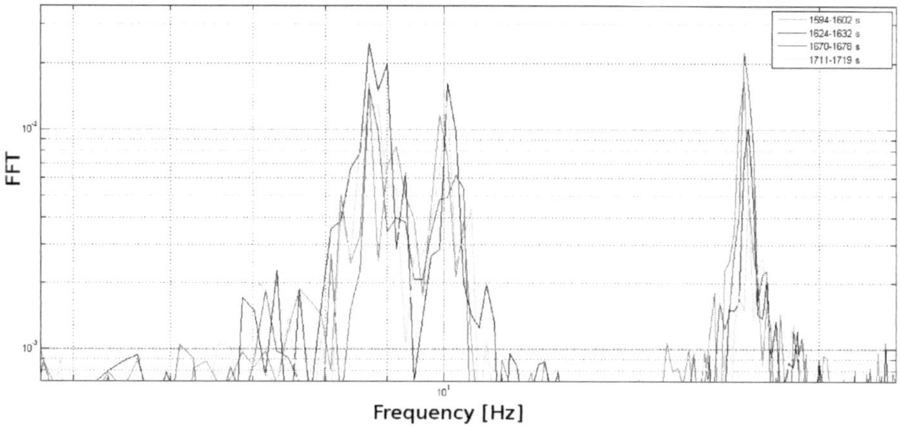

Figure 16. Fast Fourier Transform of flap signals during steady hovering flights.

Finally, with the aim of assessing the possibility of recognizing maneuvered flight phases from the strain gauges signal, an analysis on intervals of signal longer than the previous ones has been performed. Specifically, Figure 17 shows the comparison between the spectrum from a sampling period longer than 8 seconds (1638 s - 1670 s) with two 8-seconds-long periods (1638 s - 1646 s and 1646 s - 1654 s), previously analyzed. As expected, the spectra are very rich as the related periods contains frequencies associated to both control input and vehicle response. As noted before, in the range of frequencies characteristic of aeroelastic phenomena ($\approx$ 10 Hz - 100 Hz), the spectra are significantly similar. Moreover, the increase of the sampling period (black line) reveals a low frequency peak associated to the maneuver. This fact is of great interest in the development of a structural health monitoring system. Indeed, these are based on the detection of the discrepancies between actual signals and nominal ones. Hence, one of the main challenges in the definition of reliable structural health monitoring criteria concerns the difficulty of distinguishing between blade response changes due to a damage and those caused by the voluntary input of the pilot.

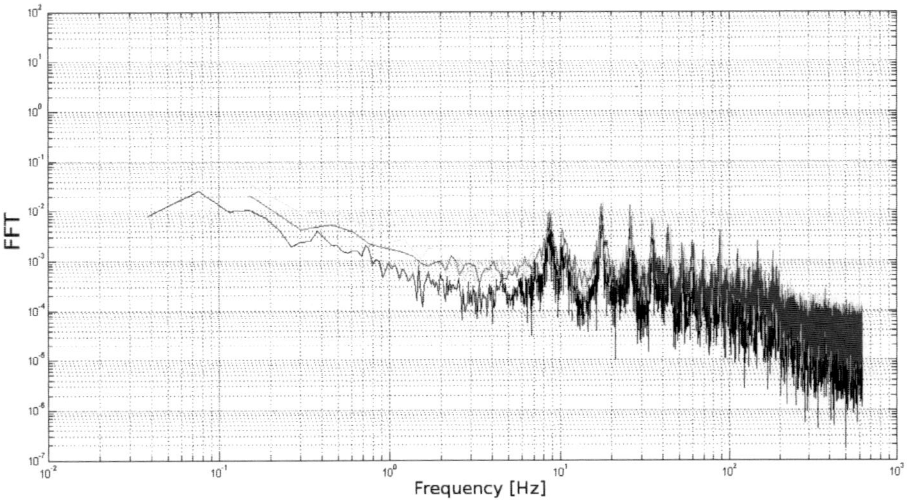

Figure 17. Fast Fourier Transform of flap signals during flight, detail. —: 1638 s - 1646 s; —: 1646 s - 1654 s; —: 1638 s - 1670 s.

At the end of this section, it is worth noting that an other important output of the flight test campaign has been the fact that the pilot did not report anomalous rotorcraft behavior due to the presence of the measuring system (interrogator and fiber optic cable) on the main rotor.

5. Conclusions and future work

The measurement system has been installed on a helicopter rotor without the need of slip rings or optical joint, which would significantly increase the complexity of the system and compromise its reliability. During the flight test no problems have been reported by the pilot related to the presence of the system. In this work the measuring system has been used for the dynamic characterization of helicopter blades both in rotating and non-rotating conditions, showing its potentialities as blades quality check system. Flight tests demonstrated that the system is able to give information on blade lead-lag, flap and torsion deformations with good accuracy up to 250 Hz, as well as the possibility of distinguishing signals associated to the helicopter rigid body motion from those pertaining the blades aeroelastic behavior. These characteristics suggest this measuring device as starting point for the development of structural health monitoring and rotor state feedback systems, which are the most interesting applications envisaged by the authors for the next future. From the technological point of view, the development of an autonomous power source not based on batteries (*e.g.,* a small electrical generator coaxial with the mast) is mandatory for real-life health monitoring applications, as well as wireless transmission capabilities to integrate the output of the monitoring system into the avionics.

References

[1] Khan A, Zafar S, Khan N and Mehmood Z 2014 *J. Space Technol.* **4** 67–74
[2] Mix P E 2005 *Introduction to nondestructive testing: a training guide* (John Wiley & Sons)
[3] Staszewski W, Boller C and Tomlinson G R 2004 *Health monitoring of aerospace structures: smart sensor technologies and signal processing* (John Wiley & Sons)
[4] Chang F K, Markmiller J F, Yang J and Kim Y 2011 *System Health Management: With Aerospace Applications* 419–428
[5] Bernardini G, Porcelli R, Serafini J and Masarati P 2017 *AHS 73nd Annual Forum & Technology Display*

[6] Enei C, Bernardini G, Serafini J, Mattioni L, Ficuciello C and Vezzari V 2015 *Journal of Physics: Conference Series* vol 658 (IOP Publishing) p 012003
[7] Bernardini G, Serafini J, Enei C, Mattioni L, Ficuciello C and Vezzari V 2016 *Measurement Science and Technology* **27** 065401
[8] Bernardini G, Serafini J, Colella M M and Gennaretti M 2011 *Proceedings of 37th European Rotorcraft Forum* (Gallarate (Italy))
[9] Bernardini G, Serafini J, Molica Colella M and Gennaretti M 2013 *Aerospace Science and Technology* **29** 175–184 ISSN 12709638
[10] Amrani A O 1991 *Journal of Guidance, Control, and Dynamics* **14** 208–210
[11] Sohn H, Farrar C R, Hemez F M, Shunk D D, Stinemates D W, Nadler B R and Czarnecki J J 2003 *Los Alamos National Laboratory*

IOP Conf. Series: Journal of Physics: Conf. Series **882** (2017) 012015 doi:10.1088/1742-6596/882/1/012015

Statistical analysis on experimental calibration data for flowmeters in pressure pipes

Alessandro Lazzarin, Enrico Orsi, Umberto Sanfilippo

Dipartimento di Ingegneria Civile e Ambientale (DICA), Politecnico di Milano
Piazza Leonardo da Vinci 32, 20133 Milano, Italy, settore-portate@polimi.it

Abstract. This paper shows a statistical analysis on experimental calibration data for flowmeters (i.e.: electromagnetic, ultrasonic, turbine flowmeters) in pressure pipes. The experimental calibration data set consists of the whole archive of the calibration tests carried out on 246 flowmeters from January 2001 to October 2015 at *Settore Portate* of *Laboratorio di Idraulica "G. Fantoli"* of *Politecnico di Milano*, that is accredited as LAT 104 for a flow range between 3 l/s and 80 l/s, with a certified Calibration and Measurement Capability (CMC) - formerly known as Best Measurement Capability (BMC) - equal to 0.2%. The data set is split into three subsets, respectively consisting in: 94 electromagnetic, 83 ultrasonic and 69 turbine flowmeters; each subset is analysed separately from the others, but then a final comparison is carried out. In particular, the main focus of the statistical analysis is the correction C, that is the difference between the flow rate Q measured by the calibration facility (through the accredited procedures and the certified reference specimen) minus the flow rate Q_M contemporarily recorded by the flowmeter under calibration, expressed as a percentage of the same Q_M.

1. Introduction

For any hydraulic application, like in every engineering field, it is very important to develop and to adopt reliable procedures and devices to measure the physical quantities of interest (i.e.: water depths, pressures, velocities, flows and so on). Moreover, the performances of each single measuring device should be monitored and ensured by means of proper maintenance and periodical calibration. This requirement is valid in general, for both laboratory researches, industrial plants and field surveys.

About calibration, it could be interesting to have also an overview about the probability distributions of the corrections that accredited calibration laboratories provide by tests results for specific kinds of measuring devices, of course hiding customer name and device brand.

To this aim, the present paper shows a statistical analysis on experimental calibration data for flowmeters (i.e.: electromagnetic, ultrasonic, turbine flowmeters) in pressure pipes. The experimental calibration data set consists of the whole archive of the calibration tests carried out on 246 flowmeters from January 2001 to October 2015 at *Settore Portate* of *Laboratorio di Idraulica "G. Fantoli"* of *Politecnico di Milano*, that is accredited as *LAT 104* for a flow range between 3 l/s and 80 l/s, with a certified Calibration and Measurement Capability (CMC) - formerly known as Best Measurement Capability (BMC) - equal to 0.2% [9].

The study presented in this paper extends the analyses carried out by the Authors and previously published about the part of the current database that was already available in 2008 [1]. Moreover, now the topic is issued with an enhanced method of analysis.

IOP Conf. Series: Journal of Physics: Conf. Series **882** (2017) 012015 doi:10.1088/1742-6596/882/1/012015

2. Flowmeters calibration test facility

Figure 1 represents the scheme of the calibration facility at *Settore Portate* of *Laboratorio di Idraulica "G. Fantoli"* of *Politecnico di Milano* (*LAT 104*), which has been accredited since 2002 [4] by the Italian branch (formerly SIT, now ACCREDIA) of the European Accreditation chain [7], [8].

The adopted calibration method is quite simple, as it is based on the comparison between on the one hand the discharge Q_M measured by the instrument and on the other hand the discharge Q that is really flowing in the pipe on which the instrument is installed.

The real discharge Q is obtained measuring the volume stored, during a fixed time interval, in a prismatic calibrated tank having a capacity of 9.37 m³ and a maximum water depth of 1.15 m. This sample tank is referred to a certified primary specimen of unitary volume (that is 1 m³).

The calibration test phases are the following, according to the scheme reported by Figure 1:

- the discharge that is wanted to flow in the pressure pipe on which the instrument is installed, is diverted from the supplying constant-level tanks;
- the discharge flowing in the pressure pipe is adjusted by a valve placed downstream of the flow meter;
- moving the "start" lever, the flow-diverter starts to divert the flow towards the calibrated tank; before this operation, the discharge went directly to the recirculation system;
- at the same time of moving the "start" lever, a chronometer starts;
- various readings of the discharge indicated by the flowmeter display are carried out;
- at the end of a fixed time interval, the "stop" lever of the flow diverter is moved, in the way that the flow is directed towards the recirculation system; the chronometer stops automatically;
- the tank level raising, due to the water volume flowed into the tank during the fixed time interval, is measured through a staff gauge placed into an appropriate stilling well, and, knowing the tank surface, the real discharge Q is calculated.

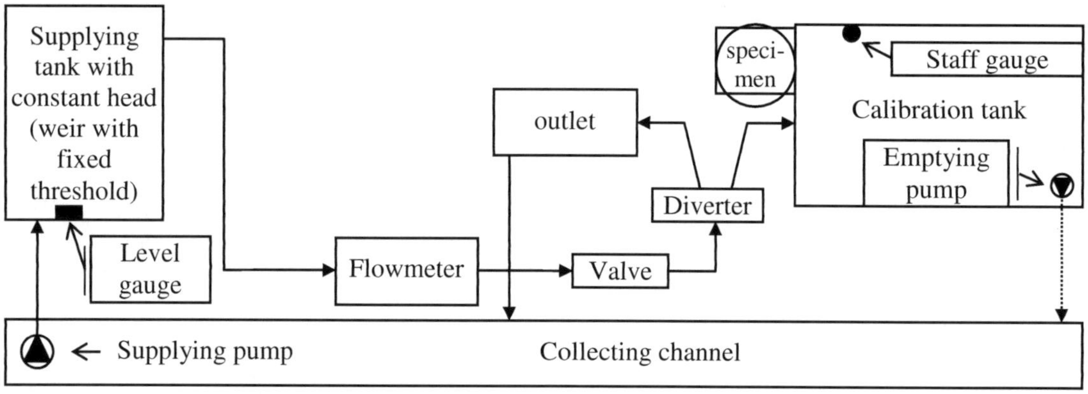

Figure 1. Scheme of the calibration facility at *Settore Portate* of *Laboratorio di Idraulica "G. Fantoli"* of *Politecnico di Milano*.

The test facility makes use of certificated instruments (tank, flow-diverter, chronometer, staff gauge, thermometer), that are subjected to periodic checks and calibrations.

IOP Conf. Series: Journal of Physics: Conf. Series **882** (2017) 012015 doi:10.1088/1742-6596/882/1/012015

During each flowmeter calibration, the performance of the gauge under test is usually evaluated for 5 different discharge values at least, comparing the Q_M value showed by the flowmeter with the correspondent real Q evaluated by the calibration tank. So, the correction $C = Q - Q_M$, is calculated.

Positive corrections ($C > 0$) indicate that the discharge measured by the instrument underestimates the real discharge, while negative corrections ($C < 0$) indicate a flow overestimation.

Each calibration certificate indicates flowmeter data and calibration environmental conditions, the values of the discharges Q_M and Q and of the correction C that testify the various calibration points and, last but not least, the adimensional ratio C/Q_M and the extended uncertainties [2] for the corrections C, both absolute $U(C)$ and adimensional $U(C/Q_M)$.

3. Experimental database

As already said, the experimental calibration data set consists of the whole archive of the calibration tests carried out on 246 flowmeters from January 2001 to October 2015 at *Settore Portate* of *Laboratorio di Idraulica "G. Fantoli"* of *Politecnico di Milano*, that is accredited as *LAT 104* for a flow range between 3 l/s and 80 l/s.

Just few of those flowmeters sustained a calibration test more than once from January 2001 to October 2015, but anyway all of their tests were considered exactly like the others, under the hypothesis that random alterations of their performances might have occurred because of the generally quite long period of time passed from one test to another.

Such a data sample have been split into three subsets, one per each of the three different kind of devices normally tested at that laboratory, respectively consisting in: 94 electromagnetic, 83 ultrasonic and 69 turbine flowmeters; then, each subset has been analysed separately from the others. Indeed, no other kind of flow meters (such as differential pressure flowmeters, Coriolis flowmeters, etc.) has been tested so far at *LAT 104* [6].

4. Data analysis method

For each single calibrated device, the following analyses on the test results have been considered about the values of the correction C, that is - as already said - the difference between the flow rate Q measured by the calibration facility (through the accredited procedures and the certified reference specimen) minus the flow rate Q_M contemporarily recorded by the flowmeter under calibration), expressed as the adimensional ratio C/Q_M:

- calculation of the mean value of the adimensional correction (with sign) $m_i(C/Q_M)$ for each generic i-th tested flowmeter, as the arithmetic mean of all the values of the adimensional correction C/Q_M for the test points (usually 5 points, that is 5 different flow rates for each tested flowmeter);
- screening and filtering of the mean values of the adimensional correction (with sign) $m_i(C/Q_M)$ through the Chauvenet criterion [5], in order to discard anomalous flowmeters of each one of the three subsets;
- calculation of both the mean value and the standard deviation of the mean values of the adimensional correction (with sign) $m(m_i(C/Q_M))$ and $s(m_i(C/Q_M))$ just for not discarded flowmeters;
- frequency distribution plots about the mean values of the adimensional correction (with sign) $m(m_i(C/Q_M))$ for not discarded flowmeters;
- implementation of the above described procedure also to the values of the adimensional correction expressed as absolute value $|C|/Q_M$ (that is without sign).

IOP Conf. Series: Journal of Physics: Conf. Series **882** (2017) 012015 doi:10.1088/1742-6596/882/1/012015

In particular, the Chauvenet criterion has been applied, for each one of the three data subsets separately (respectively about: electromagnetic, ultrasonic and turbines), as follows:

- a first estimation of the mean $m(m_i(C/Q_M))$ and the standard deviation $s(m_i(C/Q_M))$ of the N values of the elements $m_1(C/Q_M)$, $m_2(C/Q_M)$, …, $m_i(C/Q_M)$, …, $m_N(C/Q_M)$, where – as already said – each $m_i(C/Q_M)$ represents the global mean results of a specific calibration certificate;
- individuation of the maximum absolute difference between $m(m_i(C/Q_M))$ and each single value of $m_i(C/Q_M)$, that is $t = max\{|m_i(C/Q_M) - m(m_i(C/Q_M))|\}$;
- calculation of the probability value $P'(t)$, that is equal to the cumulated probability of occurrence of t under the hypothesis that the quantity $(m_i(C/Q_M) - m(m_i(C/Q_M)))\,/\,s(m_i(C/Q_M)$ is just a standard Gaussian random variable;
- calling $P(t) = (100\% - P'(t))$, the datum of $m_i(C/Q_M)$, that is the calibration certificate, which corresponds to $max\{|m_i(C/Q_M) - m(m_i(C/Q_M))|\}$ must be discarded from the sample if $N{\cdot}P < 0.5$;
- if this happens, then such a datum is discarded from the data sample, that therefore becomes made of just $N - 1$ left elements; so, new values of mean $m(m_i(C/Q_M))$ and standard deviation $s(m_i(C/Q_M))$ must be recalculated.
- such a procedure is iterated untill no further datum has to be rejected (maybe even the first turn could be already the right one, i.e. when no datum has to be discarded);
- after that, the same Chauvent filtering criterion is applied also to each one of the three data subsets separately (respectively about: electromagnetic, ultrasonic and turbines) of the adimensional correction expressed as absolute value $|C|/Q_M$ (that is without sign), in a similar way.

5. Results

Table 1, Table 2 and Table 3 show the summary of the results for the experimental data analysis carried out on the calibration certificates for respectively electromagnetic flowmeters, ultrasonic flowmeters and turbine flowmeters, for both the adimensional correction expressed with sign C/Q_M and the adimensional correction expressed as absolute value $|C|/Q_M$. In particular:

- the first column reports the number of actually considered certificates N, as the difference between the total number of certificates T minus the number of the discarded certificates D because of the Chauvenet filtering;
- the second column reports the mean value and the standard deviation of the mean values of the adimensional correction:
  - with sign: $m(m_i(C/Q_M))$ and $s(m_i(C/Q_M))$;
  - as absolute value: $m(m_i(|C|/Q_M))$ and $s(m_i(|C|/Q_M))$;
- the third column reports, in addition, also the mean value of the standard deviation (i.e: standard deviation s_i of the outcomes of the adimensional correction in a generic i-th certificate) and the standard deviation of the standard deviation:
  - with sign: $m(s_i(C/Q_M))$ and $s(s_i(C/Q_M))$;
  - as absolute value: $m(s_i(|C|/Q_M))$ and $s(s_i(|C|/Q_M))$;

Moreover, Figure 2, Figure 2 and Figure 2 show the summary of frequency distribution plots about the statistical analysis of the mean m_i of the adimensional corrections in the N actually considered certificates.

IOP Conf. Series: Journal of Physics: Conf. Series **882** (2017) 012015 doi:10.1088/1742-6596/882/1/012015

Table 1. Summary of the results for the experimental data analysis carried out on the calibration certificates for the electromagnetic flowmeters tested by *Settore Portate* of *Laboratorio di Idraulica "G. Fantoli"* of *Politecnico di Milano* (*LAT 104*).

Electromagnetic Flowmeters Tested by *Settore Portate* of *LAT 104* from January 2001 to October 2015						
Number of actually considered certificates $N = T - D$	Statistical analysis of the mean m_i of the adimensional corrections in the N actually considered certificates			Statistical analysis of the st. dev. s_i of the adimensional corrections in the N actually considered certificates		
88 = 94 – 6	Mean	$m(m_i(C/Q_M)) =$	-0.18%	Mean	$m(s_i(C/Q_M)) =$	0.66%
	St. Dev.	$s(m_i(C/Q_M)) =$	1.79%	St. Dev.	$s(s_i(C/Q_M)) =$	0.68%
87 = 94 – 7	Mean	$m(m_i(\|C\|/Q_M)) =$	1.24%	Mean	$m(s_i(\|C\|/Q_M)) =$	0.54%
	St. Dev.	$s(m_i(\|C\|/Q_M)) =$	1.19%	St. Dev.	$s(s_i(\|C\|/Q_M)) =$	0.55%

Table 2. Summary of the results for the experimental data analysis carried out on the calibration certificates for the ultrasonic flowmeters tested by *Settore Portate* of *Laboratorio di Idraulica "G. Fantoli"* of *Politecnico di Milano* (*LAT 104*).

Ultrasonic Flowmeters Tested by *Settore Portate* of *LAT 104* from January 2001 to October 2015						
Number of actually considered certificates $N = T - D$	Statistical analysis of the mean m_i of the adimensional corrections in the N actually considered certificates			Statistical analysis of the st. dev. s_i of the adimensional corrections in the N actually considered certificates		
82 = 83 – 1	Mean	$m(m_i(C/Q_M)) =$	+1.09%	Mean	$m(s_i(C/Q_M)) =$	1.65%
	St. Dev.	$s(m_i(C/Q_M)) =$	4.97%	St. Dev.	$s(s_i(C/Q_M)) =$	1.50%
81 = 83 – 2	Mean	$m(m_i(\|C\|/Q_M)) =$	3.68%	Mean	$m(s_i(\|C\|/Q_M)) =$	1.41%
	St. Dev.	$s(m_i(\|C\|/Q_M)) =$	3.04%	St. Dev.	$s(s_i(\|C\|/Q_M)) =$	1.36%

Table 3. Summary of the results for the experimental data analysis carried out on the calibration certificates for the turbine flowmeters tested by *Settore Portate* of *Laboratorio di Idraulica "G. Fantoli"* of *Politecnico di Milano* (*LAT 104*).

Turbine Flowmeters Tested by *Settore Portate* of *LAT 104* from January 2001 to October 2015						
Number of actually considered certificates $N = T - D$	Statistical analysis of the mean m_i of the adimensional corrections in the N actually considered certificates			Statistical analysis of the st. dev. s_i of the adimensional corrections in the N actually considered certificates		
69 = 69 – 0	Mean	$m(m_i(C/Q_M)) =$	-0.11%	Mean	$m(s_i(C/Q_M)) =$	0.70%
	St. Dev.	$s(m_i(C/Q_M)) =$	2.41%	St. Dev.	$s(s_i(C/Q_M)) =$	0.74%
69 = 69 – 0	Mean	$m(m_i(\|C\|/Q_M)) =$	1.73%	Mean	$m(s_i(\|C\|/Q_M)) =$	0.57%
	St. Dev.	$s(m_i(\|C\|/Q_M)) =$	1.75%	St. Dev.	$s(s_i(\|C\|/Q_M)) =$	0.63%

a)

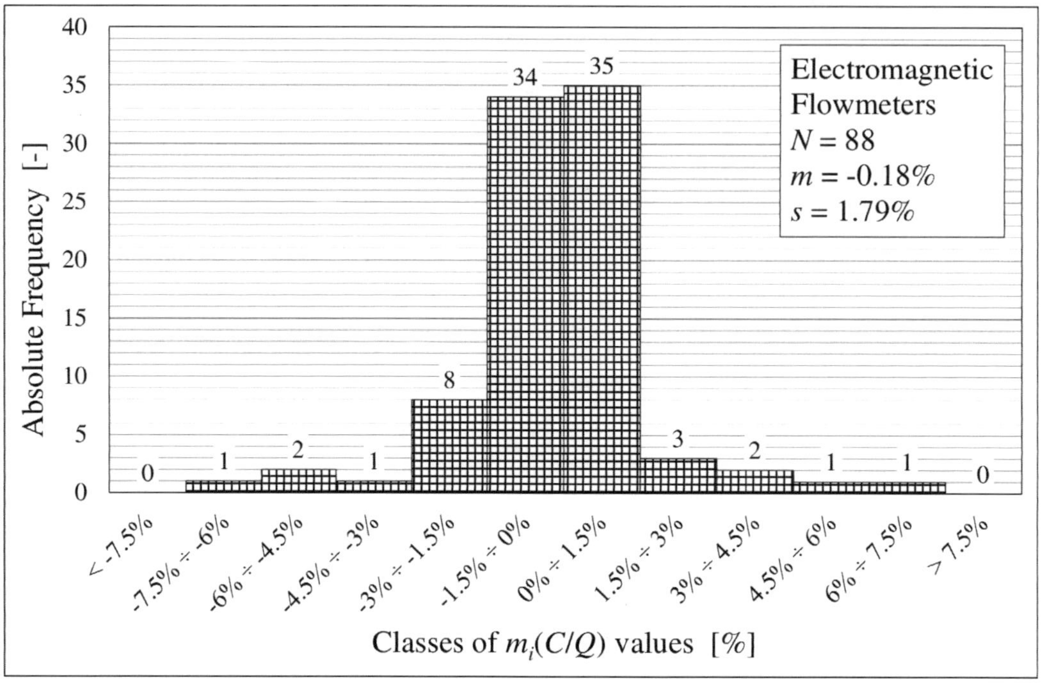

b)

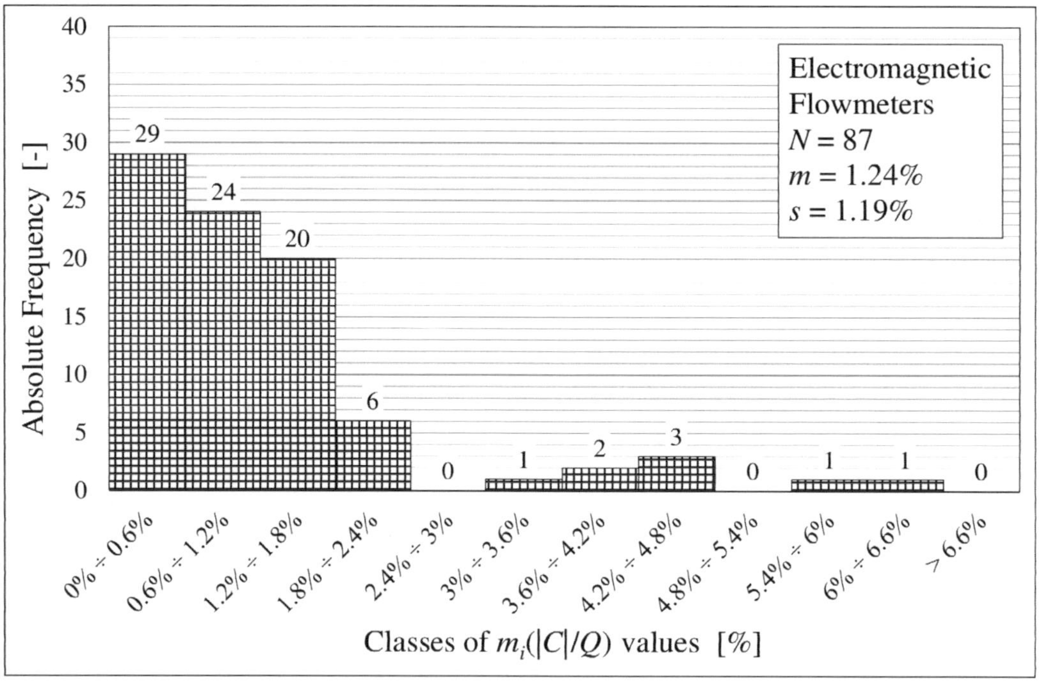

Figure 2. Frequency distribution plots about the results for the experimental data analysis carried out on the calibration certificates for the electromagnetic flowmeters tested by *Settore Portate* of *Laboratorio di Idraulica "G. Fantoli"* of *Politecnico di Milano* (*LAT 104*) from January 2001 to October 2015:
a) for adimensional values with sign C/Q_M;
b) for adimensional absolute values $|C|/Q_M$.

a)

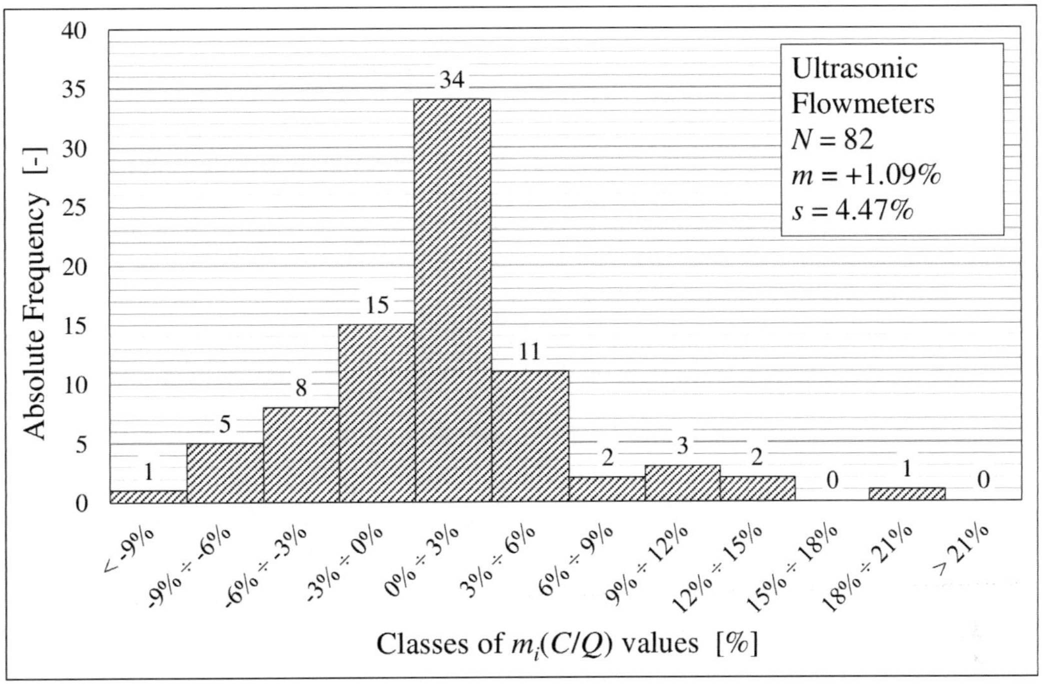

b)

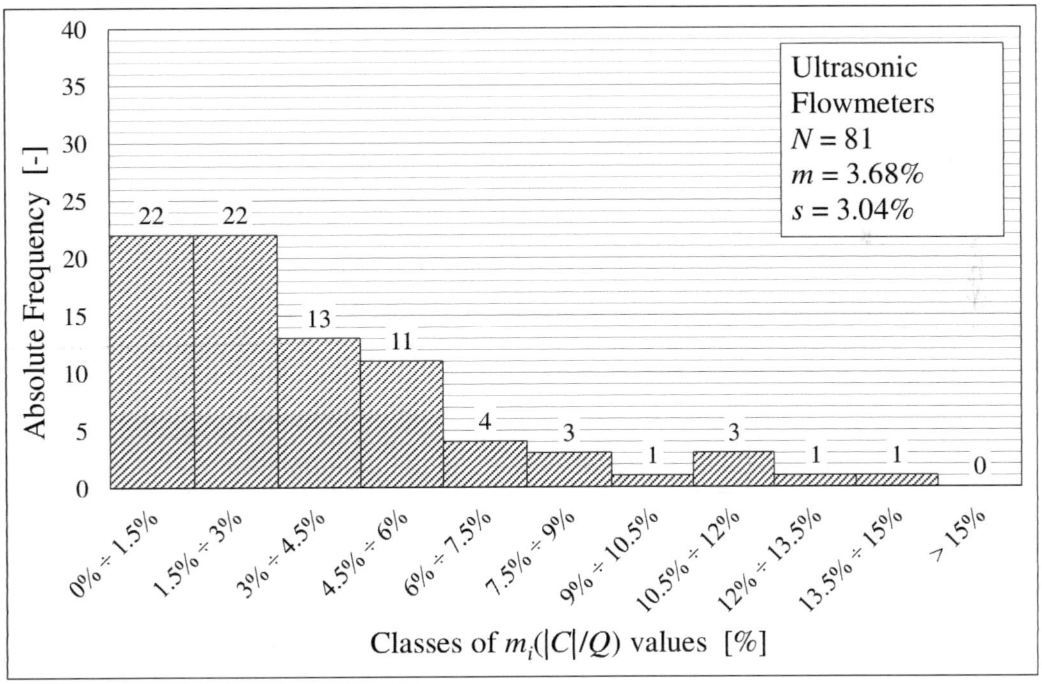

Figure 3. Frequency distribution plots about the results for the experimental data analysis carried out on the calibration certificates for the ultrasonic flowmeters tested by *Settore Portate* of *Laboratorio di Idraulica "G. Fantoli"* of *Politecnico di Milano* (*LAT 104*) from January 2001 to October 2015:
a) for adimensional values with sign C/Q_M;
b) for adimensional absolute values $|C|/Q_M$.

IOP Conf. Series: Journal of Physics: Conf. Series **882** (2017) 012015 doi:10.1088/1742-6596/882/1/012015

a)

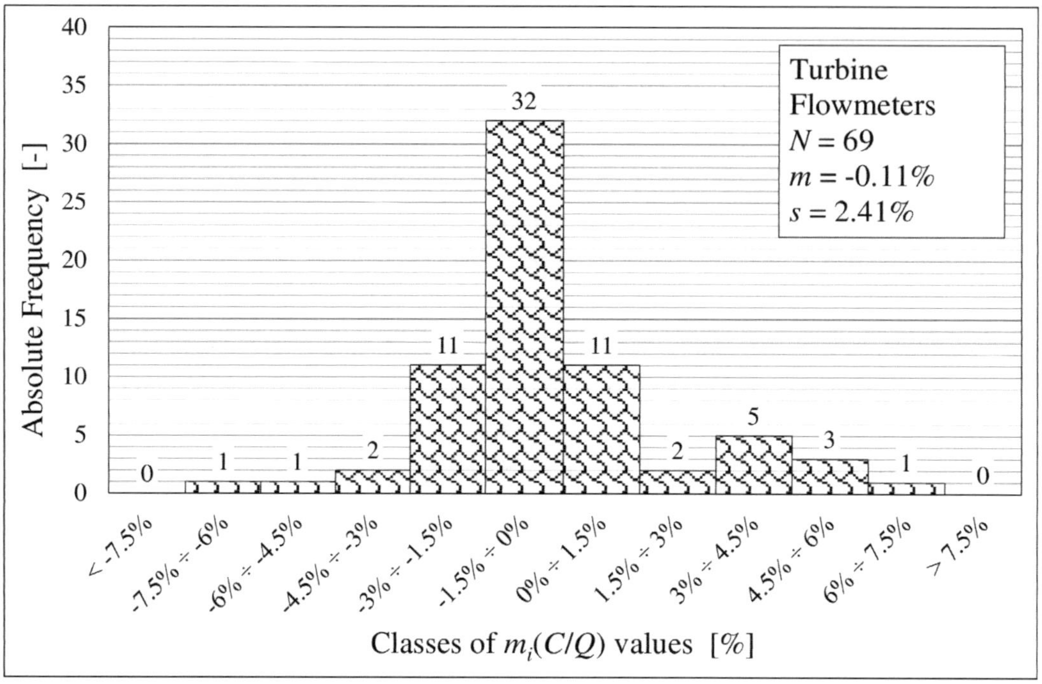

b)

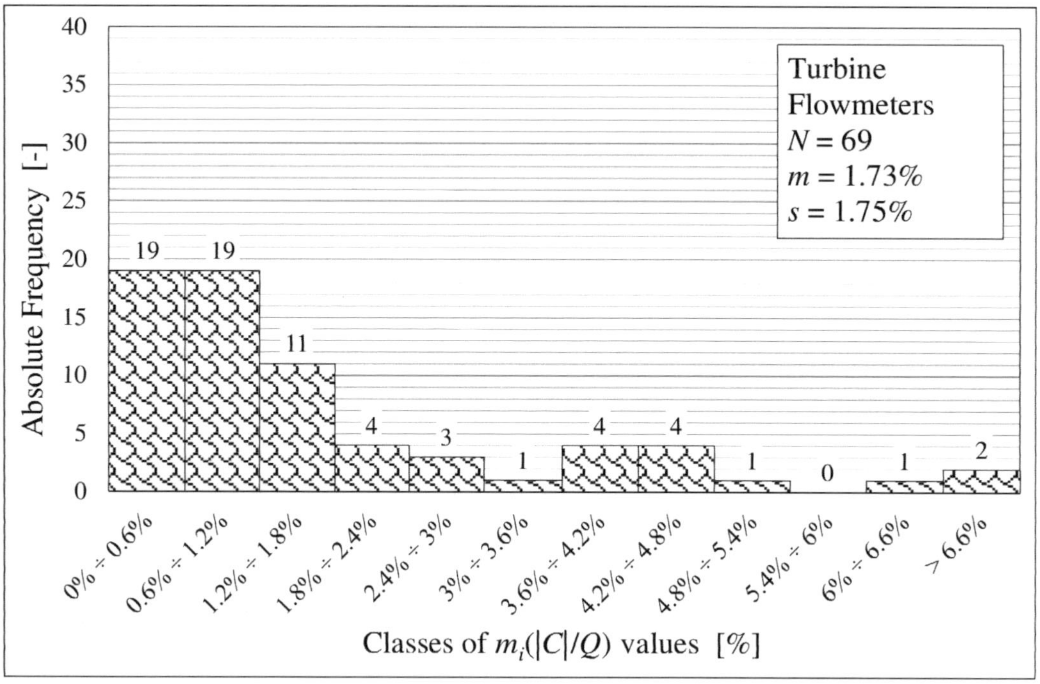

Figure 4. Frequency distribution plots about the results for the experimental data analysis carried out on the calibration certificates for the turbine flowmeters tested by *Settore Portate* of *Laboratorio di Idraulica "G. Fantoli"* of *Politecnico di Milano* (*LAT 104*) from January 2001 to October 2015:

a) for adimensional values with sign C/Q_M;

b) for adimensional absolute values $|C|/Q_M$.

IOP Conf. Series: Journal of Physics: Conf. Series **882** (2017) 012015 doi:10.1088/1742-6596/882/1/012015

6. Conclusions

The following conclusions can be drawn from the results of the statistical analysis presented here.

- The Chauvenet filtering criterion has discarded a significant number of certificates for electromagnetic flowmeters (6 out of 94 for C/Q_M and 7 out of 94 for $|C|/Q_M$), very few certificates for ultrasonic flowmeters (1 out of 83 for C/Q_M and 2 out of 83 for $|C|/Q_M$) and none for turbine flowmeters (0 out of 69 for both C/Q_M and $|C|/Q_M$); this implies that the performances among the different electromagnetic flowmeters that have been tested look less homogeneous in comparison to the other two kind of devices.
- Nevertheless, once the worst certificates have been the discarded by the Chauvenet filtering criterion, so just considering the remaining data, the standard deviation of the mean values for the tested electromagnetic flowmeters (1.79% for C/Q_M and 1.19% for $|C|/Q_M$) is slightly lower than the one for the tested turbine flowmeters (2.41% for C/Q_M and 1.75% for $|C|/Q_M$), while the highest standard deviation of the mean values is by far for the tested ultrasonic flowmeters (4.97% for C/Q_M and 3.04% for $|C|/Q_M$); this implies that the performances among just the electromagnetic flowmeters of the not discarded certificates look more homogeneous in comparison to what happens to the other two kind of devices (indeed especially in comparison to ultrasonic flowmeters).
- The mean value of the mean values of the adimensional correction (with sign) C/Q_M, again considering just not discarded flowmeters, comes out actually (as expected) close to 0% for the tested electromagnetic flowmeters (-0.18%) and turbine flowmeters (-0.11%), but not negligible for the tested ultrasonic flowmeters (+1.09%); but this crucial issue should be investigated with a larger data set than the available one, because it could be caused either by the unlucky contribute of some bad performing tested devices (although not discarded by the Chauvenet filtering criterion as the standard deviation for ultrasonic flowmeters is quite high) or by a really systematic problem.
- Frequency distribution plots about the mean values of the adimensional correction (with sign) C/Q_M, just for not discarded flowmeters, looks Gaussian for each one of the three subsets of tested flowmeters.
- In terms of the mean values of the absolute values of the corrections (that is $|C|/Q_M$), the best average performances seem to be the one of the tested electromagnetic flowmeters (1.24%), followed by the tested turbine flowmeters (1.73%) and finally by the tested ultrasonic flowmeters (3.68%).
- Frequency distribution plots about the mean values of the adimensional correction (as absolute values) $|C|/Q_M$, just for not discarded flowmeters, looks exponentially decreasing for each one of the three subsets of tested flowmeters.
- Both the mean of the standard deviation and the standard deviation of the standard deviation are quite low for the tested electromagnetic flowmeters (the mean of the standard deviation is 0.66% for C/Q_M and 0.54% for $|C|/Q_M$, while the standard deviation of the standard deviation is 0.68% for C/Q_M and 0.55% for $|C|/Q_M$) and for the tested turbine flowmeters as well (the mean of the standard deviation is 0.70% for C/Q_M and 0.57% for $|C|/Q_M$, while the standard deviation of the standard deviation is 0.74% for C/Q_M and 0.63% for $|C|/Q_M$); but they are significantly higher for the tested ultrasonic flowmeters (the mean of the standard deviation is 1.65% for C/Q_M and 1.41% for $|C|/Q_M$, while the standard deviation of the standard deviation is 1.50% for C/Q_M and 1.36% for $|C|/Q_M$); this suggest that the performances of each one of the tested ultrasonic flowmeters are more variable through the different test points (that is for the different flow rates) in comparison to the tested electromagnetic and turbine flowmeters.

Acknowledgments
The Authors wish to thank the Politecnico di Milano students Franco Gorla and Marco Trentin, who carried out most of the numerical computations for the statistical analyses on the experimental data [3].

References

[1] Camnasio E., Lazzarin A., Orsi E., Sanfilippo U., "Experimental Data on the Reliability of Flowmeters for Pressure Pipes", International Conference "IGHEM 2008", Innovation on Hydraulic Efficiency Measurement, 3-6 September 2008, Milan, Italy.

[2] European Accreditation, "Evaluation of the Uncertainty of Measurement in Calibration", EA-4/02 M: 2013.

[3] Gorla F., Trentin M., Analisi statistica su dati di taratura relativi a misuratori di portata per correnti in pressione, Thesis, Politecnico di Milano, 2016.

[4] Orsi E., Sanfilippo U., Taratura di misuratori di portata con metodo volumetrico e certificazione di qualità, in "Il Giornale dell'Ingegnere", N. 19, p. 11, Italy, 15 nov 2002.

[5] Taylor J.R., An Introduction to Error Analysis – The Study of Uncertainties in Physical Measurements, Second Edition, University Science Books, Sausalito, CA, United States,1997.

[6] https://en.wikipedia.org/wiki/Flow_measurement – Summary of flowmeters technologies.

[7] www.accredia.it – ACCREDIA official website.

[8] www.european-accreditation.org – European Accreditation official website.

[9] www.qualita.polimi.it/index.php?option=com_sppagebuilder&view=page&id=33 – Settore Portate of Politecnico di Milano LAT 104 official website.

IOP Conf. Series: Journal of Physics: Conf. Series **882** (2017) 012016 doi:10.1088/1742-6596/882/1/012016

Measuring the human psychophysiological conditions without contact

L Scalise, L Casacanditella, G Cosoli
Department of Industrial Engineering and Mathematical Sciences, Università Politecnica delle Marche, via Brecce Bianche, 60033 Ancona, Italy

E-mail: l.scalise@univpm.it

Abstract. Heart Rate Variability, HRV, studies the variations of cardiac rhythm caused by the autonomic regulation. HRV analysis can be applied to the study of the effects of mental or physical stressors on the psychophysiological conditions. The present work is a pilot study performed on a 23-year-old healthy subject.
The measurement of HRV was performed by means of two sensors, that is an electrocardiograph and a Laser Doppler Vibrometer, which is a non-contact device able to detect the skin vibrations related to the cardiac activity.
The present study aims to evaluate the effects of a physical task on HRV parameters (in both time and frequency domain), and consequently on the autonomic regulation, and the capability of Laser Doppler Vibrometry in correctly detecting the effects of stress on the Heart Variability.
The results show a significant reduction of HRV parameters caused by the execution of the physical task (i.e. variations of 25-40% for parameters in time domain, also higher in frequency domain); this is consistent with the fact that stress causes a reduced capability of the organism in varying the Heart Rate (and, consequently, a limited HRV).
LDV was able to correctly detect this phenomenon in the time domain, while the parameters in the frequency domain show significant deviations with respect to the gold standard technique (i.e. ECG). This may be due to the movement artefacts that have consistently modified the shape of the vibration signal measured by means of LDV, after having performed the physical task. In the future, in order to avoid this drawback, the LDV technique could be used to evaluate the effects of a mental task on HRV signals (i.e. the evaluation of mental stress).

1. Introduction

Several studies state that there is a significant relationship between autonomic nervous system and cardiovascular diseases, showing an association between arrhythmias and variations in the sympathetic and parasympathetic activities (1–5).
One of the most used markers of the relationship between cardiac activity and nervous system is Heart Rate Variability, HRV (6–9). It is the physiological phenomenon related to cardiac rhythm variability due to autonomic regulation; it is important not only for the prediction of cardiovascular risk (8), but also for the evaluation of different conditions, such as mental and physical stress (9–12), hypertension (13), diabetes (14) and pregnancy (10,15–17). The analysis of HRV allows to characterize the health and psychophysiological conditions of a subject, which cause changes in the balance between sympathetic and parasympathetic systems (18,19).
In the past, HRV has already been used to characterize the effects of physical or psychological stressors on the balance of the autonomic nervous system (20). In fact, stressors can be associated to

IOP Conf. Series: Journal of Physics: Conf. Series **882** (2017) 012016 doi:10.1088/1742-6596/882/1/012016

an increase in sympathetic cardiac control and/or to a decrease in parasympathetic control. In particular, several studies have focused on the effects of a physical effort on HRV (21–24). In (25), an inverse proportion between the intensity of the physical task and the variability of the heart rate has been reported. Luft et al. (22) have investigated the effects of physical efforts on the execution of cognitive tasks and the correspondent HRV variations; they have found no overall difference in cognitive performance pre- and post- test condition, and the revealed differences in HRV suggest an increased stimulation of both the sympathetic and parasympathetic systems. On the other hand, according to (26), the subject, after the execution of a physical task (i.e. in stress conditions), is expected to show an increased Heart Rate; in addition, the High Frequency (HF) component of HRV signal is supposed to decrease.

The present work is a pilot study (carried out on a participant) with the aim of evaluating the effects of a physical task on HRV. The measurement procedure involved the use of an electrocardiograph (ECG) and a Laser Doppler Vibrometer (LDV) for the non-invasive and non-contact assessment of the cardiac activity; LDV technique has already been demonstrated as an accurate and reliable means to assess physiological parameters (16,17,27–30).

2. Materials and methods

2.1 Measurement setup

The experimental setup included the following instruments:

- Laser Doppler Vibrometer (PDV 100, Polytec (31), sensitivity of 0.2V/(mm/s));
- Electrocardiograph (ECG) (MLA2540, ADInstruments (32), 5 Lead, Shielded Bio Amp Cable).

A sketch of the measurement setup is reported in 'figure 1'. LDV and ECG signals were acquired simultaneously using a proper A/D board (PowerLab 4/25T, ADInstruments (33), 12-bit).

At first, a 23-year-old subject (male) was measured in rest conditions by means of electrocardiography and Laser Doppler Vibrometry, whose main advantage is being a non-contact sensor (27,30,34)); the signal (i.e. a velocity signal) had a time duration of 6 minutes, which is suitable for HRV analysis (20). Then, the subject was asked to make a 2-minutes run. ECG and VCG signals were recorded after having performed the task. During both the acquisition periods, the subject was asked to sit and stay relaxed. The LDV was placed on a tripod at distance of 0.5 m from the subject, with the laser beam directed perpendicularly to the skin surface, in correspondence of the carotid artery, so as to record the vibrations transmitted by the passage of the blood pulse in the underlying vessel.

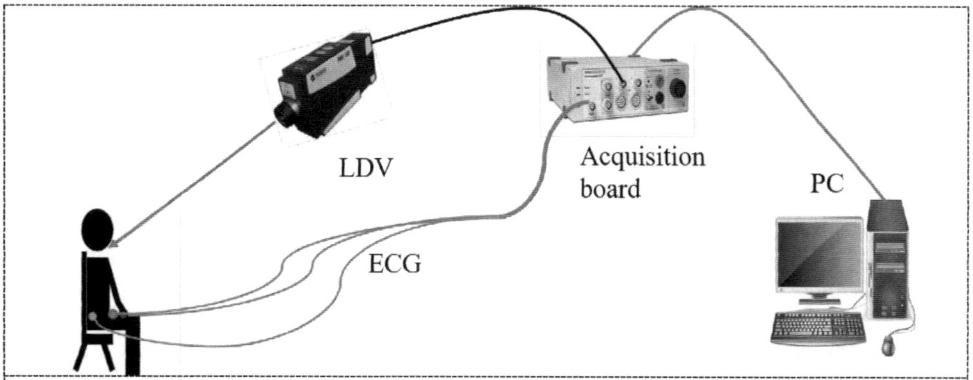

Figure 1. Sketch of the measurement setup

IOP Conf. Series: Journal of Physics: Conf. Series **882** (2017) 012016 doi:10.1088/1742-6596/882/1/012016

2.2 Data processing

The first step of data processing consisted in the detection of the main feature of ECG and LDV signals, that is the R-peak and the V-peak respectively, as shown in 'figure 2'.

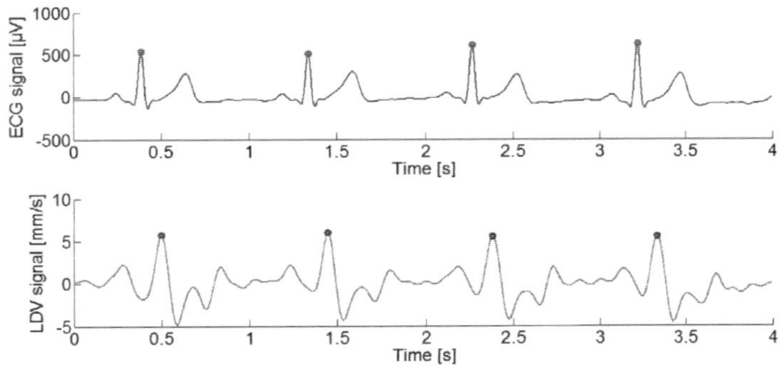

Figure 2. Example of main feature detection in ECG (top) and LDV (bottom)

The applied detection algorithm was developed by the authors on the basis of (35) and is based on the geometrical shape of the signals (36). In particular, it performs the computation of two straight lines on a small subset of points, by means of a sliding window. Then, the angle included between the two straight lines is computed; by detecting local minima of the angle, it is possible to locate the main peaks of the two considered signals (i.e. ECG and LDV).
The time location of ECG and LDV main peaks allowed to obtain the tachograms, i.e. the series of time intervals between two main features in a cardiac-related signal (37). Then, the HRV analysis was carried out by means of an open access software, named Kubios (38), considering both time and frequency domains.
Among all the obtainable parameters (30), in this work the authors considered the following features in time domain (13):

- RRavg [ms]: the average cardiac period;
- RMSSD [ms]: the square root of the mean squared differences of successive intervals between two main peaks;
- S^2 [ms^2]: the area of the ellipse obtained by means of Poincaré plot (14) (an example of this kind of representation is reported in 'figure 3').
-

On the other hand, with regard to the frequency domain, the Power Spectrum Density (PSD) of the tachogram was computed, assuming equidistant sampling by means of interpolating techniques (39). In particular, the authors focused the attention on the following indices:

- LF (Low Frequency): the percentage of the tachogram spectral content in the range 0.04-0.15 Hz; it is commonly related to the activation of the sympathetic nervous system;
- HF (High Frequency): the percentage of the tachogram spectral content in the range 0.15-0.40 Hz; it can be considered characteristic of the parasympathetic nervous system activation;
- LF/HF: the ratio of LF to HF, describing the balance between sympathetic and parasympathetic branches of the autonomic nervous system.

IOP Conf. Series: Journal of Physics: Conf. Series **882** (2017) 012016 doi:10.1088/1742-6596/882/1/012016

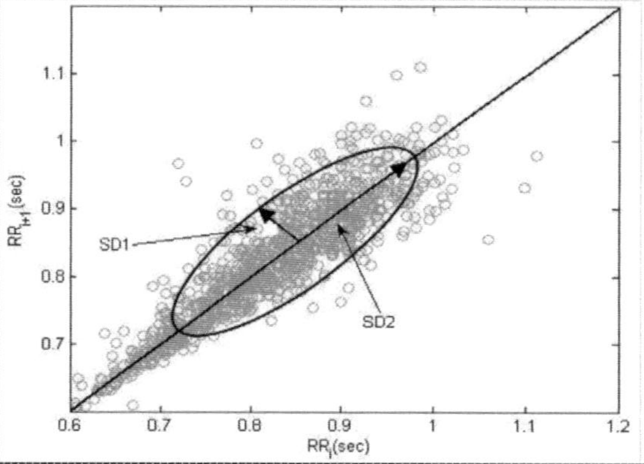

Figure 3. Example of Poincaré plot

3. Results

The chosen parameters (i.e. RRavg, RMSSD, S^2, LF, HF and LF/HF) were computed for both the signals, ECG and LDV, in both the studied conditions, i.e. at rest and after performing the physical task. This aimed to verify the effects of the physical task on the Heart Rate Variability, and to evaluate the capability of LDV technique in HRV analysis considering different operative conditions. In 'table 1' there are reported the results from ECG and LDV in time domain, while in 'table 2' the results in frequency domain are shown.

Table 1. HRV parameters in time domain, at rest and after the execution of the physical task

	Rest			After physical task		
	RRavg [ms]	**RMSSD [ms]**	**S^2 [ms^2]**	**RRavg [ms]**	**RMSSD [ms]**	**S^2 [ms^2]**
ECG	953	28	988	710	17	638
LDV	954	27	900	702	17	615

Table 2. HRV parameters in frequency domain, at rest and after the execution of the physical task

	Rest			After physical task		
	LF [%]	**HF [%]**	**LF/HF [ ]**	**LF [%]**	**HF [%]**	**LF/HF [ ]**
ECG	56	28	2.0	8	2	3.4
LDV	56	27	2.1	12	30	0.4

IOP Conf. Series: Journal of Physics: Conf. Series **882** (2017) 012016 doi:10.1088/1742-6596/882/1/012016

4. Discussion and conclusions

This pilot study aimed to evaluate the effects of the stress caused by a physical task on the heart rate variability and the capability of Laser Doppler Vibrometry, a fully non-contact technique, in assessing HRV in different conditions. The results from both the used sensors confirm the initial statements; in fact, after having performed the physical task, the subject showed a relevant percentage increase of the Heart Rate (i.e. 26 %) and a very significant decrease of HF content (i.e. 85 %) was computed in the PSD of the tachogram signal.
The measurement, performed on a 23-year-old healthy subject, at rest and after the execution of a physical task, showed variations in HRV parameters, in accordance with the literature (22,26).
Indeed, the comparison of HRV parameters derived from ECG signal in the two considered measurement conditions has put in evidence:

- a decrease of RRavg, RMSSD and S^2, which is consistent the reduction of HRV caused by the execution of the physical task;
- a decrease of HF content and an increase of LF/HF ratio, which is related to the reduced capability of the subject in varying the heart rate in response to the execution of the physical task.

LDV measurements provided results comparable with the gold standard method (i.e. ECG) in the time domain, while there are significant deviations in spectral parameters.
The aforementioned considerations allow the authors to state that HRV analysis in stress condition can be accurately performed in time domain by means of both ECG and LDV sensors. Anyway, the stress due to the physical task cannot be correctly quantified by considering the analysis of HRV in the frequency domain performed by means of LDV signals. This may be due to the movement artefacts caused by the strain of the subject after the execution of the physical task (in this case, the subject can be laboured, so that laser beam is not directed in a stable point).
However, it is of utmost importance to consider the LDV advantage of being a non-contact technique; in fact, this can be spent on carrying out measurements on particular subjects, such as (preterm) neonatal patients, who would be upset by the application of ECG electrodes.
Therefore, LDV technique should be further explored in the evaluation of stress through HRV analysis, trying to avoid the effects of movement artefacts; in this direction, another possible field of application could be the study of the effects of a mental task on HRV by means of the considered non-contact technique, so avoiding the problem linked to movement.

References

1. Lown B, Verrier RL. Neural activity and ventricular fibrillation. N Engl J Med. 1976 May 20;294(21):1165–70.
2. Du X-J, Dart AM. Role of sympathoadrenergic mechanisms in arrhythmogenesis. Cardiovasc Res. 1999 Sep 1;43(4):832–4.
3. Natali JES, Starzynski PN, Chaui-Berlinck JG. Oscillatory patterns in heart rate variability and complexity: A meta-analysis. Biomed Signal Process Control. 2017;33:66–71.
4. Cysarz D, Leeuwen PV, Edelhäuser F, Montano N, Somers VK, Porta A. Symbolic transformations of heart rate variability preserve information about cardiac autonomic control. Physiol Meas. 2015;36(4):643.
5. Akselrod S, Amitay Y, Lang RM, Mor-Avi V, Keselbrener L. Spectral analysis of left ventricular area variability as a tool to improve the understanding of cardiac autonomic control. Physiol Meas. 2000;21(2):319.
6. Kamath MV, Watanabe M, Upton A. Heart Rate Variability (HRV) Signal Analysis: Clinical Applications. CRC Press; 2012. 535 p.

IOP Conf. Series: Journal of Physics: Conf. Series **882** (2017) 012016 doi:10.1088/1742-6596/882/1/012016

7. Cerutti S, Baselli G, Civardi S, Furlan R, Lombardi F, Malliani A, et al. SPECTRAL ANALYSIS OF HEART RATE AND ARTERIAL BLOOD PRESSURE VARIABILITY SIGNALS FOR PHYSIOLOGICAL AND CLINICAL PURPOSES. In 1987. p. 435–8.
8. van Ravenswaaij-Arts CMA, Kollee LAA, Hopman JCW, Stoelinga GBA, van Geijn HP. Heart Rate Variability. Ann Intern Med. 1993 Mar 15;118(6):436–47.
9. Gang Y, Malik M. Heart rate variability in critical care medicine. Curr Opin Crit Care. 2002;8(5):371–5.
10. Lange S, van Leeuwen P, Geue D, Hatzmann W, Grönemeyer D. Influence of gestational age, heart rate, gender and time of day on fetal heart rate variability. Med Biol Eng Comput. 2005;43(4):481–6.
11. Calvo M, Gomis P, Romero D, Rolle VL, Béhar N, Mabo P, et al. Heart rate complexity analysis in Brugada syndrome during physical stress testing. Physiol Meas. 2017;38(2):387.
12. Visnovcova Z, Mestanik M, Javorka M, Mokra D, Gala M, Jurko A, et al. Complexity and time asymmetry of heart rate variability are altered in acute mental stress. Physiol Meas. 2014;35(7):1319.
13. Cardiology T force of the ES of, Pacing the NAS of, Electrophysiology. Heart rate variability standards of measurement, physiological interpretation, and clinical use. Eur Heart J. 1996;17:354–81.
14. Kamen PW, Tonkin AM. Application of the Poincaré plot to heart rate variability: a new measure of functional status in heart failure. Aust N Z J Med. 1995 Feb;25(1):18–26.
15. Carpenter RE, D'Silva LA, Emery SJ, Uzun O, Rassi D, Lewis MJ. Changes in heart rate variability and QT variability during the first trimester of pregnancy. Physiol Meas. 2015;36(3):531.
16. Scalise L, Morbiducci U. Non-contact cardiac monitoring from carotid artery using optical vibrocardiography. Med Eng Phys. 2008;30(4):490–497.
17. Morbiducci U, Scalise L, Melis MD, Grigioni M. Optical Vibrocardiography: A Novel Tool for the Optical Monitoring of Cardiac Activity. Ann Biomed Eng. 2007 Jan 1;35(1):45–58.
18. Friedman BH, Thayer JF. Autonomic balance revisited: Panic anxiety and heart rate variability. J Psychosom Res. 1998 Jan;44(1):133–51.
19. Miu AC, Heilman RM, Miclea M. Reduced heart rate variability and vagal tone in anxiety: Trait versus state, and the effects of autogenic training. Auton Neurosci. 2009 Jan 28;145(1–2):99–103.
20. Wiley: Dynamic Electrocardiography - Marek Malik, A. John Camm [Internet]. [cited 2017 Feb 24]. Available from: http://www.wiley.com/WileyCDA/WileyTitle/productCd-1405119608.html
21. Soares-Miranda L, Sattelmair J, Chaves P, Duncan G, Siscovick DS, Stein PK, et al. Physical Activity and Heart Rate Variability in Older Adults: The Cardiovascular Health Study. Circulation. 2014 May 5;CIRCULATIONAHA.113.005361.
22. Luft CDB, Takase E, Darby D. Heart rate variability and cognitive function: effects of physical effort. Biol Psychol. 2009 Oct;82(2):164–8.
23. Hottenrott K, Hoos O, Esperer HD. [Heart rate variability and physical exercise. Current status]. Herz. 2006 Sep;31(6):544–52.
24. Tonello L, Rodrigues FB, Souza JWS, Campbell CSG, Leicht AS, Boullosa DA. The role of physical activity and heart rate variability for the control of work related stress. Front Physiol [Internet]. 2014 Feb 21 [cited 2017 Feb 24];5. Available from: http://www.ncbi.nlm.nih.gov/pmc/articles/PMC3931195/
25. Heart Rate Variability Dynamics During Early Recovery After Different Endurance Exercises [Internet]. PubMed Journals. [cited 2017 Feb 24]. Available from: https://ncbi.nlm.nih.gov/labs/articles/17899162/
26. Vuksanović V, Gal V. Heart rate variability in mental stress aloud. Med Eng Phys. 2007 Apr;29(3):344–9.

27. Pinotti M, Paone N, Santos FA, Tomasini EP. Carotid artery pulse wave measured by a laser vibrometer. In: Third International Conference on Vibration Measurements by Laser Techniques: Advances and Applications [Internet]. International Society for Optics and Photonics; 1998 [cited 2015 Oct 12]. p. 611–616. Available from: http://proceedings.spiedigitallibrary.org/proceeding.aspx?articleid=954179
28. Cosoli G, Casacanditella L, Tomasini EP, Scalise L. Evaluation of Heart Rate Variability by means of Laser Doppler Vibrometry measurements. J Phys Conf Ser. 2015;658(1):12002.
29. Casaccia S, Sirevaag EJ, Richter E, O'Sullivan JA, Scalise L, Rohrbaugh JW. Decoding carotid pressure waveforms recorded by laser Doppler vibrometry: Effects of rebreathing. In: AIP Conference Proceedings [Internet]. AIP Publishing; 2014 [cited 2015 Oct 28]. p. 298–312. Available from: http://scitation.aip.org/content/aip/proceeding/aipcp/10.1063/1.4879596
30. Cosoli G, Casacanditella L, Tomasini EP, Scalise L. The non-contact measure of the heart rate variability by laser Doppler vibrometry: comparison with electrocardiography. Meas Sci Technol. 2016;27(6):65701.
31. GmbH P. Polytec: PDV-100 Portable Digital Vibrometer [Internet]. [cited 2016 Nov 30]. Available from: http://www.polytec.com/it/products/vibration-sensors/single-point-vibrometers/complete-systems/pdv-100-portable-digital-vibrometer/
32. Bio Amp Cables [Internet]. ADInstruments. [cited 2017 Feb 24]. Available from: https://www.adinstruments.com/products/bio-amp-cables
33. PowerLab [Internet]. ADInstruments. [cited 2017 Feb 24]. Available from: https://www.adinstruments.com/products/powerlab
34. Castellini P, Martarelli M, Tomasini EP. Laser Doppler Vibrometry: Development of advanced solutions answering to technology's needs. Mech Syst Signal Process. 2006 Aug;20(6):1265–85.
35. Hu X, Liu J, Wang J, Xiao Z, Yao J. Automatic detection of onset and offset of QRS complexes independent of isoelectric segments. Measurement. 2014 May;51:53–62.
36. Cosoli G, Casacanditella L, Pietroni F, Calvaresi A, Revel GM, Scalise L. A novel approach for features extraction in physiological signals. In: 2015 IEEE International Symposium on Medical Measurements and Applications (MeMeA). 2015. p. 380–5.
37. Engineering Approaches to Study Cardiovascular Physiology: Modeling, Estimation, and Signal Processing. Frontiers E-books; 136 p.
38. Biosignal Analysis and Medical Imaging Group - Kubios HRV [Internet]. 09:22:52.849189 [cited 2015 Dec 1]. Available from: http://kubios.uef.fi/
39. Merri M, Farden DC, Mottley JG, Titlebaum EL. Sampling frequency of the electrocardiogram for spectral analysis of the heart rate variability. IEEE Trans Biomed Eng. 1990 Jan;37(1):99–106.

IOP Conf. Series: Journal of Physics: Conf. Series **882** (2017) 012017 doi:10.1088/1742-6596/882/1/012017

Heavy Class Helicopter Fuselage Model Drag Reduction by Active Flow Control Systems

F De Gregorio

Italian Aerospace Research Centre, via Maiorise, 81043 Capua (CE), Italy

f.degregorio@cira.it

Abstract. A comprehensive experimental investigation of helicopter blunt fuselage drag reduction using active flow control is being carried out within the European Clean Sky program. The objective is to demonstrate the capability of several active flow technologies to decrease fuselage drag by alleviating the flow separation occurring in the rear area of some helicopters. The work is performed on a simplified blunt fuselage at model-scale. Two different flow control actuators are considered for evaluation: steady blowing, unsteady blowing (or pulsed jets). Laboratory tests of each individual actuator are first performed to assess their performance and properties. The fuselage model is then equipped with these actuators distributed in 3 slots located on the ramp bottom edge. This paper addresses the promising results obtained during the wind-tunnel campaign, since significant drag reductions are achieved for a wide range of fuselage angles of attack and yaw angles without detriment of the other aerodynamic characteristics.

1. Introduction

The performance of heavy transport helicopters, having a large, and almost flat, loading ramp suffer from poor aerodynamics of the aft body.

In cruise flight condition, the component of the rotor power due to counter balance the helicopters parasite force is of the order of 45% to 55% of the total requirements depending by the helicopter class, as observed in 1975 by Stroub and Rabbot [1] and later by Gatard et al in 1997 [2]. More recent fuselage and rotor head drag breakdown studies [3], [4], indicate that about 69% of the total parasite drag can be ascribe to the fuselage and the remaining 31% is due to the rotor head for heavy class helicopters. Further detailed drag breakdown (Figure 1) indicates that the basic fuselage contributes with about 19%, engine cowls by 11%, sponson about 5%, empennage with 7%, tail rotor head, cooling system and Aerials respectively for about 5%, 6% and 16%.

Consequently, fuselage drag reduction is one of the main objectives in order to improve helicopter performance, to reduce fuel consumption and environmental impact. Heavy helicopter fuselages are often characterized by a rear loading ramp that significantly affects the aerodynamic performance. The transport helicopter fuselage can be assimilated to a blunt body and the aft region is typically characterized by flow detachment at the lower corner, generating a separation bubble and a system of streamwise vortices at the sides. These vortical structures present some similarity to those separated from hatch-back cars [5]. The flow separation and the longitudinal vortex always cohabit on the aft region, when the flow separation is predominant the flow is named "eddy flow" otherwise it is called "vortex flow." The fuselage drag magnitude is strictly influenced by the hatch flow topology, being related with the fuselage incidence angle (α) and with the loading door upsweep angle (ϕ) as widely discussed by Seddon in 1990 [6]. Furthermore, a point not to be neglected is that the flow separation interacting with the fuselage tail induces strong vibration and submits the airframe to fatigue cycles.

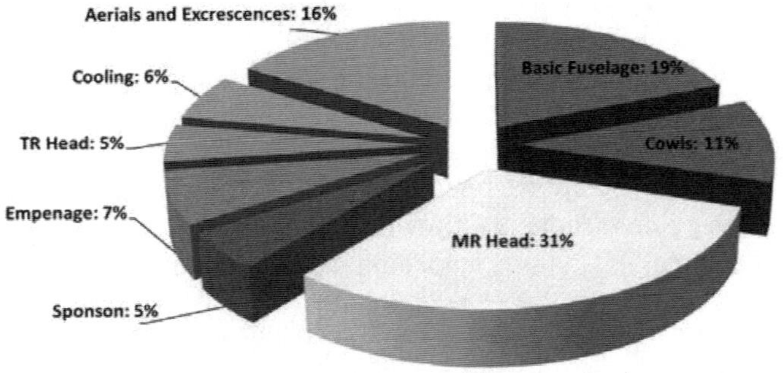

Figure 1. Heavy Class Helicopter parasite drag breakdown

Fuselage drag reduction can be obtained by means of an optimization of the aerodynamic design [7] - [8] and by improving the streamlined geometry [9](for example fishtailed geometry characterized by small upsweep angles ϕ=15° instead of blunt body fuselage with large upsweep angles ranging between 30° to 35°). In alternative, fuselage drag reduction can be pursued by using flow control systems [9], when the operative requirements prevent substantial modifications of the fuselage geometry or the retrofit of existing rotorcraft is required.

During the last years several research teams investigated the possibility to reduce helicopter fuselage parasite drag by means of active flow control (AFC) systems. Large interest was driven by synthetic jets (SJ) actuators. In 2005 Martin et al [11] investigated numerically and experimentally the influence of 12 SJ actuators on a helicopter fuselage obtaining drag reduction in the range between 6 and 10% and estimating the possibility to reduce by almost 40% the lift download. Analogous results were obtained by Ben-Hamau et al in 2007 obtaining drag reductions in the range between 3 to 11% for different attitudes and yaw angles [12] using blowing coefficient values ranging between $C\mu$: 0.025 to 0.05. In 2010 a NASA and ONERA collaboration [13] investigated numerically and experimentally the behavior of different AFC systems (steady blowing and SJ), obtaining remarkable results in drag reduction, with the steady blowing system able to reduce up to 35% the fuselage drag at a $C\mu$ of 0.06 and the synthetic jets inducing a decrement up to 26% operated at a $C\mu$ of 0.038. A contribution to understanding the ACF influence on the helicopter fuselage has been provided by Le Pape et al in 2013 [15] with their comprehensive work investigating its effect on fuselage drag and lift download. A further promising control system is the COMPACT (Combustion Powered actuation) [16], a novel technology which exploits the chemical energy of gaseous fuel/oxidizer mixture to create a high pressure burst and subsequent high momentum jet of exhaust products. In 2011 the chemical powered actuators were investigated by George et al [17] and by Woo et al [18] on the ROBIN fuselage model obtaining a drag reduction of the order of 12 to 17 % but also a significant increment of the lift download. The authors claimed that the COMPACT actuator is the possible solution to overcome the shortage of momentum of the SJ based on piezoelectric membranes. Another interesting actuator is the fluid oscillators investigated by Martin et al [19] on the ROBIN fuselage equipped of powered rotor. The results indicated, in some cases, a reduction of the total drag of the order of the 20% with respect to the baseline configuration.

Although many efforts have been made in the past, there is still a large interest and different flow control systems are being investigated in order to increase the fuselage performance. Additional efforts are necessary in order to understand the interaction between the flow topologies and the selected actuators. Previous experiments duly investigated the fuselage behavior for different attitude angles but many presented some shortcomings regarding the influence of the yaw angle on the effectiveness of the flow actuators and few discussed the effect on the lift download and on the pitching moment.

This research was driven by the interest to understand the flow topology of a heavy transport helicopter fuselage and in what manner it could be possible to obtain a drag reduction by using steady and unsteady blowing (or pulsed jet) actuators without penalizing the other aerodynamic characteristics. As initial activities, a comprehensive experimental and CFD investigation was conducted at laboratory level on a small simplified fuselage model equipped by steady blowing and pulsed jets at a Reynolds number based on the fuselage length of about 1 million [20]. The investigation provided a clear picture of the flow characteristics in the region of the loading ramp for different values of the incidence angle. The Wind Tunnel Test (WTT) delivered a better understanding of the effect of the steady and pulsed jets in alleviating the pressure drag inducing flow reattachment and deflecting the longitudinal vortex path. Furthermore, fundamental indications regarding the position and the direction of the jet slot were gathered.

This research foresees the experimental assessment of the flow control systems on a heavy class helicopter fuselage model at larger Reynolds number (Re: $8.2*10^6$). The paper first describes the simplified helicopter model and the wind-tunnel test set-up. A second part is dedicated to the presentation of the different actuators and their property and performance evaluation during specific laboratory tests. Wind-tunnel tests measurements that include aerodynamic loads acting on the fuselage, static pressure taps and PIV measurements are presented and discussed. An overall analysis of the results including actuators comparison with respect to achieved drag reduction is then presented.

2. Experimental Set up

An overview of the investigated test article, wind tunnel, experimental set up, model measurement instrumentation, flow control system and flow velocity measurement system is provided in this section.

2.1. Wind tunnel description

The test campaign was conducted at the RUAG LWTE wind tunnel, an atmospheric closed loop wind tunnel with closed test section.

Test section main sizes are: height of 5 m, width of 7 m and length equal to 11 m. The test section sizes assure negligible wall interferences on the load measurements (*Figure 6*). The ratio between model cross section and the tunnel test section is 0.45%, much smaller than the value of 5% where the wall interference becomes significant and wall corrections are necessary [21].

Maximum achievable flow speed is 70 m/s with a turbulence level of 0.3% on longitudinal velocity and 0.15% on the lateral velocity at 65 m/s. The full test campaign was carried out at constant speed of V_∞=50 m/s and Reynolds number, based on the fuselage length, of $8.2*10^6$. The fuselage was investigated varying the incidence angle in sweep mode between -12° to +16° at fixed yaw angle of β=0° and β=-5°, and sweeping the yaw angle β between -15° to +15° at incidence angle of α=0 and at cruise flight incidence angle of α=-3°. The model was mounted in the up-right position and supported by a dorsal strut in order to minimize disturbances in the regions of interest.

2.2. Simplified fuselage Model

The investigation was carried out on the basic fuselage model representing the main characteristics of a well-known heavy class helicopter. The term “basic fuselage” indicates that sponsons, cowls, empennage, landing gears, rotor head and aerial excrescences have been removed. The tested model was geometrically scaled 1 to 7 with respect to the full scale vehicle and it is characterised by a rear loading ramp having an up-sweep angle of ϕ=32°. The model is composed of an internal structure on which are mounted the external fuselage surfaces and the measurement instrumentations.

The model is equipped by 135 pressure taps (Figure 2). The pressure taps are located on the model nose, on the plane of symmetry, on several waterlines (y/H= 0, 0.26 and 0.54 where H is the fuselage height) and on the aft region at different stations (x/L=0.64, 0.68. 0.72 and 0.75 where L is the fuselage length) along the loading ramp and the tail boom. A single transition strip was placed on the nose cone to facilitate the laminar to turbulent transition on the fuselage.

IOP Conf. Series: Journal of Physics: Conf. Series **882** (2017) 012017 doi:10.1088/1742-6596/882/1/012017

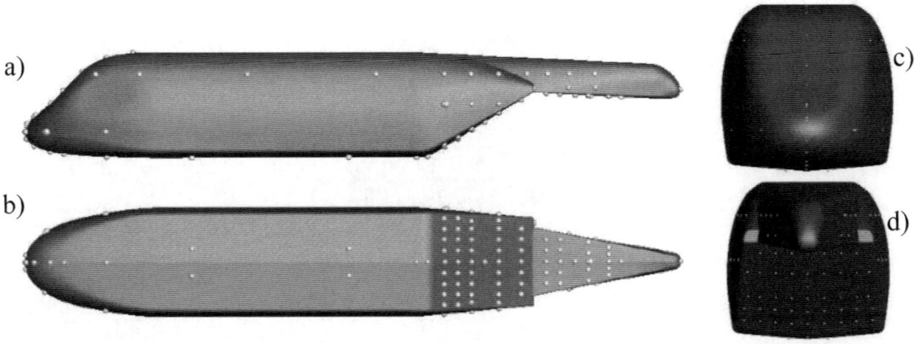

Figure 2: Model PTS location: lateral view (a), bottom view (b), front view(c) and rear view (d).

2.3. Model Instrumentation

The aerodynamic loads were measured by an internal six components balance, whose main characteristics in terms of full scale and accuracy are summarised in *Table 1*. The balance was installed in the model central part on a steel plate as close as possible to the mass centre.

Table 1: Balance type 192 RUAG main characteristics

	Fx [N]	Fy [N]	Fz [N]	Mx [Nm]	My [Nm]	Mz [Nm]
Design loads	350	250	1200	100	120	130
Accuracy	0.4	0.7	0.9	0.02	0.02	0.06

Two inclinometers with accuracy of 0.03° degree were installed to measure the pitch and roll attitude of the wind tunnel model. The yaw angle position was provided by the rotating model support. In addition, an electrolevel was used to reliably and quickly set the model to the zero reference pitch attitude and a further MEMS inclinometer was placed as backup. All these instrumentations were mounted on the aluminum plate available on the front of the fuselage (*Figure 3*).

A total of three pressure transducers modules with 64 ports each (with full scale of 1 psi and accuracy of 0.03 % FS) were used for measuring the 135 pressure ports. One pressure transducer module was mounted on the front part of the model to cover the nose pressure taps while the remaining two were mounted in proximity of the model aft where the PTS are concentrated.

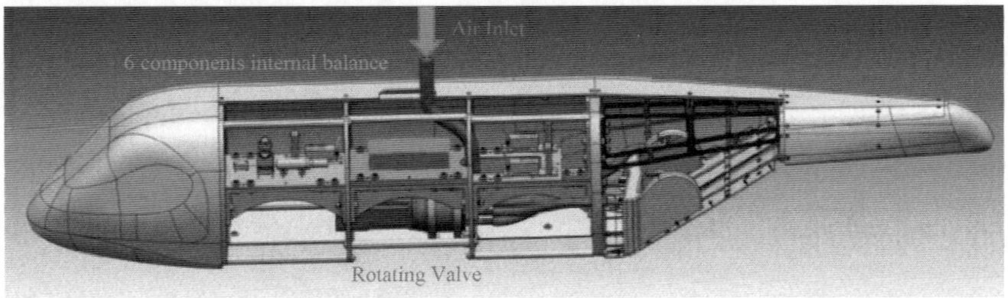

Figure 3*:* Model with instrumentation and flow control actuators iso-view

2.4. Flow control system description

The simplified fuselage model foresees three actuator slots, crosswise directed and located on the bottom of the ramp. Each blowing slot is characterised by length and width respectively of L_j=105 mm and W_j= 1.8 mm and directed with respect to the ramp surface of a jet angle of α_j=-37°. Each slot was connected to a dedicated pneumatic circuit feed by pressurized air. Up-stream of each slot a resonant cavity is present in order to level the flow coming out from the slots. Particular care in designing the pneumatic circuit was given in order to minimize the pressure losses.

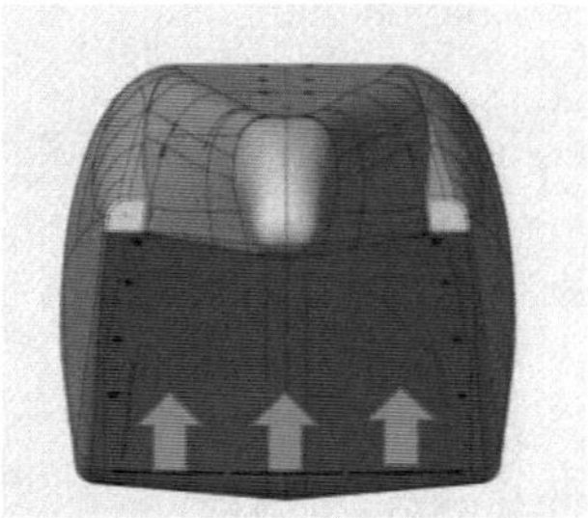

Figure 4: AFC operating scheme

Two different AFC systems were considered: steady blowing and pulsed jet operating from the rear bottom slots. The steady blowing jet was obtained feeding the cavity by external air supply through the model allowing several flow rates and blowing velocities to be tested. The pulsed jet was achieved modulating the supplied mass flow by means of a rotating valve. The valve consists of two concentric cylinders. The inner cylinder (or rotor) rotates around its axis of revolution driven by an electrical stepper motor. The Outer cylinder (or stator) is fixed. The inner cylinder contained 7 apertures with diamond shape and equally angular spaced on the same circumference. The outer cylinder contains 3 circular apertures along the circumference and in correspondence of the rotor apertures. The air transfer is obtained when the rotor apertures align with the stator apertures. The pulse frequency f_j is calculated multiplying the rotating speed by the number of rotor apertures $f_j=\omega*Ns$. Varying the rotating speed and the flow rate was possible to change the jet frequency and the blowing speed. *Figure 3* shows the rotating valve, inlet and outlet, tube routing and the plenum chambers mounted inside the model. The rotating valve supply the pressurised air to the three slots at the same time. A strong Coanda effect induces the jet flow attachment to the rear ramp (*Figure 4*).

2.5. Flow Measurements
The flow field characteristics downstream of the fuselage aft region were investigated by two PIV measurement systems: a standard two components PIV (2C-PIV) system and a stereo PIV (S-PIV) system. The stereo measurements were carried out at three different vertical cross planes at different distances from the model nose (x/L=0.75, 0.81, and 0.86) and respectively named PIV1, PIV2, PIV3. The two components measurements were performed on the symmetry plane (y/W=0) downstream the loading ramp and indicated as PIV4 (*Figure 5*).

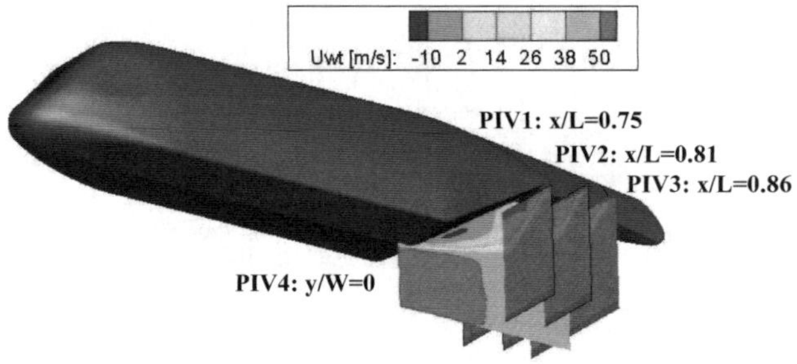

Figure 5: PIV recording region

The S-PIV system was composed of two Nd-Yag resonator heads providing a laser beam of about 320 mJ each at 532 nm and by two double frame CCD cameras (2048x2048 pixels) whereas the 2C-PIV system was composed of two Nd-Yag resonator heads (E=200 mJ at 532 nm) and by a single PIV sCMOS sensor camera (2560x2160 pixels). Particles of about 1 μm of diameter, composed of DEHS oil, were used as seeding. The seeding was injected downstream of the test section in order to obtain uniform seeding concentration of the full circuit. The lasers were located under the test section. The laser light sheet was projected upward into the test section trough an acrylic window installed in the test section floor. The light sheet optics were mounted on a linear traversing system remotely controlled in order to translate the light sheet along the wind tunnel longitudinal axis. Each recording camera belonging to the S-PIV was mounted on a 2D linear traversing system and located outside of the test section, inside the door frames of the side wall rear doors, downstream of the model. The traversing systems allowed to rigidly translate the cameras and light sheet plane without the need for additional calibrations of the stereo set up. Each camera was equipped with a motorised Scheimpflug support, 200 mm Canon EOS lens and lens remote control. The viewing angle between the stereo cameras was about 82°, close to the optimum values of 90°. The 2C-PIV camera equipped with an 80mm Canon lens which was mounted in a fixed position outside the left lateral wall behind a windows (*Figure 6*). The camera was directed towards the model loading ramp region.

Figure 6: Recording lay-out in WT test section.

3. Active Flow Control laboratory characterization.

Before the wind tunnel test campaign, particular care was taken for characterising the steady blowing and pulsed jet pneumatic system. A dedicated laboratory test campaign was carried out in order to obtain the transfer function relating the flow volume rate to the mean and maximum jet velocity for each single slot.

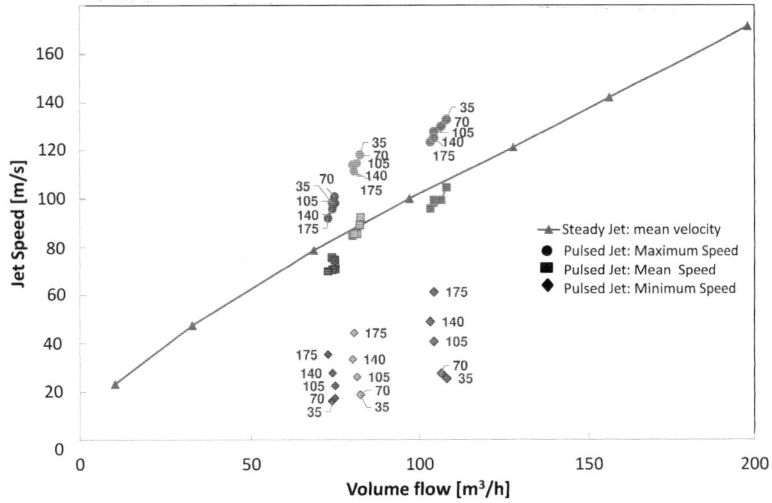

Figure 7: Steady and Pulsed Jet velocity vs the flow rate. Steady jet velocity are indicated by a triangle symbol, the pulsed jet maximum, mean and minimum speed are indicated respectively by circle, square and diamond symbols. The jet frequency is indicated next to the symbol.

IOP Conf. Series: Journal of Physics: Conf. Series **882** (2017) 012017 doi:10.1088/1742-6596/882/1/012017

The flow rate was measured by a flowmeter model SD8000 with measuring range between 0.25 to 225 m^3/h and accuracy equal to the 3% of the FS. The flow velocity at the different slot exit was measured by means of an IFA300 constant temperature anemometer system using single and double wire sensors.

The steady blowing jet speed was measured varying the volume flow rate. Analogously the velocity time history was measured varying the jet frequency and the volume flow rate for the pulsed jet actuators. The results indicate an almost linear behaviour of the mean jet speed varying the flow rate for the case of the steady blowing actuators (*Figure 7*). The pulsed jet characterization presents similar behaviour for the mean velocity while an influence of the jet frequency is evident on the maximum and minimum values. The results (*Figure 7*) present a reduction of the velocity amplitude increasing the jet frequency. The pulsed jet induces an increment of the maximum speed of about the 30-35% with respect to the steady jet for the same flow rate.

Once that the steady and pulsed jet were characterized in terms of mean and peak velocity in the full achievable frequency range it was possible to calculate the characteristic non dimensional quantities defined for evaluating the flow control systems [9]. In particular the jet frequency normalised as:

$$F^{+}=f_j \cdot W/V_\infty$$

where f_j is the pulse jet frequency, W the fuselage width and V_∞ the free stream velocity and the blowing coefficient c_μ, defined as the sum of the contribution of each blowing jet:

$$c_\mu = \frac{\sum_j \rho_j A_j V_J^2}{A_{CS} 0.5 \rho_\infty V_\infty^2}$$

where A_j, ρ_j and V_j are respectively the actuator slot surface, flow density and maximum speed and A_{cs} , ρ_∞ and V_∞ are the fuselage cross section, free stream density and velocity. The AFC laboratory characterization provided the achievable non dimensional parameters reported in *Table 2*.

Table 2: AFC normalised parameters

Parameters	**range**
Reduced Frequency < F^+>	0.15 – 1.4
Blowing momentum coefficient <$c\mu$>	0.02 – 0.1
Number of slots	5

4. Test Matrix

The test matrix foresaw first the investigation of the aerodynamic behaviour of the baseline model without AFC system varying the angle of attack at fixed yaw angle and for prefixed values of the angle of attack varying the yaw angle. The model incidence angle was varied in the range between -12° to +16° in sweep mode with an angle resolution of 0.1° at β=0 and β=-5°. Similarly the fuselage behaviour was investigated varying the yaw angle in sweep mode from β=-15° to β=15° at fixed attitude: null angle of attack (α=0°) and cruise condition (α=-3°).

Once that the baseline behaviour was assessed, the influence of the steady and pulsed jet was investigated varying the flow control parameters for all the selected incidence angles in the operating range reported in *Table 3*.

Table 3: Investigated AFC values.

Actuators	**Flow Volume [m³/h]**	**Blowing coefficient cμ [-]**	**Frequency [Hz]**	**F+ [-]**
Steady Blowing	50-60-80	0.005-0.007-0.012	-	-
Pulsed Jet	50-60	0.008-0.012	140	1.14

5. Results
The wind tunnel flow quality and experimental set up provided very high quality measurements, and an excellent test repeatability. The comparison between the aerodynamic forces and moments coefficient show a fairly good agreement. For example the drag coefficient discrepancy was much smaller of the CD accuracy $\pm 0.4 \cdot 10^{-5}$.

5.1. Baseline
The basic fuselage model was considered as baseline configuration and compared with the different flow control systems. The alpha sweep polar were carefully investigated (*Figure 8*). For confidentiality reason the discussion of the results is carried out normalizing the aerodynamic coefficient with respect to their values at null incidence angle (CD_0, CL_0 and CMY_0). The normalized drag coefficient indicates a reduction varying the model incidence from α=-12° to about α=-5°. Here the curve slope decreases in concomitance with the occurrance of the flow separation on the loading ramp for reaching a minimum value at about α=+3°. Further increasing the angle of attack, the drag shows a continuous increment up to α=+11.4° where an abrupt reduction is present with a minimum at about α=12.10°. At this point, the flow is fully attached. Increasing the angle of attack the drag coefficient presents a positive slope with a marked CD growth. The normalized lift coefficient shows a constant positive slope along the full sweep range, characterized by a slope reduction in the region affected by the flow separation. Similarly the flow separation region is visible also in the pitching moment behavior, where the curve slope increases in the range between α=-5° to α=12°.

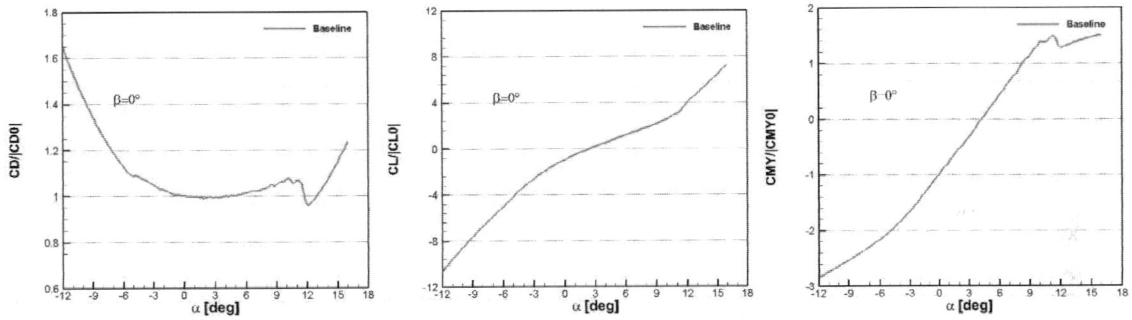

Figure 8: Baseline aerodynamic characterization: normalised drag, lift and pitching moment coefficient with respect to the corresponding values at α=0° varying the incidence angle at β=0°.

The surface pressure distribution measurements provide valuable information of the flow behavior. The longitudinal pressure distribution indicates the values of the incidence angle where the eddy flow condition occurs (left diagrams in *Figure 9*). For the cases of $\alpha \leq$-5° and $\alpha \geq$ 12°, the pressure coefficient presents a marked expansion in concomitance of the sweep angle followed by a pressure recovery along the length of the model, indicating a fully attached flow.

For the fuselage attitude interval between -5°<α<12°, immediately downstream the rear sweep angle, the pressure coefficient shows a constant value confined between x/L=0.63 to x/L=0.68 indicating separated flow on the ramp. Furthermore, the pressure data indicates that increasing the incidence angle the separated region presents an increment (right diagram in *Figure 9*).

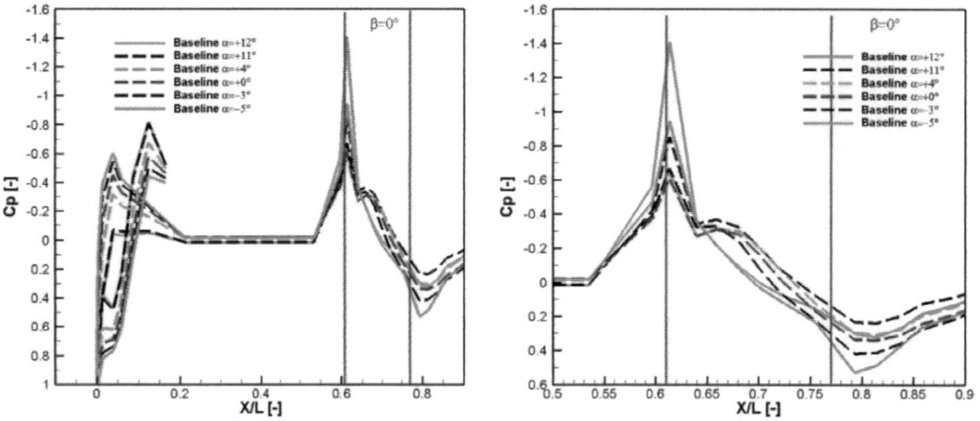

Figure 9: Baseline longitudinal pressure distribution at different angle of attach and β=0°. The boundaries of the loading ramp are indicated by the vertical red lines.

This result is confirmed by the flow velocity measurement on the rear ramp symmetry plane. At α=-3°, the ensemble average velocity field colour map shows a recirculating region extending from the bottom edge of the fuselage to about the 70% of the loading ramp. As the angle of attack is increased, the flow separation presents a growth up to reach at α=+7.3° the tail boom (*Figure 11*).

The spanwise pressure distribution (*Figure 10*), together with the cross flow measurement (*Figure 12*), provides a clear indications about the flow topology. At α=-8°, the spanwise pressure distribution at different stations (x/L=0.64, 0.68, 0.72, 0.75 and 0.86) shows a pressure recovery moving along the x direction, indicating an attached flow. Analysing the spanwise direction, an expansion toward the fuselage edge is encountered suggesting the presence of two longitudinal counter rotating vortices. At α=0°, the pressure distribution shows at the first two stations straight constant values representative of the presence of flow separation. Moving downstream the ramp, the pressure coefficient indicates a flow reattachment and the presence of longitudinal vortices. At α=12° straight spanwise distributions indicate the absence of vortex flow and the pressure recovery suggest a fully attached flow.

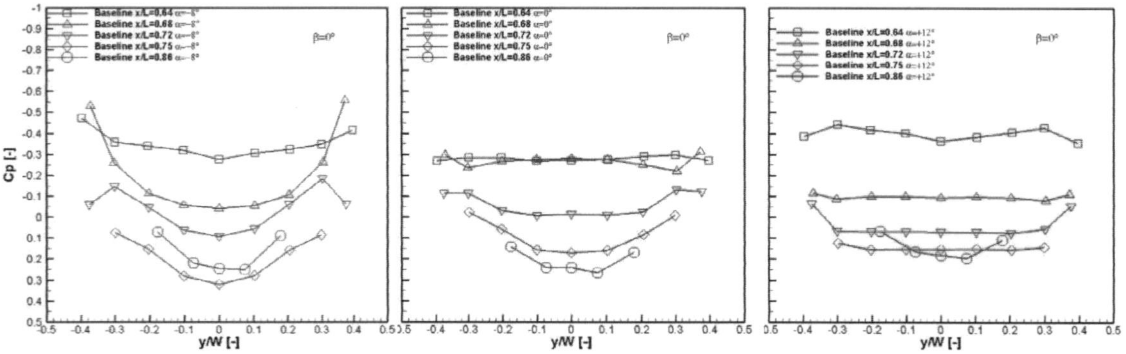

Figure 10: Baseline crosswise pressure distribution at different incidence angles. Left diagram α=-8°, central diagram α=0° and right diagram α=12°.

Also in this case the pressure indications are confirmed by the PIV results (*Figure 12*). For angle of attack of α=-3°, the cross flow velocity map indicates the presence of separated flow and two counter rotating vortices. Similar behavior is encountered at α=0° but in this case the vorticity colour maps suggests that weaker vortices are present. Flow velocity results at α=+4.5° and +7.3° indicate the presence of eddy flow and the absence of vortex flow.

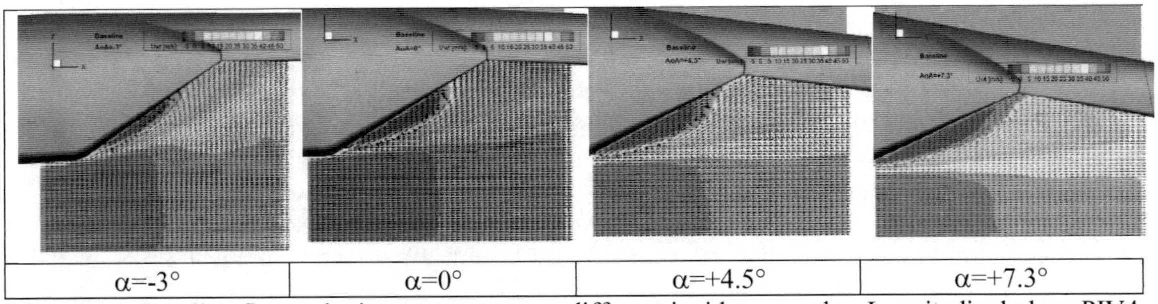

Figure 11: Baseline flow velocity contour map at different incidence angles. Longitudinal plane PIV4

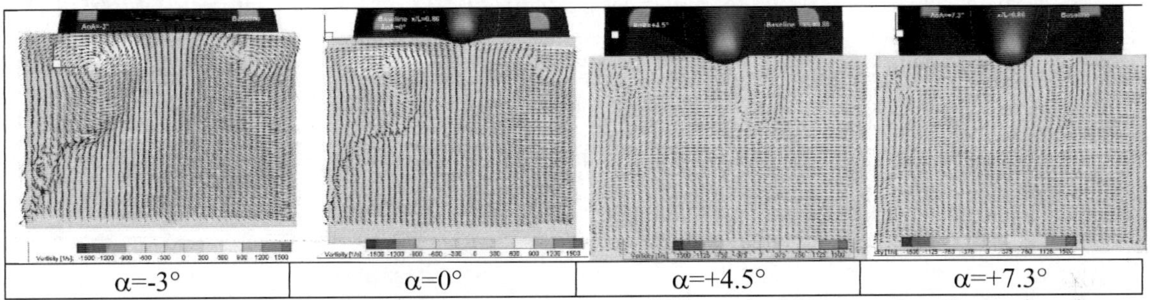

Figure 12: Baseline flow velocity field contour map at different incidence angles. Crosswise plane

Helicopters are often operated in side flow conditions. For this reason the fuselage behaviour was investigated sweeping the yaw angle between $-15° < \beta < +15°$ at constant values of the incidence angle ($\alpha=-3°$ and $\alpha=0°$). The drag coefficient presents a symmetric behaviour whereas the lift and pitching moments presents an hysteresis contribution (*Figure 13*). The side flow induces an increment of the fuselage drag as an increment of lift download.

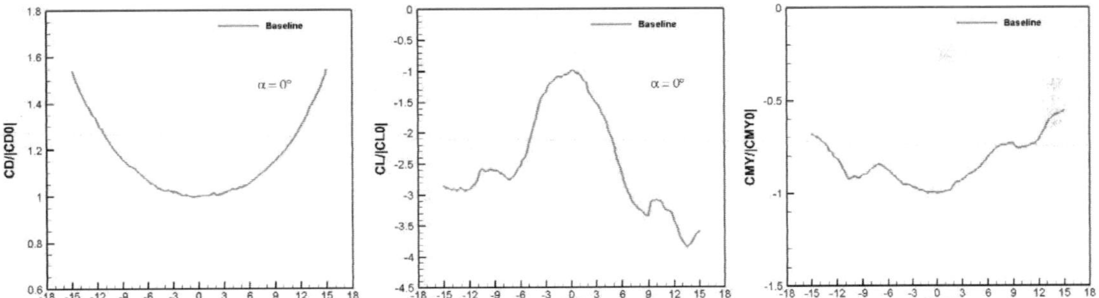

Figure 13: Baseline aerodynamic characterization: normalised drag, lift and pitching moment coefficients versus beta sweep angle at $\alpha=0°$.

5.2. Steady Blowing Actuator Results

The experimental assessment of the AFC systems in terms of drag reduction was the main scope of the project. At the same time, it was of great importance to verify that the benefits obtained by the drag alleviation were without detriments to the other quantities in particular lift and pitching moment coefficients.

Three different values of the blowing flow rates were investigated corresponding respectively to c_μ=0.005, 0.007 and 0.012. In the full incidence angle range, the diagram of the drag coefficient presents a clear benefit in terms of drag reduction varying from a minimum of 5% to a maximum of 22 % with respect to the baseline configuration (left diagram in *Figure 14*). The case characterised by a flow rate of Q_1=50 m^3/h corresponding to c_μ=0.005, shows a drag alleviation although the data still indicate that flow separation occurs in the range between α=-2° to α=+11°. The drag reduction is further enounced for Q_2=60 m^3/h (c_μ=0.007), but still limited flow separation occurs in the range between α=0° to α=+10°. The larger flow rate value (Q_3=80 m^3/h corresponding to c_μ=0.012) provides the best behaviour inducing a full flow reattachment on the fuselage.

The steady blowing system presents a positive effect on the lift coefficient as well. The steady blowing results indicate a benefit in terms of lift download reduction up to 70% for almost all the attitude angles except in the range between α=-5° to α=+1° where a maximum price of about 22% is paid in terms of download increment (central diagram in *Figure 14*).

Some consideration can also be made regarding the effect of the steady jet on the pitching moment behaviour. The steady blowing actuator generates a benefit increasing the aerodynamic stability of the fuselage. Mostly the pitching moment coefficient decreases with respect to the baseline and it presents an advantage in terms of longitudinal stability.

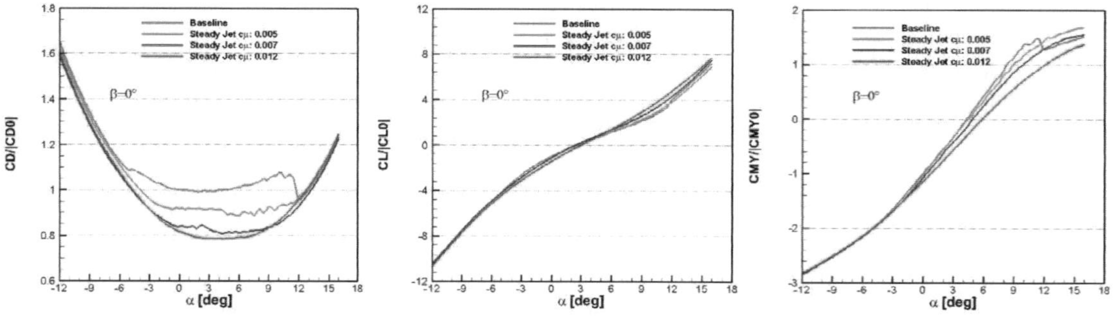

Figure 14: Steady blowing actuator contribution on the normalised aerodynamic coefficients

In detail the effect of the steady blowing actuators is described by the longitudinal and spanwise pressure distribution (*Figure 15*). The longitudinal pressure distribution shows a pressure recovery with respect to the baseline behaviour as the blowing coefficient increases. The lower value of the blowing coefficient is not sufficient to induce the flow reattachment partially obtained with c_μ=0.007 and fully reached using c_μ=0.012. The flow velocity map on the fuselage symmetry plane at α=-3° confirms this behaviour. The baseline presents a noticeable flow separation on the loading ramp, that almost disappears activating the steady blowing actuators at c_μ=0.007 (*Figure 20*). Just a limited circulation bubble is located on the conjunction between the ramp and the tail boom.

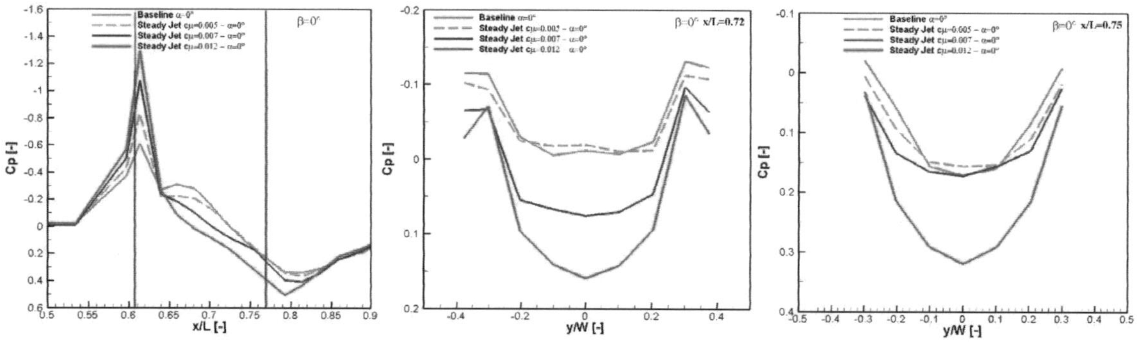

Figure 15: Steady blowing contribution to pressure distribution. Longitudinal (left diagram) and spanwise (centre and right diagrams) behaviour for different blowing coefficient.

The spanwise distribution at x/L=0.72 and 0.75 confirms the pressure recovery and the presence of two longitudinal vortices, characterised by two expansion peaks. The spanwise pressure distribution indicates a reduction of the expansion peaks due to the steady blowing actuators justifying the reduction of the fuselage pressure drag. The crosswise PIV plane at x/L=0.81 shows a fully attached flow and the presence of the two counter rotating vortices shedding by the fuselage. The results indicates an increment of the intensity of the vortices due to the steady blowing actuator and a displacement of the centre of the vortices moving apart by the symmetry plane (*Figure 21*).

The actuator effectiveness is demonstrated also on the sweep beta polar carried out at α=0°. The drag coefficient presents a reduction ranging between 6% to 16% with respect to the baseline configuration. The lift download is characterised by an increment up to a maximum value of 50% in the incidence range between α=-4° to α=+5° and a marked reduction up to 70% on the rest of the angle range. The pitching moment coefficient presents for the complete range of the yaw angle a reduction (*Figure 16*).

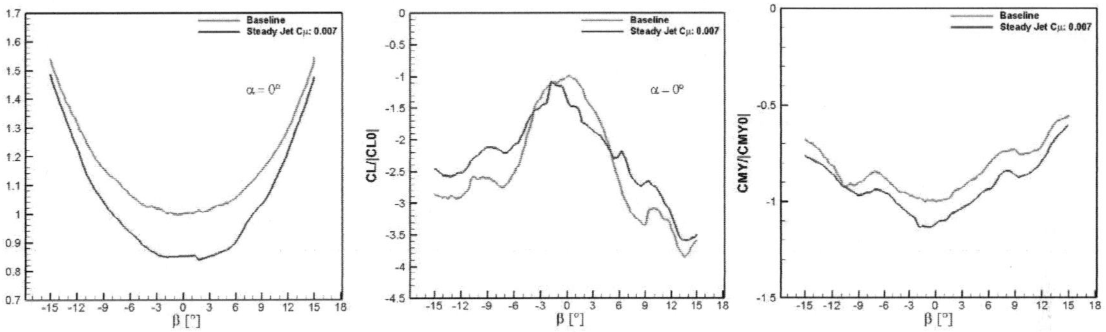

Figure 16: Steady blowing effect on the sweep yaw angle polars.

5.3. *Pulsed Jet Actuator Results*

The main characteristic of the unsteady blowing with respect with the steady blowing actuator is to provide larger momentum and blowing coefficient for equal flow rate used. In the following the discussion is centred on the pulsed jet assessment at flow rate of Q=60m^3/h and characterised by a total blowing coefficient of c_μ=0.012 versus the steady jet operated at flow rate of Q=80m^3/h and similar blowing coefficient.

The drag coefficient indicates a reduction with respect to the baseline ranging between 4% to 24% along the full incidence sweep. The pulsed jet, although fed by a flow rate quantity smaller by 25%, has almost the same behaviour if not slightly better with respect to the steady blowing system (*Figure 17*). Analogously to the steady blowing behaviour the lift coefficient presents improvements with respect to the baseline up to 100% for α<-7 and α>+6 and a loss in terms of download increment up to a maximum of 50%. The pitching moment coefficient shows a reduction of the diagram slopes indicating an increment of the longitudinal stability.

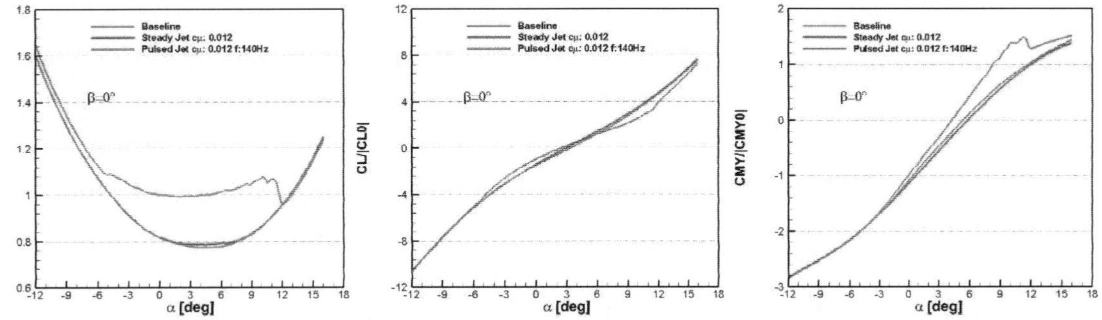

Figure 17: Pulsed jet and steady blowing contribution on the normalised aerodynamic coefficients.

The longitudinal and spanwise pressure distributions show a flow reattachment induced by the activated pulsed jets (Figure 18). The pressure track of the steady and pulsed jet are almost equivalent up to x/L=0.72. Downward the additional flow rate quantity of the steady blowing actuators further forces the pressure recovery. The wake investigation downstream the rear part of the fuselage provides similar results. The flow separation is reduced and almost removed by the pulsed jets except on the edge between the ramp and the tail boom (*Figure 20*). The longitudinal vortex flow detected on the crossplane at x/L=0.81 presents a behaviour similar to that one induced by the steady blowing. Larger vorticity intensity and core displacement moving apart from the symmetry plane (*Figure 21*).

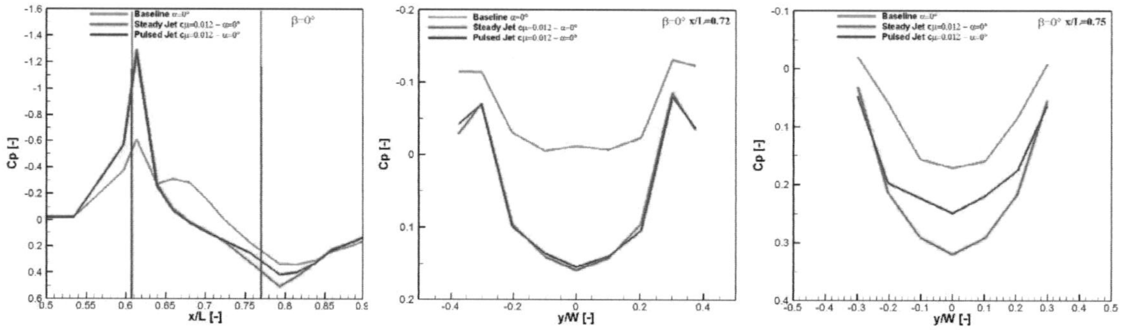

Figure 18: Pulsed Jet and Steady Blowing contribution to pressure distribution. Longitudinal (left diagram) and spanwise (x/L=0.72 on centre and x/L=0.75 on the right) behaviour at c_μ=0.012.

The pulsed jet effectiveness is demonstrated also on the sweep of the beta polar carried out at α=0°. The drag coefficient presents a reduction ranging between 8% to 20% with respect to the baseline configuration and a better behaviour regarding the steady blowing. The lift download is characterised by an increment up to a maximum value of 70% in the incidence range between α=-5° to α=+5.5° and a marked reduction up to 80% on the rest of the angle range. The pitching moment coefficient presents a reduction for the complete range of the yaw angle (*Figure 19*).

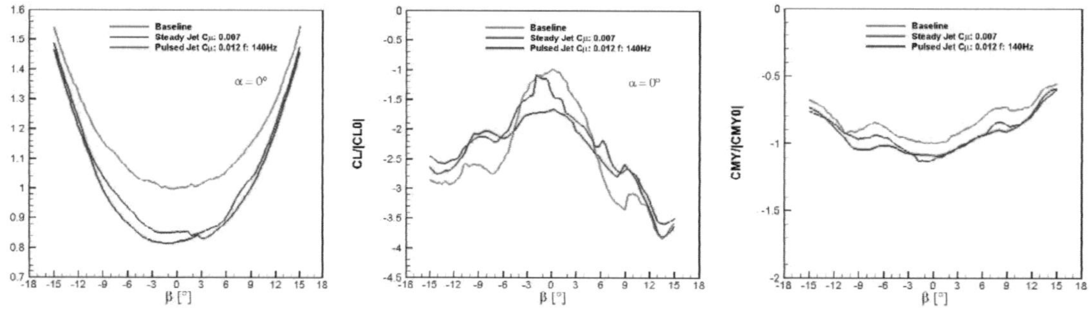

Figure 19: Pulsed Jet effect on the sweep yaw angle polar.

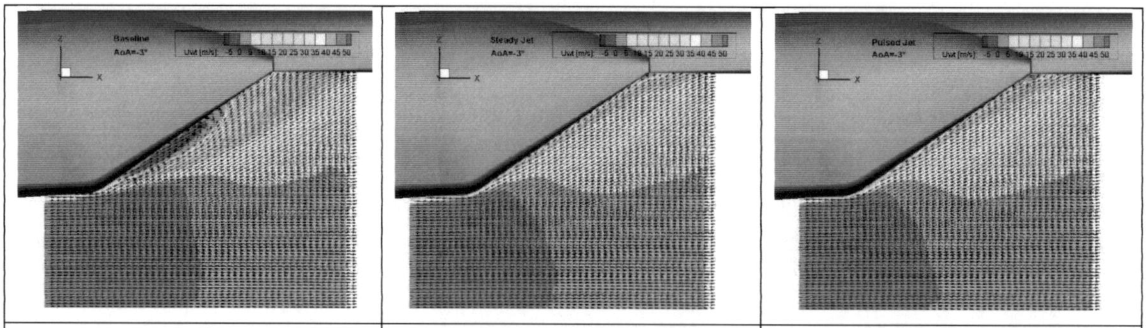

Baseline	Steady Blowing (c_μ=0.007)	Pulsed Jet (c_μ=0.012)

Figure 20: Flow axial velocity contour map comparison: Baseline, steady blowing and pulsed jet.

IOP Conf. Series: Journal of Physics: Conf. Series **882** (2017) 012017 doi:10.1088/1742-6596/882/1/012017

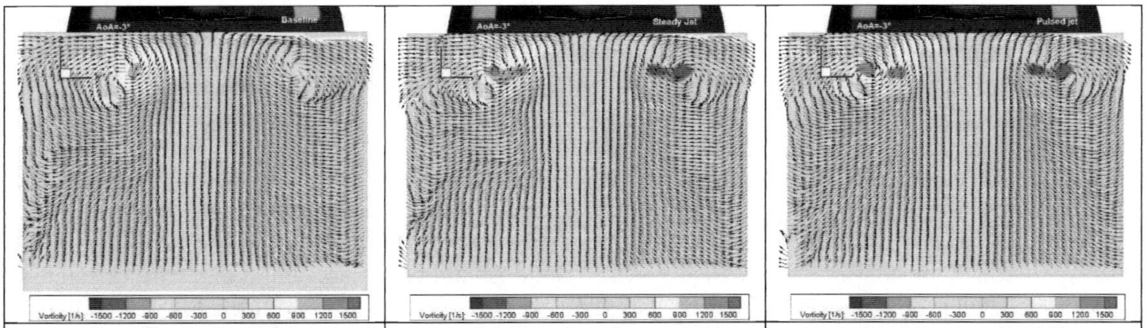

Baseline x/L=0.81	Steady Blowing (c_μ=0.007)	Pulsed Jet (c_μ=0.012)

Figure 21: Crossplane velocity components and vorticity colour map at x/L=0.81: Baseline, Steady Blowing and Pulsed Jet

6. Conclusions

A comprehensive test campaign was successfully performed at the RUAG LWTE wind tunnel aimed to investigate the possibility to reduce the helicopter fuselage passive drag by means of different flow control systems. Two types of AFC systems were designed, manufactured and tested: steady blowing and pulsed jets operating from the bottom of the rear fuselage hatch.

The comprehensive test campaign has foreseen the following activities:

- Aerodynamic model force and moments measurements;
- Model surface pressure characterization;
- Wake flow field measurements by 2C and 3C PIV measurements.

Both flow control systems provided remarkable results in terms of drag reductions without almost any detriments of the other aerodynamic quantities. The flow control systems were effective for all the investigated incidence and yaw angles.

The flow control systems were successful for values of the blowing coefficient smaller respect the values reported in the literature [13] and [15] and with respect to the values measured during the previous laboratory test [20].

The following main results are reported:

1. The steady blowing actuators, operated from the ramp bottom at different values of the blowing coefficient, induce a model drag reduction with respect to the baseline configuration from a minimum of 5% to a maximum of 23 % for the complete incidence model range.
2. The steady blowing actuator generates a benefit in terms of lift download reduction up to a maximum of 100% for the total angle of attack range except between $-5° < \alpha < +2°$ where the lift download can increases up to a maximum of 50%.
3. Some consideration may be made regarding the effect of the steady blowing on the pitching moment behaviour. The slope of the pitching moment diagram decreases with respect to the baseline and it presents an advantage in term of longitudinal stability.
4. The steady blowing system presents benefits in terms of aerodynamic coefficients also for the complete sweep of the beta polar. In particular the results show a drag reduction with respect to the baseline configuration between 6 to 16 %, benefits and some drawback are encountered for the lift coefficients and reduction of the pitching moment diagram slopes increasing the longitudinal stability.
5. The pulsed jet actuator operated at different values of the blowing coefficient and jet frequency of fj=140 Hz, induces a model drag reduction with respect to the baseline configuration between 4 to 24% for the model incidence range of $-12° < \alpha < 12°$. For values of $\alpha > 12°$ a small increment of about 1 % is present.
6. The Pulsed Jet actuator induces a remarkable benefit in terms of lift download reduction up to 100% for the total angle of attack range except between $-7° < \alpha < +6°$ where the lift download can

increases up to a maximum of 70%. With respect to the steady blowing the lift detriment is increased.

7. Analogously, the pulsed jets induce a model drag reduction with respect to the baseline configuration between 8 to 20% also for the complete sweep of the beta polar.
8. The longitudinal pressure measurements clearly indicate a flow separation reduction and an increment of the pressure recovery on the loading ramp due to the flow control systems. The spanwise pressure distribution indicates the flow topology occurring on the loading ramp, whether eddy or vortex flow and the effect induced by the AFC systems.
9. The flow field measurements indicate clear wake alleviation in terms of size and momentum loss due to the flow control systems and confirms the indication provided by the pressure measurements regarding the flow separation and topology.
10. The vortices development for the different model attitudes has been measured and vortex growth and the dissipation phenomena can be investigated.
11. A comprehensive experimental data base has been generated for future comparison with the CFD simulations.

Acknowledgement
The work has been partially founded by the European Union in the framework of the Clean Sky Joint Technology Initiative under grant agreement number: CSJU-GAM-GRC-2008-01. The authors would like to acknowledge P.L. Vitagliano, G. Ceglia, G. Spaziani, D. Steiling and the LWTE crew for the valuable support provided during the measurement campaign.

References

[1] Stroub RH and Rabbot JP 1975 Wasted Fuel another Reason for Drag Reduction, in proceddings of 31 American Helicopter Society, Washington
[2] Gatard J, Costes M, Kroll N, Renzoni P, Kokkalis A, Rocchetto A, Serr C, Larrey E, Filippone A, and Wehr D 1997 High reynolds number helicopter fuselage test in the ONERA F1 Pressurized Wind Tunnel *in 23rd European Rotorcraft Forum* Paper 167 Dresden Germany
[3] Renaud T, Le Pape A and Pèron S, 2012 Numerical analysis of hub and fuselage drag breakdown of a helicopter configuration *proc. 38th European Rotorcraft Forum*, September 4-7 Amsterdam, The Netherlands
[4] Grawunder M, Reß R, Breitsamter C and Adams N A 2012 Flow Characteristics of a Helicopter Fuselage Configuration Including a Rotating Rotor Head, *proc. 28th int. congress of the aeronautical sciences* Brisbane Australia
[5] Ahmed S R 1983 Influence of base slant on the wake structure and drag of road vehicles. *J. Fluids Eng.* **105**, 429–434
[6] Seddon J 1990 Basic helicopter aerodynamics *BSP Professional books* ISBN 0-632-02032-6.
[7] Reß R, Grawunder M and Breitsamter C 2014 Parasite drag reduction of a twin engine lightweight helicopter configuration *proc.In: German Aerospace Congress,* Augsburg, DGLRK2014–0046 pp. 1-9
[8] Breitsameter C, Grawunder M and Reß R Aerodynamic design optimization for a helicopter configuration including a rotating head *proc. 29th int. congress of the aeronautical sciences*, San Petersburg, Russia ICAS 2014_0686_paper
[9] Batrakov A S, Kusyumov A N, Mikhailov S A, Pakhov V V, Sungatullin A R, Zherekhov V V and Barakos G N 2013 A Study in helicopter fuselage drag *proc. 39th European Rotorcraft Forum* Moskow Russia
[10] Gad-el-Hak M 1996 Modern developments in flow control. *in Applied Mechanics Reviews*, vol. **49**, pp. 365–379
[11] Martin PB, Tung C, Hassan AA, Cerchie D and Roth J 2005 Active flow control measurements and CFD on a transport helicopter fuselage, *proc. 61st American Helicopter Society*, Grapevine, Texas vol. 1 pp 131-151

[12] Ben-Hamou E, Arad E and Seifert A 2007 Generic transport aft-body drag reduction using active flow control *Flow Turbulence Combust* **78** 365–382, DOI 10.1007/s10494-007-9070-x.

[13] Schaeffler N W, Allan B G, Lienard C and Le Pape A 2010 Progress towards fuselage drag reduction via active flow control: A combined CFD and experimental effort, *proc. 36th European Rotorcraft Forum,* Paris, paper 64

[14] Allan B G and Schaeffler N W 2011 Numerical investigation of rotorcraft fuselage drag reduction using active flow control *proc. 67th Annual Forum of the American Helicopter Society, Virginia Beach, VA*

[15] Le Pape A, Lienard C, Verbeke C, Pruvost M and De Coninck J L 2013 Helicopter fuselage drag reduction using active flow control: a comprehensive experimental investigation *proc. 69th AHS Annual Forum* Phoenix Arizona USA

[16] Crittenden T, Glezer A, Funk R and Parekh D 2001 Combustion-driven jet actuators for flow control AIAA Paper 2001-2768

[17] Woo GTK, Glezer A, Bain J and Sankar L 2011 Rotorcraft fuselage drag reduction using combustion powered actuators *proc. 49th AIAA Aerospace Sciences Meeting including the New Horizons Forum and Aerospace Exposition*, Orlando, Florida

[18] Woo GTK, Glezer A and Crittenden TM 2011 Transitory separation control on a ROBIN fuselage using pulsed actuation *proc. 37th European Rotorcraft Forum* Gallarate, Italy

[19] Martin PB, Overmeyer AD, Tanner PE, Wilson JS and Jenkins LN 2014 Helicopter fuselage active flow control in the presence of a rotor *proc. 70th American Helicopter Society,* Montreal, Quebec, Canada

[20] De Gregorio F and Vitagliano PL 2014 Experimental and numerical investigation of helicopter fuselage model with active flow control *proc. 40th European Rotorcraft Forum* Southampton UK

[21] Barlow J W Rae WH and Pope A 1999 Wind Tunnel Testing 3rd edition, John Wiley & Sons, Inc.

Fiber suspension investigation in a backward-facing step by PIV

A Capone[1], G P Romano[2], A Soldati[3]

[1]CNR-INSEAN Via di Vallerano 139, Rome Italy
[2]Dipartimento Ingegneria Meccanica e Aeronautica, Università La Sapienza, Via Eudossiana 18, Rome Italy
[3]Dipartimento Politecnico, Università di Udine, Via delle Scienze 208, Udine Italy

alessandro.capone@cnr.insean.it

Abstract. A dilute suspension (volume fraction 0.05%) of rod-like particles in a turbulent backward-facing step flow at Reynolds number Re_H=14900, is investigated by means of Particle Image Velocimetry. Two-way interactions between fluid and dispersed phase are analyzed by exploiting the high spatial resolution of the acquisitions. Mutual interactions between phases can be investigated by considering flow turbulence modulations and phenomena related to preferential concentration and orientation of fibers. Slight turbulence enhancement is reported in the laden flow and concentration data show a moderate tendency of fibers to accumulate at the channel centreline. Orientation data display a strong preferential orientation of fibers. Local fiber orientation is correlated to the direction of maximum shear showing a high level of correlation also in the flow regions featuring strong gradients.

1. Introduction

In many industrial and environmental applications such as chemical reactors, cyclone separators and devices for pollution abatement, multi-phase turbulent flows in which a carrier fluid is laden with a solid, dispersed phase are commonly encountered.

The understanding of these peculiar suspensions is strongly connected to the investigation of the dynamic interactions between fluid and dispersed phase. An important direction of investigation for engineering applications and of interest for turbulence fundamentals is represented by the study of the impact of the dispersed phase on the flow dynamics, specifically on turbulence modulation. Many efforts have been put forth in this direction ([1], [2] and [3] among others) and a comprehensive review is given in [4]. In addition to the mutual interaction between phases, the distribution and concentration of the dispersed phase is particularly relevant for the study and control of combustion and pollutant dispersion. Several works focused also on this aspect: [5] investigated a forced turbulent jet; homogeneous and isotropic turbulence was investigated by [6]; [7] studied thouroghly preferential concentration of particles induced by turbulence in several configurations. In an unifying perspective, [8] shows how the Stokes number St, defined as the ratio of the particle response time to a characteristic flow time scale, in combination with the particles volume fraction can describe the interactions in multiphase flows.

Although often the dispersed phase is effectively modeled via spherical particles, there exist many applications where non-isotropic, non-spherical particles are of interest. A particular class of such particles is represented by rod-like particles, characterized by high (>5) aspect ratios.

The dynamics of a single rod-like fiber in shear flow under creeping flow conditions have been theoretically predicted by [9] showing that a fiber tends to align mostly with the mean flow following a periodic revolving motion. Many works have been published on this subject: low-Reynolds number conditions were investigated experimentally by [10] in planar shear flow and by [11] on an inclined plate; contracting channels were investigated by [12]. Turbulent flows at high Reynolds number were studied experimentally by [13], [14] for different contraction shapes, while simulations in channel flow are presented in the works by [15] and [16][17]. To the best of authors' knowledge no experimental work was carried out on rod-like particle laden flows in a turbulent backward-facing step channel.

In this work Particle Image Velocimetry (PIV) is employed to investigate the modification of flow structure and turbulence modulation effects by fibers. Furthermore, fiber concentration and orientation data is extracted and correlated to fluid velocity field so as to shed light on their mutual interactions.

The use of an optical technique as PIV requires a non-trivial image pre-processing step in order to achieve an effective separation of phases. The current experimental set-up makes it possible to employ a discrimination algorithm based on ellipse fitting on fiber images, detailed in [18].

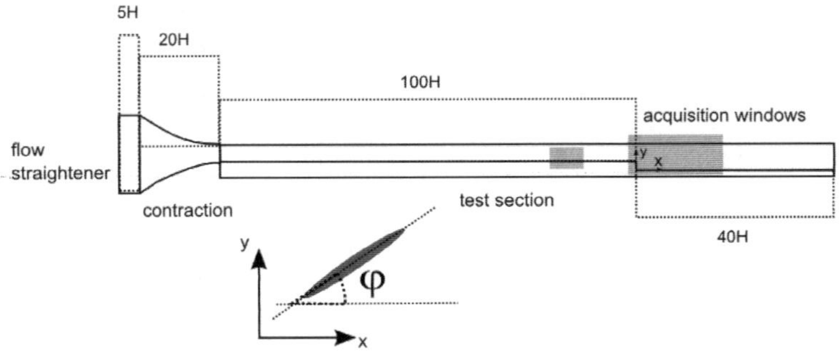

Figure 1 Experimental set-up

2. Experimental set-up

The channel and backward-facing step apparatus is shown in Figure 1. It consists of a channel section approximately 140H long, 2H high and 20H wide, where H=1cm is the step height. The channel section is preceded by a 10:1 contraction in order to reduce turbulence intensity by accelerating the mean flow and by a 5H long flow straightener to increase flow uniformity. The contraction section, which is 20H long, features a polynomial profile to avoid flow separation. At about 100H downstream of the contraction outlet, a backward-facing step is present having an expansion ratio ER=1.5. The flow is driven by a head tank (not shown).

The standard planar, two-component PIV system consists of a 14-bit BW, PCO-Pixelfly CCD camera (1392 X 1040 pixel resolution). The lenses used for all acquisitions are a Nikon F 50 mm with maximum aperture of 1.2. Lighting is provided by a Big Sky Nd:Yag pulsed laser, 488–514 nm, a maximum power equal to 200 mJ. Laser sheet thickness is approximately equal to 1mm. Fluid is seeded with neutrally buoyant 10μm diameter hollow glass spheres (Dantec HGS-10), whereas the dispersed phase is provided by nylon fibers (Polyamide 6.6, density 1.13-1.15g/cm^3, produced by Swissflock AG) with a mean length of approximately L=320μm and a mean diameter of d=24μm so that the aspect ratio k=L/d=13.3. Images were taken in the upstream channel approximately 80H after the 10:1 contraction, thus 20H before the expansion, in a region where the flow can be reasonably deemed as developed and no disturbance from the step is present, and after the expansion, up to roughly 6H downstream. In order to increase the spatial resolution, this latter acquisition region is subdivided into smaller overlapping parts each covering an area of nearly 1.2H X 1.6H, with a spatial resolution corresponding to 0.012mm/pixel. Acquired image sets consist of 2000 image pairs with an acquisition frequency of 10Hz. A commercial PIV software, DaVis by LaVision Gmbh, has been employed for instantaneous fluid velocity field computation. The minimum window size and overlap were respectively 32x32 and 50%, leading to a vector spacing of 16 pixels corresponding to approximately 0.02H.

Investigation of multiphase flows may be approached with many techniques (a review is given in [19]). In the present case the spatial resolution allows application of the algorithm described in [18], which is based on fiber ellipse fitting.

The experimental campaign was made up of two main acquisitions: a reference case with no fibers in the flow and a fiber-laden cases with volume fraction of C=0.05%. Reynolds number Re_H based on flow centre-line velocity was 14900. The fiber suspension can be considered as dilute because $nL^3<<1$ holds, given n as the number density of fibers.

Single-phase data are validated (not shown) upon comparison to data from [20] collected with a 3D PTV technique in similar experimental conditions.

3. Two-way interaction: fibers to fluid

Based on the classification by [8], multiphase systems may present one-way, two-way and four-way coupling regimes, depending on the extent of the interactions between fluid and dispersed particles and among particles themselves. In the current experiment the volume fraction concentration falls within the 10^{-6}-10^{-3} range where two-way coupling holds. Therefore, the momentum transfer from particles towards the fluid is not negligible and turbulence modulation effects could be observed. Several works ([21], [3] and [1] among others) showed that within this regime, the Stokes number has a particular importance in predicting the quality of modulation.

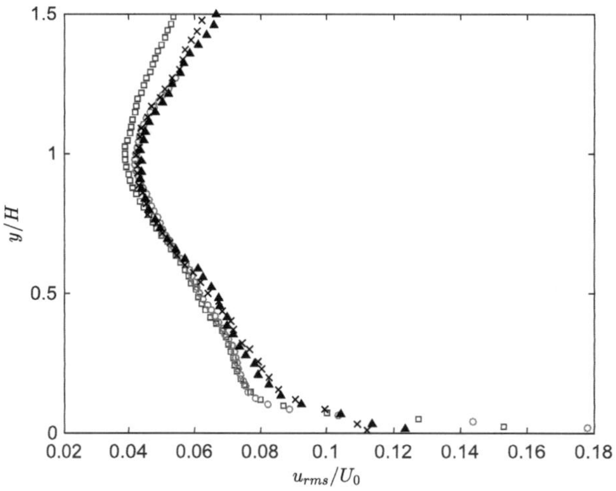

Figure 2 Fluid RMS of streamwise velocity of single phase (blue squares), and fiber-laden (red squares) for the upstream step region. Comparison to data from [22] single-phase (triangles) and multi-phase at 0.00014 volume fraction (crosses).

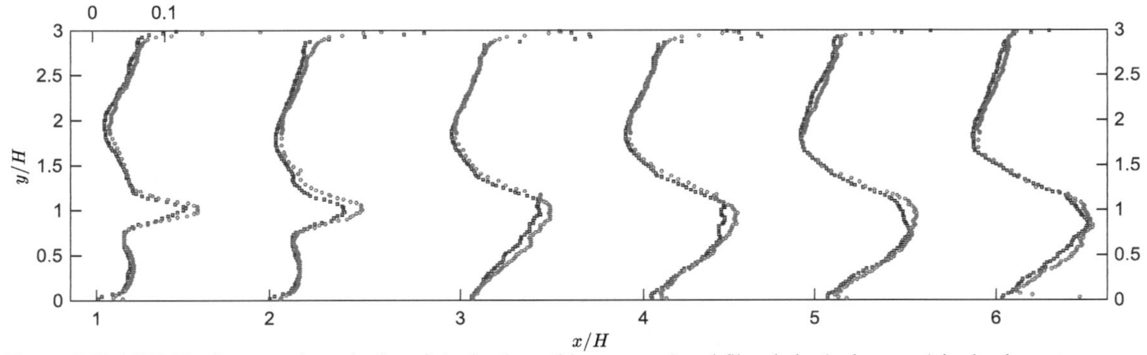

Figure 3 Fluid RMS of streamwise velocity of single phase (blue squares) and fiber-laden(red squares) in the downstream step region

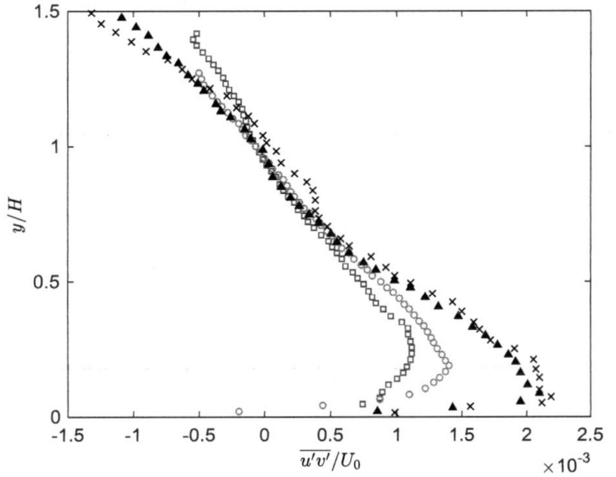

Figure 4 Fluid turbulent stress mixed term for single phase (blue squares) and fiber-laden (red squares) in the upstream step region. Comparison to data from [22] single-phase (triangles) and multi-phase at 0.00014 volume fraction (crosses).

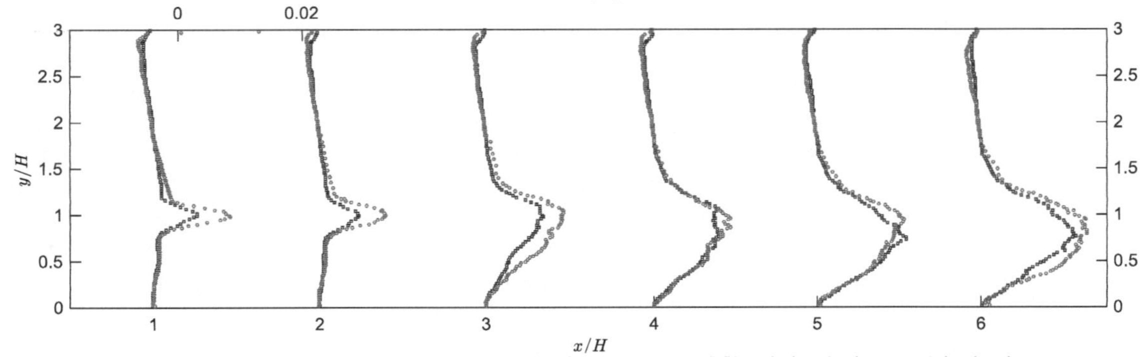

Figure 5 Fluid turbulent stress mixed term for single phase (blue squares) and fiber-laden (red squares) in the downstream step region.

The discussion about the appropriate particle response time, and consequently Stokes number, has been tackled by [17], who pointed out that when the fiber velocity induced by the hydrodynamic drag depends only on fiber characteristic length and D, and if the latter is nearly equal to one, the fiber timescale may be computed, upon choice of fiber length as a characteristic size, as

$$\tau_p = \frac{L}{U_0}$$

In the current conditions, assuming isotropic fiber orientation and adopting as flow timescale τ_f=P/U_0, where P represents the flow integral length scale, it descends that the Stokes number St ranges from 0.2 to 0.5. Specifically, St is larger inside recirculation regions, where on average the flow undergoes steep changes in direction and in these zones strong particle-fluid interactions take place. Outside these regions the Stokes number lies in the lower end and particle inertia effects are weak.

Turbulent intensity profiles are shown in Figure 2 and Figure 3, upstream and downstream the step respectively. Upstream, a slight turbulence enhancement for the laden cases is reported around the channel centre-line as opposed to the data from [22]. Downstream of the step an increase of turbulence in the shear layer up to x/H=2 is observed for the laden conditions. Higher levels are also observed inside the main recirculation bubble, i.e. x/H>2, which is the region where the Stokes number takes on the maximum value. An 1% error introduced by the phase discrimination procedure, should be accounted for.

Reynolds turbulent stress profiles for the upstream and downstream step regions are provided in Figure 4 and Figure 5 and compared to data by [22]. The results in the upstream section, in agreement to the findings of [22] show that the Reynolds stresses (absolute value) are enhanced in the near-wall region and to a larger extent when fibers are present. The extent of this increase is larger downstream of the step and it reaches a 35% increase for x/H=1.

4. Two-way interaction: fluid to fibers

Literature works have shown that conditions characterized by St≈1 correspond to situations where the preferential concentration by turbulence is the strongest ([4], [7]). This considered [7] point out that in water flows, due to the usual small difference between particle and fluid density, preferential concentration does not take place due to the same mechanisms. In fact rod-like particles' dynamics are strongly dependent on orientation and in wall-bounded flows the interactions with boundaries also has an impact on concentration. In the near-wall region, elongated particles are subject to the "pole vaulting" effect described in [10] and [23]. According to the description of this phenomenon, fibers close to the wall are "pole vaulted" away off the wall towards flow regions where they can freely rotate, with this process affecting the equilibrium concentration profile. In Figure 6, the relative local fiber concentration in the channel is compared to results from [24] obtained in a channel at concentration nL^3 =3.6. The concentration distribution in the wall-normal direction is normalized so that the integral of the distribution is equal to the total number of expected fibers divided by nL^3. A marked gradient is reported in the near-wall region, with the relative concentration reaching 90% of the centerline value at approximately 0.2H from the wall. Starting from roughly 0.3H the concentration levels at a nearly constant value. These findings confirm that in the upstream channel the relationships between phases are fully developed and bear resemblance to the findings presented in [25], where fibers were reported to concentrate in core region of a fiber-laden turbulent jet. Insight into fiber orientation dependence on the wall distance is provided by probability density functions (PDF) of fiber orientation computed at different distances from the wall, expressed in terms of the fiber length L. Results are summarised in Figure 7, and confirm that close to the wall, fibers are forced to align along the streamwise direction due to wall physical constraint, whereas as this constraint is relaxed, i.e. for y>L, the PDFs widen.

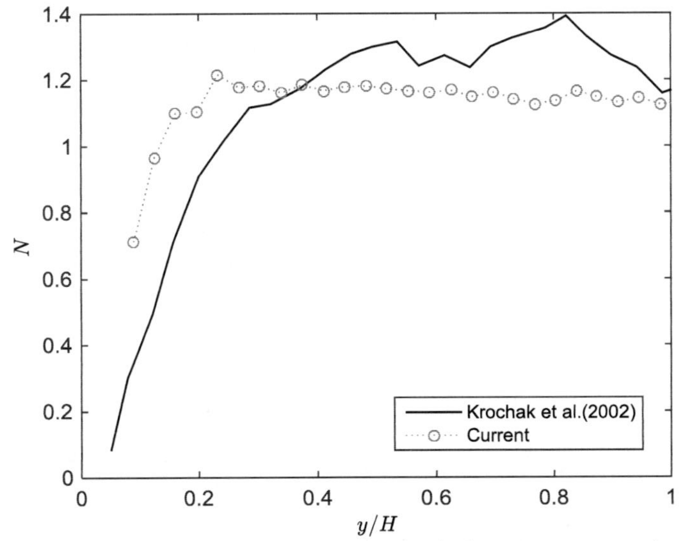

Figure 6 Relative fiber local concentration in the upstream step region.

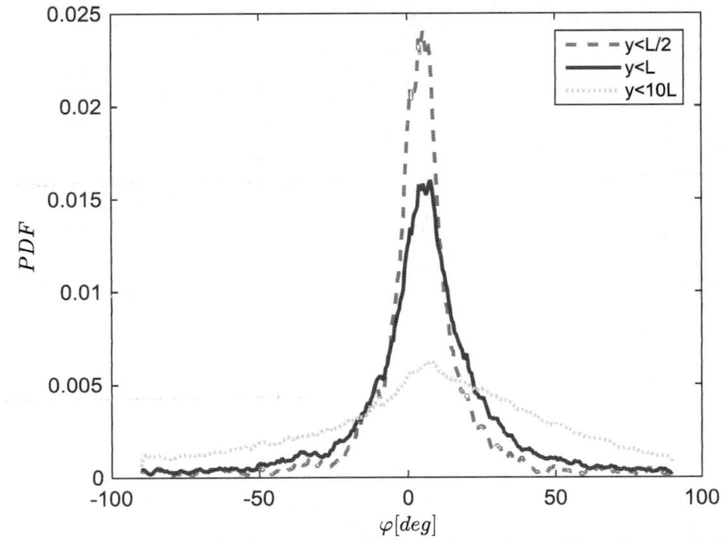

Figure 7 PDF of fiber orientation angle (in degrees as defined in Figure 1 conditioned on wall distance y in the upstream step region.

Several efforts have been carried out to establish a link between fiber mean orientation and some feature of the flow. [10] and [12] [24] focused on low-Reynolds number simple-shear flows, linearly contracting channels and channel contractions with various shapes whereas [26] investigated rods alignment with strain rate in a chaotic flow. Results on global fiber orientation downstream of the step are presented in Figure 8 in terms of mean orientation map and orientation profiles at selected locations.

The map shows the existence of a region featuring slightly negative fiber orientation towards the top of the channel and a region of positive orientations close to the step. This latter region expands towards the bottom when moving downstream, as also the point of sign inversion. Inside the shear layer, fibers take on the largest inclination value, i.e. with an angle roughly equal to 15 degrees, this orientation persisting up to x=5H. Orientation data shown can be effectively associated to turbulence modulation findings reported in the preceding section. The flow areas where most of the turbulence and turbulent stress modulation effects take place match those where fibers take on the highest orientation angle. A reasonable assumption is that as φ increases, local vortex shedding induced in the surrounding flow is enhanced and turbulence increased. Furthermore, near the bottom side of the step, i.e. for y/H<0.5 x/H<2, a negative orientation region is visible, thus suggesting a relation between orientation and mean flow direction in the lower side of the recirculation bubble.

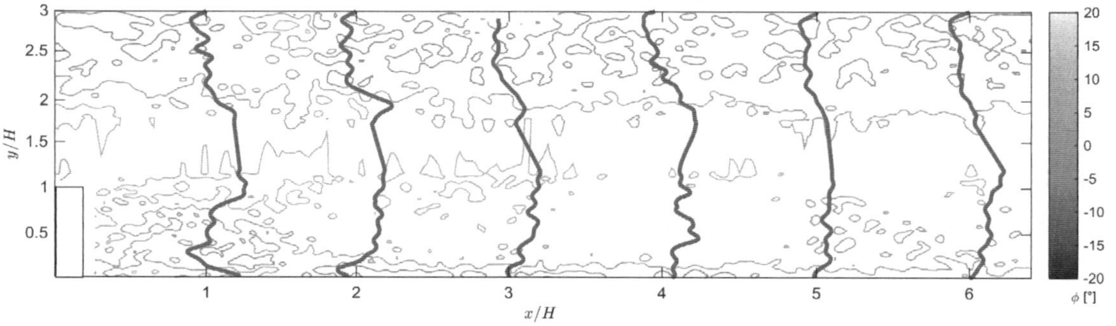

Figure 8 Mean fiber orientation angle map and profiles in downstream step region.

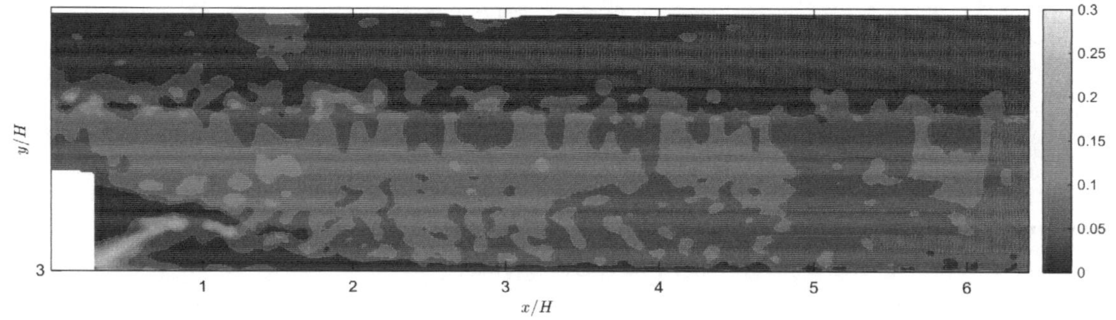

Figure 9 Ratio of angle between fiber and direction of maximum shear to angle of fiber orientation in the downstream step region.

Global alignment of fiber to mean flow direction was assessed based on instantaneous fiber and fluid data. Data were obtained considering instantaneous flow velocity and fibers orientation and obtaining flow velocity at fibers location by tri-linear interpolation of gridded velocity data and averaging the result for each slot, taking into account the resolution/convergence trade-off . Results (not shown) indicate that fibers tend to align to the mean flow with the exception of the flow regions where velocity gradients are stronger. This observation suggests the use of a different parameter rather than local mean flow direction to correlate to fiber orientation. Therefore, fiber orientation has been compared to different directions defined by the flow. In particular, alignment of fibers to strain-rate showed to be weak. Thus, we propose to correlate fibers orientation to the direction of the maximum shear defined as the angle θ for which the following holds

$$\tan 2\theta = -\frac{(\sigma_x - \sigma_y)}{\tau_{xy}}$$

Where σ_x, σ_y and τ_{xy} are respectively the normal stresses along x and y directions and shear stress in the x-y plane. The map of the ratio S = 1-|cosθ|/cosφ is shown Figure 9.

The results show that fibers align to the direction of maximum shear also in the shear layer. Indeed, in the flow region below the main recirculation bubble, S performs very well, with values below 0.05. The stretched thin region between x/H<1 and y/H<0.5 is the one characterized by the poorest performance of the S index with an average value of 0.17 especially due to developing turbulence. For y/H>1 and on the upper side of the main recirculation bubble fibers alignment to maximum shear plane is still relevant with an average value of S=0.1.

5. Conclusions

The upstream and downstream flow developing in a backward-facing step laden with rod-like particles at dilute concentration is investigated by means of Particle Image Velocimetry. Mean flow structures were reported to be basically unchanged when fibers are present. Moderate turbulence enhancement was reported in the near wall region, in the shear layer and in the recirculation region, as also increments of Reynolds stress cross term in the measurement plane in comparison to unladen conditions. Concentration data show fiber accumulation at the centreline, supposedly due to the combined effect of flow shear and wall physical constraints. Evidence of a marked preferential orientation of fibers is presented. In the channel section, fibers tend to align symmetrically with respect to the centreline. Correlation of fiber orientation angle to the direction of the maximum shear is shown to be strong in general, reaching high levels also in flow region where strong gradients are reported.

References

[1] Parthasarathy RN and Faeth GM 1990 *J. Fluid Mech.* **220** 485-514
[2] Rashidi M, Hetsroni G, Banerjee S 1990 *Int. J. Multiph. Flow* **16** 935–949
[3] Paris AD and Eaton JK 2001 Rep. TSD-137 Dep. Mech. Eng., Stanford Univ.
[4] Balachandar S, Eaton JK 2010 *Annu Rev. Fluid Mech.* **42** 111-133
[5] Longmire EK and Eaton JK 1992 *J. Fluid Mech.* **36** 217-257
[6] Wang LP and Maxey RP 1993 *J. Fluid Mech.* **256** 27-68
[7] Eaton JK and Fessler JR 1994 *Int. J. Multiphase Flow* **20** 169-209
[8] Elghobashi L 1994 *App. Sci. Resh.* **52** 309-329
[9] Jeffery GB 1922 *Fluid Proc. R. Soc. London Ser. A* **102** 161-179
[10] Moses K, Advani S, Reinhardt A 2001 *Rheol. Acta* **40** 296–396
[11] Holm R, Soederberg D 2007 *Rheol. Acta* **46** 721–729
[12] Krochak P, Olson J, Martinez D 2008 *Phys. Fluids* **20** 073303
[13] Parsheh M, Brown M, Aidun C 2005 *J. Fluid Mech.* **545** 245–269
[14] Parsheh M, Brown M, Aidun C 2006 *Int. J. Multiph. Flow* **32** 1354–1369
[15] Marchioli C, Fantoni M, Soldati 2010 A *Phys. Fluids* **22** 033301
[16] Mortensen P, Andersson H, Gillissen J, Boersma B 2008 *Phys. Fluids* **20** 093302
[17] Mortensen P, Andersson H, Gillissen J, Boersma B 2008 *Int. J. Multiph. Flow* **34** 678–683
[18] Dearing S, Campolo M, Capone A, Soldati A 2013 *Exp. Fluids* **54** 1419
[19] Cheng, Y, Pothos, S,Diez FJ 2010 *Exp. Fluids* **49** 1375-1391
[20] Kasagi N, Matsunaga A 1995 *Int. J. Heat and Fluid Flow* **16** 477-485
[21] Kulick JD, Fessler JR, Eaton JK 1994 *J. Fluid Mech.* **277** 109-134
[22] Van Hout R 2011 *Int. J. Multiphase Flow* **37** 346–357
[23] Stover CA, Cohen C 1990 *Rheol. Acta* **29** 192-203
[24] Krochak P, Olson J, Martinez D 2010 *J. Fluid Mech.* **653** 431-462
[25] Capone A, Romano GP, Soldati A 2014 *Exp. Fluids* **56** 1
[26] Parsa S, Guasto J, Kishore K, Ouellette N, Gollub J, Voth G 2011 *Phys. Fluids* **23** 043302